LUMINOUS CREATURES

LUMINOUS CREATURES

The History and Science of Light Production in Living Organisms

Michel Anctil

McGill-Queen's University Press
Montreal & Kingston • London • Chicago

ISBN 978-0-7735-5312-5 (cloth)
ISBN 978-0-7735-5409-2 (ePDF)
ISBN 978-0-7735-5410-8 (ePUB)

Legal deposit second quarter 2018
Bibliothèque nationale du Québec

Printed in Canada on acid-free paper that is 100% ancient forest free (100% post-consumer recycled), processed chlorine free

Funded by the Government of Canada | Financé par le gouvernement du Canada | Canada | Conseil des arts du Canada | Canada Council for the Arts

We acknowledge the support of the Canada Council for the Arts, which last year invested $153 million to bring the arts to Canadians throughout the country. Nous remercions le Conseil des arts du Canada de son soutien. L'an dernier, le Conseil a investi 153 millions de dollars pour mettre de l'art dans la vie des Canadiennes et des Canadiens de tout le pays.

Library and Archives Canada Cataloguing in Publication

Anctil, Michel, 1945–, author
Luminous creatures : the history and science of light production in living organisms / Michel Anctil.

Includes bibliographical references and index.
Issued in print and electronic formats.
ISBN 978-0-7735-5312-5 (cloth).–ISBN 978-0-7735-5409-2 (ePDF).–
ISBN 978-0-7735-5410-8 (ePUB)

1. Bioluminescence – History. I. Title.

QH641.A53 2018 572'.4358 C2018-900897-0
C2018-900898-9

This book was typeset and designed by studio oneonone in Adobe Garamond 11/13.5.

Contents

Acknowledgments

As an investigator in the field of bioluminescence for many years, I collected many scientific books and articles, as well as correspondence with colleagues and memorabilia, that were helpful in putting this book together. But the book would have suffered without help from outside.

I wish to thank the American Philosophical Society, Philadelphia, for access to the Papers of E. Newton Harvey. Harvey is probably the central figure in this book, and the mine of information found in his papers, particularly his correspondence, has greatly enriched it. For the papers of another iconic figure in this book, Raphaël Dubois, I owe a great debt to Christian and Renée Bange, who produced precious documents from the archives of the French investigator. For information on the scientific production of Yata Haneda, I relied on Namika Fukuhara of the Harano Agricultural Museum, Amami Culture Foundation, Kagoshima, Japan. She also kindly provided a photograph of Yata Haneda. Further information on the scientific life of Yata Haneda was graciously provided by his student Dr Nobuyoshi Ohba and, for the war years, John K. Corner of Melbourne, Australia. Maura Jess of Portland, Oregon, sent me the unpublished memoirs of James F. Case. From the Special Collections of the Scripps Library in La Jolla, California, I enjoyed the expert assistance of Heather Smidberg in sifting through the archives of John Buck and the *Alpha Helix* expeditions and making available the documents I needed. I also thank Heidi Stover, of the Smithsonian Institution Archives, Washington, DC, for making accessible the photographs of Paolo Panceri and Albert Günther.

I thank James G. Morin of Cornell University for his invaluable criticism of the science in the manuscript. His constructive comments have saved me from several blunders and significantly improved the final product. My editors at MQUP, Jacqueline Mason and Kathleen Kearns, took my project to heart and steered it smoothly through the various stages of its early life. Finally, my copy editor, Jane McWhinney, did her usual utmost to make this book an enjoyable reading experience.

Prologue

In the many versions of the theoretical hollow earth proposed by imaginative or deluded minds over the centuries, little direct attention has been paid to the question of lighting. And yet it is there. Whatever form the hollow spaces take deep inside the earth – concentric galleries, labyrinthine corridors, vast grottos – the underground creatures are somehow lit by some undefined inner ambient light. And sometimes mention is made of bioluminescence as a light source. In Mark J.P. Wolf's book *Building Imaginary Worlds: The Theory and History of Subcreation*, we learn that Edgar Rice Burroughs, the science fiction writer of *Tarzan* fame, imagined a *Pellucidar* "lit by subterraneous suns," and that Baron Ludvig Holberg (1845) evoked the planet Nazar, with "individual subterranean homes lit by luminous creatures." We also learn that the underground cavern of the D'ni people imagined by computer game creators Rand and Robyn Miller is lit by a lake inhabited by bioluminescent plankton. Fast forward to 2009 and we find in James Cameron's movie *Avatar* that almost all vegetation growing on the fictional planet Pandora is bioluminescent, as are many of the grotesque animal creatures roaming the Pandora world.

In the real world, there is a parallel with this evolution of fantastical representations of bioluminescence. To creatures dwelling on the surface of the earth, the outer lights – sunlight, moonlight, artificial light – are always associated with the immediacy of everyday living. They were, and still are, embedded in the consciousness of humans, if only for their critical value to survival, in a way that living lights, which are largely hidden from view (except for fireflies), could not be. In recent times, the ubiquitous bioluminescent creatures populating the vast spaces of the conquered oceanic underworld have captured the attention of many in the general public. This book follows the trajectory that brought humans from their lack of acquaintance with, and ignorance of living lights, to the current scientific understanding of bioluminescence and its attendant glare of publicity. Its evolution from an object of bewildered observation to an object of analysis, from the obscurantism of uncritical interpretation to the scientific insights that led to a Nobel Prize and turned the field into an indispensable tool for medical research, molecular biology, and biotechnology, is the object of this book.

The history of bioluminescence has been dealt with only once before in book form. E. Newton Harvey, who is discussed in chapter 10, published *A History of Luminescence* in 1957 at the end of his career as the most prominent and distinguished investigator of bioluminescence in the scientific world. How, sixty years later, does the present book differ in style and approach from Harvey's book? His book covered other physical light emissions as well as bioluminescence: fluorescence, electroluminescence, chemiluminescence, and so forth. I will limit my coverage to bioluminescence, which a more general readership can better identify with. Harvey's book was more akin to a comprehensive compendium than to a historical narrative, whereas the present book aspires to a contextual narration in which the scientific literature is abundantly cited, but not exhaustively so. Choices dictated by my perception of the main historical currents driving the field of bioluminescence have determined the quantity and selectivity of the cited literature. (If contributors to the field feel left out or neglected in this book, I can only say that any such omissions were not intentional.) Harvey's literature review, in contrast, was all-inclusive.

Finally, Harvey's coverage only reached the end of the nineteenth century. The present book, therefore, exposes for the first time the history of the field throughout the twentieth century and into the burgeoning twenty-first. The century missed by Harvey saw key scientific achievements regarding the chemistry, physiology, and ecology of bioluminescence, as well as its evolutionary origins and usefulness as a tool in medical research, biotechnology, and agriculture. Of course, I refer to Harvey's massive opus several times for perspective and to point out the few bibliographic errors or missed historical events that I unearthed during my research.

Recently, Aldo Roda of the University of Bologna summarized the history of chemiluminescence and bioluminescence from Chinese antiquity to the present (Roda, 2011). Roda's account, which largely ignores non-chemical aspects of bioluminescence, provides new information not found in Harvey's book, especially regarding the earliest mentions of luminous organisms and the contributions of the twentieth century. Several short popular books have appeared over the years with a view to explaining bioluminescence in simple language, some of which have targeted children. None have provided significant historical background. The authors were not insiders; they were reporters, or freelance or science writers, but not especially practised in bioluminescence research. Two notable exceptions are recent books: *Bioluminescence: Living Lights, Lights for Living* by Thérèse Wilson and J. Woodland Hastings, a beautifully illustrated, highly readable short tome leaning heavily on the biochemistry side of the phenomenon; and *Silent Sparks*, by evolutionary ecologist Sara Lewis, whose account of the life of fireflies reads like a rhapsody as much as an up-to-date review of the science. Lewis admits to being a "firefly junky," and her reflections on the sense of wonderment fireflies summon find resonance in passages of my book.

As one who has taken part in scientific research programs devoted to bioluminescence, I have tried to strike a balance between my interest in exposing scientific research and the need to provide the human and social backgrounds from which the research sprang. Obviously, my training places me at risk of tipping the scales toward hard science. I can only leave readers to decide whether the effort at balance is successful. The book is divided into six parts, each part subdivided into three chapters and highlighting what I consider key milestones in the historical development of the field.

Part One briefly guides the reader from Antiquity (Aristotle and Pliny the Elder) to the Renaissance, the Age of Enlightenment, and the first half of the nineteenth century. This was a long period when superstition, preconceived ideas, and optical illusions led even students of natural history into imaginative worlds; when light from animated creatures was like fire, only cold; when a sparkling sea was the watery equivalent of the aurora borealis; when will-o'-the-wisps over marshes conjured devilish dancing creatures; when organic putrefaction went hand in hand with glowing light; when simple reflections from human eyes were mistaken for genuine light emissions of a portentous nature; when phosphorus was believed to be the source of animal light – hence the ancient use of the word "phosphorescence." But also a period when naturalists were slowly beginning to make sense of the phenomenon, when their circumnavigations of the globe in the wake of Captain Cook's voyage put them in contact with an increasing number of luminescent species.

Part Two chronicles the firm grounding of living lights in the scientific approach during the second half of the nineteenth century. The scientific method as practised today came about around the middle of that century, and it quickly had an impact on bioluminescence research. Especially instrumental in this outcome was the development of oceanography. The great oceanographic explorations started with the British *Challenger* expedition in 1872–74; and others, such as the German *Valdivia* expedition, followed. They surveyed the oceans using new research tools that defined physical and biological oceanography, they mapped the sea bottoms, and they discovered an unsuspected bonanza of deep-sea creatures that emitted light. Then came the realization that bioluminescence was an important player in the lives of a great many oceanic organisms. The anatomy of the light organs of the animals captured during these expeditions was intensely investigated.

The careers and scientific accomplishments of individuals who shaped the field of enquiry, particularly in relation to coastal and terrestrial bioluminescence, are the subject of Part Three. Now that the scientific study of bioluminescent organisms was on a firm footing, bioluminescence as a serious biological discipline was ushered in, thanks to numerous dedicated biologists who plied their trade from the 1870s and into the early 1900s in laboratories around Europe, Japan, and America. For the first time, research on bioluminescence could be envisaged as a career-long endeavour, as it was for Paolo Panceri, Raphäel Dubois, whose work, along with others I will scrutinize. For the importance of its discovery of the basic chemical mechanism taking place in light-emitting cells – the luciferin-luciferase reaction – Dubois's contribution cannot be overemphasized.

Around the time of the First World War, the centre stage of the story moved to the United States. Part Four recounts how, at Princeton University, E. Newton Harvey ran a busy laboratory devoted largely to bioluminescence research, picking up the thread from Dubois. A larger-than-life and indefatigable figure, Harvey dominated the field in the first half of the twentieth century, steering research toward biochemical and physiological approaches, and uncovering some of the luciferin-luciferase systems in nature. His numerous students in turn launched their own careers in the field and in turn again begot Harvey's academic grandchildren – among the latter Nobel recipient Osamu Shimomura. All this progeny contributed to the triumph of the biochemical approach to the study of bioluminescence. In parallel to Harvey's academic career, another larger-than-life American, William Beebe, played the part of the old-school naturalist and entrepreneur, whose major contribution was to peer at luminescent creatures in their own element, deep in the ocean. In the early 1930s he and his team devised a "bathysphere" able to be dropped to great depths from a cable winch to watch live animals and their spectacular luminescent displays through a porthole.

In Part Five I examine what went on beyond the Princeton circle during and after Harvey's lifetime. I include a portrait of the important and enigmatic figure of Japan-based biologist Yata Haneda, emphasizing the tumultuous political times during which his research foundered and flourished. I also look at three other figures who distinguished themselves in sub-disciplines of bioluminescence other than biochemistry: the American John Buck, the Canadian J.A. Colin Nicol, and Jean-Marie Bassot of France. An innovative way of doing research – assembling a team of bioluminescence specialists to do research together in Papua New Guinea – was spearheaded by John Buck, and its ups and downs are chronicled.

The chapters in the final section (Part Six) round out the field as it has unfolded in the past forty years. The thrust of bioluminescence research, edging to the present, has focused on marine bioluminescence, physiological control mechanisms, and the molecular biology of chemiluminescent reactants – luciferins, luciferases, and photoproteins. Of particular interest is the role of technological developments in the spectacular advancement of the field. Oceanic bioluminescence research benefitted from the arrival of field instruments – sophisticated bathyphotometers, autonomous submersibles – developed in cooperation with or under the sponsorship of the US Navy. The dynamics of luminescent sources in light organs and luminous cells became understood thanks to advances in low-light image intensifiers and fluorescence microscopy. As the tools of molecular genetics were developed, the chemistry of bioluminescence made further great strides. It led not only to a better understanding of chemical mechanisms in bioluminescence but also to the development of tools using bioluminescent molecular trackers for pharmacological and cancer research, to name only two.

In the course of this narrative, it becomes increasingly clear that the terminology used to identify living lights in the old texts has evolved over time. In early works the word "phosphorescence" comes up often. Originally, it described light purportedly emitted by phosphors and, as many early observers assumed that all living lights had a phosphor as their source, the phenomenon in living organisms became categorized as phosphorescence. Later, in the nineteenth century, when physicists and chemists came to understand the properties that differentiate luminous bodies, the umbrella term "luminescence" was applied to all forms of light emission. Subcategories of luminescence were created to account for light emissions triggered by specific chemical or physical means: chemiluminescence and fluorescence, among others. The currently used term for the phenomenon in living organisms, "bioluminescence," only appeared early in the twentieth century. All these forms of light emission are generally detectable by human eyes adapted to the dark. The use of the word "bioluminescence" separates the phenomenon from a form of luminescence generally undetectable by human eyes and considered an attribute of all living cells: ultraweak photon emissions (Devaraj, Usa, and Inaba, 1997). The latter, because it is outside the experience of the observers recorded in this book, is not treated here.

As this story unfolds, it also becomes clear how appreciation for the importance of bioluminescence grew dramatically as the phenomenon became understood as a signal detected and decoded for biological purposes: defence, predation, reproduction, kin recognition – in short, for survival. Defence can be accomplished by deterring or escaping predators using luminescence for camouflage, to startle or decoy, and to issue warning signals that particular luminescent prey are unpalatable. For predation, luminescent organisms can startle or attract prey, or use luminescence as a visual aid to locate prey. For reproduction, luminescence can be used in courtship or in mating; kin recognition involves light organs arranged in specific patterns recognized by conspecifics.

This book demonstrates how far the field has progressed, not only in deepening our overall understanding of the forms and functions of the phenomenon, but also by unearthing evolutionary processes – natural selection, sexual selection – that shaped these forms and functions. Many bioluminescent species use their light for more than one function. Firefly "femme fatales," for instance, use their light for reproduction by responding to a male of their own species with the appropriately timed flash signal. But they also use their light for predation by mimicking the flash signal of females from other species so as to attract foreign males and eat them. This example also illustrates the plasticity of the phenomenon; a slight variation in flash timing by the female shifts the context in which her bioluminescence is used to her advantage.

If living lights are to serve a signalling function, they must be detected by another organism for whom the signal may carry meaning. The alternative is, of course, that the light-emitting organism detects its own light in order to see its way around, a role of bioluminescence that was all too often invoked in the past to the detriment of other functions. The receiver of the light emission is, of course, a photoreceptive structure – dermal photoreceptor cells or ocular organs (eyes). Pioneers of the nineteenth and early twentieth centuries struggled to understand the interrelationship between vision and bioluminescence, especially in the oceanic world, but only in the past few decades have we made significant headway on this difficult topic. The field of study embraces an entire cast of characters: the bioluminescent organism, the organism perceiving the bioluminescence, and the ambient natural light in the ecosystem where these organisms dwell. The modern field of visual ecology made it a mission to discover visual adaptations that had evolved to enhance an organism's perception of living lights while retaining its ability to visually navigate under ambient light conditions.

A striking example of such visual adaptations relates to deep-sea fishes. The eye sensitivity of most deep-sea fishes is attuned to the blue light selectively transmitted at these depths from sunlight, allowing them to see their surroundings as well as the blue light emission from ventral and lateral light organs of their congeners or other species. But there are a few deep-sea dragonfish (family Stomiidae) which, in addition to blue bioluminescence, also possess light organs below the eyes that produce bioluminescence in the far-red range. Their eyes have evolved visual pigments sensitive to the red emission of their own suborbital organs. As Julian Partridge and Ron Douglas

(1995) explain, "the red photophore of *Aristostomias* could be used both for private, intraspecific communication, invisible to most deep-sea animals, and for covert illumination of prey at distances about ten times greater than the range of lateral line senses." The prey, which cannot see the red emission, is in a way visually ambushed.

What dawns most pregnantly on modern students of the field is the recognition of the multiplicity of bioluminescent species on this planet. From the few species recorded in Antiquity, we have now reached huge numbers. Thirty years ago, the British bioluminescence expert Peter J. Herring attempted to draw up a census of known luminescent species (Herring, 1987), and still more recently an updated figure of luminescent genera (or assemblages of species) has been produced (Oba and Schultz, 2014). Over seven hundred genera are compiled, and many of them include several species. Fireflies alone tally up close to two thousand species worldwide, according to Sara Lewis (2016), and luminescent fishes are also approaching comparable numbers.

Since Herring's exercise, it has seemed futile to try and keep up with the numbers, as new species keep being discovered. But one must look beyond the numbers and make sense of the way luminescent species are distributed among the various phyletic groups: bacteria, dinoflagellates, fungi, radiolarians, cnidarians, ctenophores, nemertean worms, molluscs, annelids, arthropods (pycnogonid, crustaceans, insects, chilopods and diplopods), bryozoans, chaetognaths, echinoderms, hemichordates, and chordates (tunicates and fishes). This distribution reflects the well-recognized fact that bioluminescence is overwhelmingly associated with marine environments as opposed to terrestrial and, to a much lesser extent, freshwater environments. But more important, phylogenetic and molecular analyses have led experts to propose that bioluminescence arose independently multiple times in different groups of organisms in the course of evolution.

It is in the context of these major currents in the field that I highlight the historical milestones of discovery that populate the following chapters. My hope is that the reader, as he or she follows the travails, triumphs, and setbacks of precursors and pioneers in the field of bioluminescence, will appreciate the challenges they were up against and the dizzying speed of the progress of recent decades.

Even for me as an insider, researching this book has been a voyage of discovery, an enthralling journey through time from the superstitions, misconceptions, and wonderment of the early observers of living lights to the incredibly successful science of bioluminescence that prevails today. Like most quests for knowledge, it unravels in twists and turns against a backdrop of evolving science and human foibles.

PART ONE

GROPING IN THE DARK

1

Glows, Flashes, and Marvel-Mongers

> Strange: I stand upon the shore and see the sea seething with
> the light of organisms brought to break down an atmosphere
> a man, or even a plant, couldn't breathe.
> –Phillip Ellis ("Bioluminescence," 2009)

From the deepest of times, observers of nature must have encountered phenomena of luminous displays by living organisms. And, as with large-scale celestial and atmospheric light displays, humans have resorted to mythologies to manage their puzzlements or frights, explaining living lights as manifestations of powers wielded by the gods or ethereal spirits from whom they sought appeasement. One myth that persisted over millennia told of marine bioluminescence originating in stars falling from the sky and being engulfed by the sea. Victor Benno Meyer-Rochow (2009) and Aldo Roda (2011), in their brief historical surveys of bioluminescence, recount the mention of fireflies in the Chinese poetry classic *Shi Jing*, written three thousand years ago. A millennium later, fireflies are mentioned again, this time in a text from the Upanishads in India (Meyer-Rochow, 2009).

Roda (2011) writes that "folklore-based, artistic works, including paintings and poems, document awareness of bioluminescent phenomena among the people of Japan, India, Africa as well as Central and South America." In the New World, ancient Indigenous cultures spun legends about fireflies, as Gene Kritsky and Ron Cherry illustrate in their book *Insect Mythology* (2000):

> In a story reminiscent of tales told in Homer's *Odyssey*, Mayan twins must keep two cigars lit all night. The twins appear to complete this Herculean task by substituting fireflies at their cigar tips, thus tricking their captors … Among the Aztecs of Mexico … people believed that

> fireflies were witches or wizards that went around at night throwing out fire from their head or mouth … According to the Jicarilla Apaches of New Mexico, fire came from a mythical campfire ignited by fireflies.

These myths were transmitted by oral tradition over generations, and the comforting reassurance of mythical interpretation put off any unseemly curiosity about the physical phenomenon itself and its place in nature. But in the Greco-Roman period, when "civilization" came to imply a drive for the empowerment of knowledge and an awareness of the cultural value of the written word, personal encounters with living lights began to be permanently documented. Since E. Newton Harvey (1957) compiled detailed records of numerous bioluminescence sightings during this period, however, I will comment only on the central figures, Aristotle (384–322 BCE) and Pliny the Elder (AD 23–79).

Before we compare what Aristotle and Pliny wrote about luminous animals, it is important to place them in the context of the scholarly enterprise of their times. Reading the Ancients' accounts of nature, it seems preposterous to construe their descriptions as scientific reports. According to the science historian Roger K. French, the word "science" "has now connotations of purpose and methods that are quite out of place when describing the entirely different enterprises of the ancient world" (French, 1994). The Ancients' own depiction of what they were practising is best encapsulated in the words "Natural History." French interprets these words to mean for the Greeks "an enquiry into what was remarkable." But it was more than just a collection of showpieces to marvel at. As French explained, "It was research, and the Greeks who travelled and interviewed people about historiae looked down their noses at those who confined themselves to libraries" (French, 1994). It was akin to the kind of research an investigative journalist practises today.

It was tempting for Aristotle's Greek contemporaries to represent luminous animals as exotic oddities that inspired awe, especially as they were considered rare and very few luminous species were then known. But for Aristotle their investigation was all part of a rigorous analytical examination of the natural world. French emphasizes this when he proposes that "some historiae were selected for their strangeness, but for example the collaborative 'history of animals' put together in the Lyceum under Aristotle was systematic and impartial. Modern science gains credibility by the same device of presentation, and this is why Aristotle's historiae have been called scientific" (French, 1994). However, sharing a "device of presentation" with modern science does not a scientist make.

Aristotle's depictions of luminous organisms are few, cursory, and scattered in several of his writings. One that E. Newton Harvey highlights in his *History of Luminescence* is the following:

> Some things, indeed, are not seen in daylight, though they produce sensation in the dark: as for example the things of fiery and glittering appearance for which there is no distinguishing name, like fungus (*mukes*) horn (*keras*) and the head scales and eyes of fishes. But in no one of these cases is the proper color seen. Why these objects are seen must be discussed elsewhere. (*De anima*)

The recurrence of the metaphor of fire in the pre-scientific period clearly reflects the Ancients' lack of understanding of how that light is produced. The struggle to come to terms with the phenomenon is further revealed in another of Aristotle's writings: "Some things, though they are not in their nature fire nor any species of fire, yet seem to produce light" (*De Coloribus* of the *Opuscula*). Animals produce light because it is their nature: "It is the nature of smooth things to shine in the dark, [as, for example] the heads of certain fishes and the juice of the cuttle-fish" (*De Sensu*). French remarked that Aristotle was not content to provide materialistic answers to questions on the nature of living organisms, to describe them as composed of basic components, but sought rather to discover their purpose – some teleological goal-directedness. To put it plainly, this meant that animals "had senses to perceive and follow what was good and to avoid what was bad. It meant that the individual animal could preserve itself by feeding" (French, 1994).

To Aristotle, the concept that animals have a purposeful nature leads to the notion of their possessing a "soul" that keeps them animated; when the Aristotelian soul is gone, what is left is the perishable and the purposeless. The soul provides for animal activity, including light emission, and the soul ensures that the animal is greater than the sum of its parts. As Andrea Falcon put it, for Aristotle, "the job of the student of nature is to investigate the soul in so far as it is a principle of motion and rest – in so far as it animates a body of a specific type" (Falcon, 2005).

Aristotle never specifies that light emission by the animals he observed – or heard about from his sources – is a dynamic process worthy of the soul, although it is implied in some of his passages. But for the most part he does not pretend to know specifically the purpose of the light in the animals possessing it; nor does he realize that some of the phenomena he takes as intrinsic light emission are actually plays of light such as light reflections from fish scales or eyes, or misleading sights such as morbid flesh infected by luminous bacteria, as Harvey (1957) infers. Some of the accounts that Aristotle mentions, however, appear to refer to genuine cases of light-emitting organisms: those of fungi, squids, glow-worms, and fireflies. Of the glow-worm Aristotle writes: "From a certain small, black and hairy caterpillar comes first a wingless glow-worm; and this creature again suffers a metamorphosis, and transforms into a winged insect named the bostrychus" (*Historia animalium*: Thompson, 1910). This excerpt is typical of most of Aristotle's offerings on luminous animals, in that he concentrates on the life cycle or the classification of the animal but does not describe the external characteristics of the light emission.

Pliny the Elder, in contrast, comes forward with more physical descriptions. Four centuries after Aristotle, the "natural history" of the Roman is quite a different enterprise. First of all, Pliny collated his material into one large book, the *Naturalis historiae*, which included all his observations on luminous organisms. But he also departs from the Aristotelian model in "his lack of systematic approach and scientific [read 'analytic'] judgement" (Enenkel, 2014). Pliny focused on the uniqueness of the animal species, asking himself what makes these species such miraculous creations of God. And, given the inherent show value of luminescent phenomena in nature, it is no surprise that "the *mirabile* (information that causes admiration and astonishment) is of the greatest importance for Pliny's zoology" (Enenkel, 2014). Enenkel goes on to argue that Pliny's zoological outlook was rooted in the ideology of the Roman Empire, which emphasized man's – and especially Rome's – domination of nature. French (1994) also emphasizes this point when he writes that Pliny's "book has much more the air of a survey of the material resources available in the natural world for the use of man –

Roman man." This makes the tone of his descriptions of bioluminescence and other zoological topics sound quite anthropomorphic.

While acknowledging that Pliny was known primarily as a chronicler, Harvey (1957) stresses that his descriptions of bioluminescence are nonetheless more detailed than Aristotle's: "Although fundamentally a reader and compiler, Pliny's military career took him to all parts of the ancient world and his travels afforded the opportunity for observation and anecdote. He has been maligned by many writers but his descriptions of luminescence were often quite specific and complete." Pliny's accounts reveal the same confusions as Aristotle's did – mistaking light reflections from silvery surfaces, and luminous bacteria/fungi growing in putrid flesh or wood for genuine displays of intrinsic bioluminescence. But this shortcoming is more than made up for by his descriptions of what we know today to be true cases of bioluminescence. As Harvey did, we will use the translation by Philemon Holland (*The Historie of the World*) published in 1634, leaving intact its antiquated language.

Searching through Pliny's work, one encounters first the quaint description of a light organ on the tongue of a fish. "There is a fish comes ordinarily aboue the water, called Lucerne, for the resemblance that it hath to a light or lantern: for it lilleth forth the tongue out of the mouth, which seemeth to flame and burne like fire, and in calme and still nights giues light and shineth (Book IX, Chap. XXVII)." Harvey (1957) cites sources claiming that salps, colonial tunicates known to float near the sea surface, were the culprits. But another possibility is a lanternfish whose tongue bears photophores (Kuwabara, 1958); this discovery was made shortly after the publication of *A History of Luminescence*, so Harvey could not have been aware of it.

The next luminous species over which Pliny marvels ecstatically is the bivalve mollusc *Pholas dactylus.* The account leaves no doubt that a glandular secretion is behind the light emission:

> Of the shell fish kind are the Dactyli, so called of the likenesse of mens nailes, which they resemble. The nature of this fish is to shine as if with fire in dark places, when all other light is taken away. The more moisture they haue within them, the more light they giue: insomuch as they shine in men's mouths as they [are] chawing of them: they shine in their hands: vpon the floore on their garments, if any drops … their

> fattie liquor chance to fall by: so as it appeareth, that doubtlesse it is the very iuice & humor of the fish which is of that nature, which we do so wonder at in the whole body. (Book IX, Chap. LXI)

Glow-worms, previously noted by Aristotle, are also admired by Pliny:

> The Glo-wormes, are named by the Greeks Lampyrides, because they shine in the night like a sparke of fire: and it is no more but the brightnes of their sides and taile: for one while as they hold open their wings, they glitter; another while when they keep them close together, they be shadowed and make no shew. These Glowbards neuer appeare before hay is ripe vpon the ground, ne yet after it is cut downe. (Book IX, Chap. XXVIII)

And Pliny's last account of animal light invites one to try his procedure to light a torch out of what Harvey (1957) interpreted as jellyfish: "Rub a piece of wood with the fish called Pulmo Marinus, it will seem as though it were on a light fire; in so much as a staffe so rubbed or besmeared with it, may serue in stead of a torch to giue light before one" (Book XXXII, Chap. X).

Throughout all these accounts Pliny stays with Aristotle's fire metaphor. It cannot be surprising. What other element in their familiar environment could they have compared animal light with? In this they echoed the reports of their contemporaries, especially tales of the mariners who witnessed burning seas and wondered if they had incurred the wrath or the approval of the gods. It was like the light emanating from fire, only cold. But Pliny's presentation differs from Aristotle's, in that his accounts rely heavily on anecdotes, on what is in animal light for man's use, and on how its manifestations are in tune with the seasons. Apart from myth and legend, this was the sum-total of the worldview of bioluminescence for these Ancients.

As in other spheres of knowledge, the recording of plant or animal light fared poorly in the ensuing Dark Ages and Middle Ages. Harvey (1957) put it succinctly: "During the Dark and Middle Ages no new discovery or new phenomenon important for the history of luminescence had been made. A few mysterious and miraculous stories appeared which were to be later refuted, but the end of the Middle Ages left luminescence information in the same condition as during classic times." All spheres of knowledge suffered. Philosopher of science Joseph Agassi (2008) explained it thus: "The little contribution in the Middle Ages to what we would today consider scientific was made futile by the confusion caused by the medieval method of reconciling all disagreements [in favour of religion] in order to reconcile the conflict between religion and science."

When fireflies were recalled in these epochs, it was generally in a literary, not an academic work. Roda (2011) makes reference "to the two most common firefly species in Japan, the *genji-botaru* and the *heike-botaru*, whose names have an interesting history: it is said that the souls of countless soldiers who were killed in the battle of Dannoura (1185) turned into fireflies. The larger and more profuse *genji-botaru* fireflies were no doubt named after the winning Genji, and the smaller *heike-botaru* after the defeated Heike." In Europe, fireflies similarly made their appearance in literature. "In Canto XXVI of 'Inferno' of *The Divine Comedy*," writes Roda, "one of the last poems he penned before his death in 1321, [Dante] wrote of peering down into the Eighth Chasm of Hell and seeing 'fireflies innumerous spangling o'er the vale.'"

In the wake of the Spanish and Portuguese explorations of the New World in the fifteenth and sixteenth centuries, at a time when curiosity about the natural world was rekindled, an upsurge of reports of luminous organisms occurred. One venue for this revival was the ambitious *Historia General y Natural de las Indias* by the historiographer Gonzalo Fernández de Oviedo (1478–1557). Born in Madrid and a member of the Asturian aristocracy, Oviedo served during his teenage years as a page to Juan, Prince of Asturias and son of Ferdinand and Isabella (Gerbi, 2010). Stints of civil service led to posts in Hispaniola (Santo Domingo) and to his appointment as historiographer of the Indies in 1523. In the latter capacity he produced the *Historia*, a massive inventory of information on the native habits and natural wonders of the New World collected at first hand. The first part was published in 1555, but it took three centuries for the complete work to become available to the reading public (1851–55). The long and eventful saga of this book is vividly recounted by Jesús Carillo (2002).

The new species of bioluminescent animals described in the first volume of the *Historia* are centipedes and insects. Oviedo's accounts are detailed and evocative. In the first quotation below, he describes a luminous centipede and then alludes to a new and exotic insect with a glowing head, later known as the railroad worm:

> There exist in this island of Hispaniola many kinds of centipedes, some of which are thin and as long as a finger, and like those in Spain, they bite and cause enough pain … There are other worms, thin and about half a finger length, with many legs, which shine at night, casting light about and showing the path of travel and which can be seen from fifty to a hundred paces away; and it is not the entire worm that shines, but only the base or joints where the legs come out of the body and the light is very bright. There are other worms that are said to be very similar in size and shape, and said to shine as just described, but they have another great difference, and that is that their heads shine as much, but the light is very vivid and red, and like a glowing ember. (Oviedo 1851, Book XV, Chap. II)

In the second passage Oviedo describes a Caribbean firefly (*luciernaga*, an ambiguous Spanish word that can signify glow-worm or firefly) and the *cucujo*, an elaterid (click) beetle with a strikingly brilliant luminescent display. The length of the following extract is justified by the vivid language Oviedo adopts to extol the beetle's usefulness to man:

> There are many flies, butterflies and beetles in all these islands [West Indies] that shine as they fly at night, like those that are called *glow-worms* and other names in Castille [fireflies] and fly during the summer … In addition there is one special kind called *cocuyo* which is noteworthy. This is a well-known animal in this Island Hispaniola and in the neighboring islands; which is a type of beetle about the size of the first joint of the thumb, or slightly smaller. It has two thick wings, and below these, two thinner ones which are kept beneath the others when not in flight. It has shining eyes like candles, in such a manner that when flying the air is illuminated as by a firelight … If placed in a dark chamber it shines so much that one can see very well reading or writing

a letter; and if four or five of these cocuyos are grouped together they serve as a lantern would in the fields or in the forest in a dark night. When waging war in Hispaniola and other islands the Christians and the Indians use these lights in order not to get lost. And particularly the Indians, as they have greater dexterity to capture these insects, use them to make collars when they wish to be seen a league away. So that in the country and in homes at night, men do whatever they see fit with these cocuyos as neither water nor strong winds will affect their luminosity nor hide their light to find their way. When on an assault party at night, their leader wears a "cocuyo" on his head as a guide for those following. The light that these insects have in their eyes is also present in their body and when opening their wings for flight it gives forth more light from under its wings, as well as from its eyes, joining one light with the other, thus giving forth greater luminosity when on flight. They keep these insects caged for the services of homes for use at night without the need of other lights. The Christians did the same in the past to save money on the oil they had to buy for their lamps, which was either lacking or expensive, at that time. When they saw that the cocuyo was getting thin, or because of the stress of confinement their faculty of producing light was weakening, they turned them loose and got others for the following days. The Indians also stained their hands and faces with a paste made from these cocuyos and when in their celebrations they wanted to have fun, they scared others that were not watching and knew not what it was, by the great light given by the matter stained by these cocuyos. As this insect loses weight or dies, thus its light diminishes and finally disappears. (Oviedo 1851, Book XV, Chap. VIII)

Although Oviedo's *Historia* can be considered partly a work of natural history, he was little more than a dilettante of the discipline. And yet, his book was first published at an important historical juncture – the middle of the sixteenth century – when professional natural historians were emerging as a learned community in Europe. In the words of Brian Ogilvie (2006), it was the time when naturalists came to think of themselves as "practitioners of a discipline that, though related to medicine and natural philosophy, was distinct from both." The two practitioners who best epitomize the naturalists of that period, Conrad Gesner (1516–1565) and Ulisse Aldrovandi (1522–1605), are of special interest to our topic.

Born in Zurich, Gesner earned a medical degree and practised medicine in his native city, where he died of the plague at the age of forty-nine (Gudger, 1934; Pettitt, 2015). As a polymath, he spent his free time gathering and collating information of a broad nature on animals and plants, among other things. In so doing he was regarded as the Swiss Pliny. His lifelong interest in natural history was spurred by an early interest in medicinal plants, and his study of plants led to his authoring the first full monograph dedicated to luminescence. Written in Latin, the language of the Renaissance scholars, and published in 1555, it had an unwieldy title: *A short commentary on rare and marvelous plants that are called lunar either because they shine at night or for other reasons; and also on other things that shine in darkness* (translation in Harvey, 1957). It compiled past and contemporary records of luminous plants for the most part. Gesner was himself skeptical of many of these sightings, rightly suspecting that the apparent light emission resulted from natural light reflected by the plant surfaces. In this monograph and in his monumental work, *Historiae animalium* (1551–56), he also reported on records of luminous animals from Aristotle to his times. Harvey (1957) astutely commented: "It is unfortunate that there was so much repetition of the statements of others, and so little original observation." Gesner even carried over the mistaken notion that the eyes of animals actually emit light.

Another book devoted solely to luminescence, *De luce animalium*, published by Thomas Bartholin, a Dane, in 1647, was no improvement on Gesner's; it was a hodgepodge of religion-fuelled fantasies about human and animal luminescence as well as a rehash of beliefs held by the scholars of Antiquity. The only part of the book deserving of rescue is of great historical importance. In Book II, Chap. II, Bartholin reports on a letter from Sicilian poet, mathematician, and archaeologist Carlo Maria Ventimiglia (1570–1667) to a third party in which mention is made of "a miracle of nature, which denied the [firefly] females wings, but endowed them with a more vigorous light in order that they could call the males at night with their shine" (Harvey, 1957). This is the first account attributing to luminescence a role in firefly courtship.

Ulisse Aldrovandi, the Italian colleague and competitor of Gesner, shared with the latter the encyclopaedic gourmandise that drove their ambitions.

Aldrovandi was born in Bologna to a wealthy family and was educated at the universities of Bologna and Padua (Findlen, 1996). Immediately after earning his degree in medicine and philosophy in 1553, he began his teaching career at the University of Bologna, first in logic and soon afterward in natural history. Besides collecting numerous animals and plants for his innovative museum enterprise, what Findlen (1996) calls his "repository of natural wonders," Aldrovandi made it his goal to do one better than Gesner by writing as comprehensive a compendium of knowledge about the natural world as feasible at the time. But, as Findlen explained, he managed to publish only a small fraction of his encyclopaedic project before his death because "the burden of becoming the latest in a succession of new Aristotles was simply too much for one lifetime."

Aldrovandi did succeed, however, in publishing a monograph on insects, *De animalibus insectis* (1602), in which he recorded instances of luminous species. Like Gesner, Aldrovandi had intended to produce a book wholly dedicated to luminescence but, as he explained: "We, too, have drafted a book on things that shine at night, and will publish it if we live long enough, for we must first finish the ones that are more necessary" (quoted by Harvey, 1957). That manuscript joined the pile of his other unpublished books. The quotation pointedly illustrates how low a priority luminous organisms commanded.

In chapter 7 of the fourth book of *De animalibus insectis*, Aldrovandi starts off writing about the cucujo of the New World, relying on the sources at his disposal (Oviedo and Petrus Martyr). In chapter 8 he expounds at greater length on cicindelae (fireflies), but his text is mostly a rehash of the accounts of Aristotle, Pliny the Elder, and other writers of Antiquity. His only originality is the inclusion of drawings of glow-worms and fireflies, and especially the cucujo (page 495). He speculated on the function of the lantern of the firefly: "Nature has given this light to the cicindelae on their hind part so that they can see in their nightly flight. To us, I admit, it seems very small, but it is sufficient for them to show them their way over the fields in their flight at night" (Harvey's translation).

The great legacy that naturalists of the likes of Gesner and Aldrovandi bequeathed to subsequent generations was to provide them with reliable guideposts as to where the field then stood and what had not yet been investigated (Kraemer and Zedelmaier, 2014). Any future researcher of luminous organisms would hark back to these bibliographers. Two other naturalists of the 1600s worth mentioning, who built on the heritage of the Renaissance duo, are Athanasius Kircher (1602–1680) and Robert Boyle (1627–1691). Kircher, a German-born Jesuit scholar who eventually lived and worked in Rome, was a polymath and polyglot who, in Findlen's words, "had become a bookmaking, knowledge-regurgitating machine" (2004). Although Kircher's great encyclopaedic enterprise was considered a sham by some, what he had to say about luminous animals has the ring of authenticity and genuinely personal interpretation. His observations on the behaviour of the firefly and the functional role of its luminescence from his book *Ars magna lucis et umbrae* (1671) are quoted in Harvey's *A History of Luminescence*:

> I spent some time at Malta, where I found a great multitude of them [fireflies] shining at night, and I collected a large number in order that I might both observe their nature and investigate rather deeply the origin of this kind of living light; and I noted that the animalcule voluntarily, as I might say, at one time drew back and at another put forth that shining matter, as it sensed the presence of a friend or foe; for when it was pinched or moved, it drew back the matter and, after a little while, it brought it forth again. But especially when rather many cincindelae or lampades were put together, then most of all did it display the proud ornament of its shining liquid, as if it were exulting in the ostentatious glory of light. You would say that it was walking around in order to be seen.

Similarly, in his discourse on the jellyfish – probably a scyphomedusa – Kircher speculates that its luminescence serves as an antidote to the darkness of the deep sea in order to ensure that life goes on:

> Mention should be made here of another marvel of the sea, which, although it is nearly the lowliest and most despised of blood-containing animals, yet has not a little nobility by virtue of its innate light. Some call it the Pulmo marinus [sea lung], others Urtica [sea nettle], because its private parts in some marvelous way affect the hands with a burning

itch. I have found the humor of this animal, or zoophyte, so similar to the liquid implanted in the Dactylus that there is scarcely any effect produced by the latter that cannot be produced also by the former. But it is a marvel that the liquid of this Pulmo when rubbed on black sticks and certain other things causes them to shine in darkness no differently than fire: I discovered this by experiment, first at Aquae Martiae near Marseilles, and then again I remember having observed it at Bellonium. Twigs and sticks, when smeared, glowed at night like torches. After this, I discovered that traces of this liquid that is luminous by its own light are implanted in nearly all "fish," but especially in soft Crustacea and testacea. I think that this is the reason why nature wanted to imbue these animals with light: namely, that they should not live in perpetual darkness and seem to have been provided with eyes by nature in vain, since they live in the depths of the sea and cling to sticks, but the depths of the sea are dark and are not reached by the rays of the sun, as divers inform us. Thus, nature gave to these animals this viscous liquid imbued with counterfeit light, that by its help, as it were by a lamp born with them, they might both seek food and also easily elude the snares of foes by the voluntary emission of light and darkness, and thus they might not be destitute of those things that are necessary for their own life.

Robert Boyle was as remote from Kircher as naturalists go. Born in Ireland, he inherited wealth that allowed him the freedom to act the gentleman naturalist – as Darwin did two centuries later. Far from the scholastic tradition represented by Kircher, Boyle was "a natural philosopher who devoted his life to developing the details of a new way of knowing that he called the experimental philosophy" (Sargent, 1995). This new outlook, according to Rose-Mary Sargent, "sought to give experimental practices a rational foundation." As an experimentalist Boyle was instrumental in introducing modern chemistry: indeed, Harvey (1957) went as far as to claim that the chemical study of organic light production began with Boyle. After inventing the "pneumatic engine" (air pump) with Robert Hooke and using it to show that coal fire is extinguished in a medium devoid of air (vacuum chamber), Boyle soon wanted to repeat the same experiment with the cold light of luminous organic material. Through a serendipitous acquisition of luminous wood, fish, and veal flesh made luminous by putrefaction, he reported in two articles of the *Transactions of the Royal Society* (of which he was a co-founder) that this form of light also vanished in a vacuum but was revived after readmitting air in the chamber (Boyle, 1666, 1672).

These experiments led Boyle to conclude "that notwithstanding the coldness (at least as to sense) of fishes and other animals, there may be in the heart and blood a vital kind of fire, which needs air, as well as those fires that are sensibly hot: which may lessen the wonder, that animals should not be able to live when robbed of air" (Boyle, 1666). There was an intimation here, perhaps not acknowledged in so many words, of chemical oxidation as the process behind fire as well as organic phenomena such as light production by fungi, bacteria and animals. This formidable insight and the rational approach it entailed boded well for the kind of science to come in the enlightened period of the eighteenth century.

2

The Age of Enlightenment

> New discoveries, sound experiments, advanced machines, nothing escapes the notice of the scholars of the enlightened century in which we live.
> –Godeheu de Riville, 1754

The keen sense of observation and the embryonic experimentalism practised by Boyle in the seventeenth century was adopted by many of the great eighteenth-century naturalists. The "Age of Reason" or "Age of Enlightenment," as it became known, fostered a revolutionary spirit that profoundly transformed the sciences. As the century unfolded, natural history as it was practised – the *esprit de système* decried by the encyclopaedists Diderot and d'Alembert – gradually ceded the field to a more rigorous science that gave primacy to scientific investigation based on observation and experiment, with a view to discovering new facts and reaching a better-grounded understanding of the natural world (Mazliak, 2006). This movement culminated with the simultaneous coining of a new flag-bearer word – *biologie* – first used in its modern meaning in 1802 by Jean-Baptiste Lamarck in France and Gottfried Reinhold Treviranus in Germany. The spirit of enquiry in this new age greatly affected the approach to the study of luminous organisms and the understanding of how these organisms became – yes – enlightened.

The two prominent chemists of the eighteenth century, Antoine Lavoisier and Joseph B. Priestley, were the true heirs of Robert Boyle in that they pursued his work on combustion, studied respiration, and discovered that something in atmospheric air was specifically responsible for combustion, respiratory or otherwise. Lavoisier pinned a name on that air component – "oxygen" – and that has stuck ever since. This was a crucial discovery that paved the way for the eventual understanding of chemiluminescence as an oxydation reaction. But, as Harvey (1957) explained, both Lavoisier and

Priestley were less forthcoming than Boyle in extending the reach of combustion to light-emitting animals, let alone discoursing at all on living light. Nevertheless, other naturalists continued to explore different facets of bioluminescence.

One such naturalist of note was René-Antoine Ferchault de Réaumur (1683–1757). Born in La Rochelle, he studied and practised mathematics first, then turned to other fields, particularly the natural history of invertebrates (Torlais, 1961). He became a member of the Académie Royale des Sciences de Paris and several times served as its director. His work on the mollusc *Pholas dactylus*, which he calls *le dail*, was published in the *Mémoires de l'Académie Royale des Sciences*. Réaumur's first article describes in detail for the first time not only the anatomy of the mollusc, but also its life history in the coastal environment – one might be tempted to say its ecology (1712). In his second paper (1723), he addressed the luminescence of *Pholas*, with the stated goal of checking Pliny's observations on *dactylus* for himself. He confirmed Pliny's findings and added to them. What particularly struck Réaumur was the ubiquity of the light both outside and inside the body and the association of the light with the copious liquid covering the body inside the shell – and his unexpected way of discovering its properties:

> Their entire surface was luminous. There is no dark spot, every part seemed to shine by its own light. It is not only the external skin layers that show this property, it is shared by all the flesh of the body. Ripped or sliced, the surfaces formed by these separations are as luminous as the others were, in short their whole substance is luminous just like all the fragments of a well-lit coal … The light conferred by these fish to bodies against which they are rubbed is of short duration, it vanishes as soon as the smear on the rubbed body has dried. When I neglected to wash my rubbed fingers immediately, I saw the light intensity they had acquired fade slowly and eventually disappear. But when I got my fingers wet again to wash them, I found them as luminous as they were before.

What Réaumur did not understand, in retrospect, is that the luminous liquid was the result of a secretion by special glands. Like his contemporaries, he was poorly equipped to summon such insights. What was potentially more attainable to him was finding a way to preserve the luminous material in some form for future use, but all his attempts to achieve this were frustrated.

In the same paper Réaumur digressed about other luminous forms as examples of species like *Pholas* that emit light only in the warm season. He mentioned myriapods briefly, but discussed fireflies at greater length. The French naturalist, who seemed to be unaware of Ventimiglia's allusion to the link between the female firefly flashes and male attraction three quarters of a century earlier, offerred a personal observation of their courtship that is more explicit and detailed than the Sicilian's account:

> I was holding this glow-worm [firefly] in my hand during the night, and I was observing the vividness of its light when another insect alighted on my hand. I mistook it first for a species of scarab beetle but I soon corrected my error; it mated right away [with the firefly] and remained coupled for a while. Since then I had several occasions to host other firefly males when I held females in my hand. They also come flying around the candle, and if the latter did not attract butterflies, there would be no reason to doubt that these insects are attracted by the candle as they are by the glow of their females.

Réaumur contributed new information on species already known to be luminous. But what about reports of new luminous species in the course of the eighteenth century? This quest started off on a wrong foot early in the century when the celebrated German entomologist Maria Sybilla Merian reported the existence of a Brazilian insect with a long head process that emitted a bright flash of light strong enough to read by (Merian, 1705). The insect was a plant hopper named *Fulgora laternaria* by the famed botanist Carl Linnaeus after corresponding with Merian. On the basis of Merian's account, it became known as the lantern fly. Kim Todd's biography of Merian, however (Todd 2007), suggests that the circumstances of her discovery appeared nebulous and should have invited skepticism. After all, she was also investigating whether the lantern bug was a stage in the life cycle of a cicada! But the weight of Linnaeus's scholarly reputation helped spread the belief that the insect was luminous. In the following centuries, as increasing numbers of naturalists failed to observe the luminescence, Merian's claim was totally discredited (Brenner, 1885). This is a cautionary tale about how easily false reports of bioluminescence could spread in the absence of careful scrutiny and confirmatory support.

As the century unfolded, other naturalists were more rewardingly vindicated. Giuseppe Vianelli (died 1803), a Venetian physician and poet, wrote a short monograph in which he recorded his observations of a "marine glowworm" (Vianelli, 1749). He collected luminous sea water from the shores of his town (Chioggia) on the Adriatic coast, and by sifting it, he was able to assign the bright light to a worm which he illustrated in his monograph. Its appearance suggests a scale-worm (*Acholoe* or *Polynoe*). Of its luminescence Vianelli had this to say:

> In one respect our marine glowworms excel all their lucid brethren of the terrestrial species, for these latter emit light only at a particular spot near the tail, whereas the whole body of the former is luminous. There is also one further particular to be observed, with respect to these marine animals, which is, that they do not emit the least light so long as they are still and motionless, but the parts of their little bodies are no sooner moved and agitated, that they begin to sparkle with a very extraordinary lustre. From hence may we not conclude, that their shining depends on their motion, and is probably excited by a strong vibration of the constituent parts of their bodies since the luminous effusions, or coruscations, seem to be exactly proportionable to the briskness and vigour of their motion. (Translated in *Gentleman's Magazine* 23: 513–15, 1753)

This appears to be the first record of scale-worm bioluminescence.

Another first soon followed: the discovery of yet another cause of sea phosphorescence, the flashes of the dinoflagellate *Noctiluca*. This was recorded by Henry Baker (1698–1774), an eclectic mind like so many produced by the Age of Enlightenment and a fellow of the Royal Society (Henderson, 1885). Baker made a fortune educating the deaf mutes of the high classes, and his reputation travelled to Daniel Defoe of *Robinson Crusoe* fame, with whom he founded a weekly journal and whose youngest daughter he married. A poet at first, Baker developed an interest in natural history, particularly small species whose study required the aid of the microscope. Adept at using the instrument, he wrote two books (1743, 1753) of microscopical observations in which are found descriptions of luminous "animalcules." In his 1753 paper he recorded the description that a correspondent transmitted to him of an organism that seems to correspond to *Noctiluca*, although no illustration was appended:

> In the glass of sea water I send with this are some of the animalcules which cause the sparkling light in sea water; they may be seen by holding the vial up against the light, resembling very small bladders or air bubbles, and are in all places of it from top to bottom, but mostly towards the top, where they assemble when the water has stood still some time, unless they have been killed by keeping them too long in the vial.

The luminescence was said to be blue, a colour that matches that of *Noctiluca* and distinguishes it from that of other suspects also described by Baker which resemble his "water-insects," possibly syllid or polynoid worms known to emit a green light. An illustration found in a monograph by Martinus Slabber (1740–1835), a Dutch jack-of-all-trades who also dabbled in natural history (Kingsley, 1879), shows for the first time a recognizable *Noctiluca* meeting Baker's description (Slabber, 1778).

The ostracod *Cypridina*, another species that figures prominently later in this book, was discovered in 1754 by Godeheu de Riville. Little is known of this Frenchman – not even his first name – apart from the fact that he was a member of the French nobility and a Chevalier de Malte, a religious and civic order founded during the Crusades. We also know that he corresponded with Réaumur (Torlais, 1961). Godeheu's *Mémoire sur la mer lumineuse* (1760) gives us an unparalleled contextual description of how a naturalist of that period went about this kind of quest, so we shall give it due attention here.

During a trip abroad with his brother, Godeheu was actively looking for a luminous sea he could investigate at length. He was told that luminous seas shone particularly brightly in the region between Malabar in the state of Kerala in India and the Maldives Islands southwest of the southern tip of India and Sri Lanka. Aboard a ship sailing in the Laccadive Sea, they happened upon some vividly luminescent waters, which Godeheu likened to sparkling snow:

> The weakly agitated sea seemed covered with starlets; each wave that broke around us spread a vivid light, similar in color to a silvery cloth electrified in the dark. The waves, which seemed to blend into each other as they moved away in the distance, formed on the horizon a snow-covered plain; and the wake of our vessel, the brightness of which lasted for a long time, was of a luminous white, strewn with brilliant and bluish dots.

These dots intrigued him, so he had a sample of the seawater brought to him and he filtered it through a thin cloth into a flask. Noticing that the filtered water had little luminescence left and that luminous "fish eggs" littered the cloth, Godeheu proceeded to examine the latter with a strong magnifying glass. What he saw mesmerized him:

> I saw instantly a large number of small insects which swam rapidly and which resembled at first sight what in France we call *Puces d'eau* [water fleas]. In spite of their agility, I managed to trap one against the side of a cup with a pair of tweezers. This slight pressure was apparently too much for such a delicate insect; it was quite hurt and despite the light coming from two candles we saw a bluish luminous fluid coming out of its body and streaming linearly into the water. I maintained my grip on the insect but then released it from my tweezers, and no sooner was it placed under the microscope than it ejected again a large amount of this bluish fluid … I kept many which appeared sluggish the next day, but by changing the water they revived. The bright fluid, of which they have large supplies, was not altered, for having kept for some time at the tip of my tweezers one that was destined for microscopic examination, it broadcast a glow that lasted seven or eight seconds and which in full daylight was visible to persons two or three feet away.

Godeheu also noticed that the dried secretion of this animal failed to luminesce but that soaking it with water revived the light. (As we shall later see, this characteristic would be found useful in the future.) Godeheu enlisted an officer on board to draw specimens, and those illustrations accompany the text. The draftsman was clearly gifted, as the striking images he rendered allow an unequivocal identification of the so-called insect as the ostracod *Cypridina*, of which more later. At the end of his *Mémoire* Godeheu urged chemists to look into this phenomenon. It would take two more centuries for this wish to be fulfilled.

A few years later another grand traveller observed luminescent medusae and, for the first time, luminescent crustaceans (probably shrimps). Joseph Banks (1743–1820) needs little presentation as the botanist who accompanied Captain James Cook on the first voyage of the HMS *Endeavour* and was later instrumental in the colonization of Australia. Banks, a member of the Royal Society, was appointed at the age of twenty-five as a naturalist on the voyage of the *Endeavour* sponsored by the Royal Navy and the Royal Society (Holmes, 2009). He paid to have the Swedish-born botanist Daniel Carlsson Solander accompany him. Early in the voyage, on 29 October 1768, as the *Endeavour* was crossing the Atlantic from Madeira, the ocean made a display of lights that attracted the keen attention of Banks and Solander:

> This evening the sea appeard uncommonly bea[u]tifull, flashes of light coming from it perfectly resembling small flashes of lightning, and these so frequent that sometimes 8 or ten were visible at the same moment; the seamen were divided in their acco[u]nts some assuring us that it proceeded from fish who made the light by agitating the salt water, as they call it, in their darting at their prey, while others said that they had often seen them and knew them to be nothing but blubbers (Medusas). This made us very Eager to procure some of them, which at last we did one by the help of the landing net. They prov'd to be a species of Medusa which when brought on board appeard like metal violently heated, emitting a white light; on the surface of this animal a small Lepas was fixd exactly the colour of it, which was almost transparent not unlike thin starch in which a small quantity of blue is disolv'd. In taking these animals three or 4 species of Crabbs were taken also but very small, one of which gave light full as much as a glow-worm in England tho the Creature was not so large by 10/9ths; indeed the sea this night seemd to abound with light in an uncommon manner, as if every inhabitant of it furnishd its share, which might have

been the case tho none kept that property after being brought out of the water except these two.

It is puzzling that Banks recorded this bioluminescent event, which lasted into the next evening, so early in the voyage, but that no comparable record followed for the next three years of roving the seas of the globe. But thirty-three years later, Louis Bosc (1801) reported another such event in the Atlantic, which records the first sighting of the luminescence of a comb-jelly (ctenophore). Louis Augustin Guillaume Bosc (1759–1828) had a very eventful life, as detailed by his colleague the Baron Augustin-François de Silvestre (1829). A Parisian, the son of a medical doctor, he spent all his free time studying natural history while assuming various posts in government offices. The French Revolution appealed to him but his party (Girondins) was persecuted in 1793 and he was forced to go into hiding. During the Directoire he took consulate posts in the United States (1796–98), and it was during his Atlantic crossing in 1796 that he witnessed the luminescence of comb-jellies (*Beroe*). He wrote that the *Beroe* species "are all phosphorescent; they scintillate in the night like so many lights, and their brightness is proportional to the speed of their movements. The tentacles are more active in this regard than the body surface." He returned to France to fill various government posts in agriculture and horticulture, and died in 1828 after a long illness.

All the naturalists noted above can be credited with representing the kind of natural history practised in the Siècle des Lumières, with their more reasoned approach to the study of their material and their willingness to travel and explore the world to see for themselves the treasures it contains. But some of the *savants* of the period broke ranks even with these enlightened naturalists and became the vanguard of a new scientific approach that on the one hand created a distinction between zoology, botany, and paleontology as sources of specimen collection and knowledge (Findlen, 1996), and on the other hand saw the need to build a scientific foundation for experimental biology (Rostand, 1951). This apotheosis of the scientific development of the eighteenth century is best embodied in the person of Lazaro Spallanzani (1729–1797).

Spallanzani was born in Scandiano near Reggio Emilia in northern Italy. His early schooling took place at home under his father's guidance, followed by studies at the Jesuit College of Reggio (Rostand 1951). He completed his education at the University of Bologna, where his interest in science was aroused by the lectures in experimental physics of Laura Bassi, his cousin and the first woman to hold an academic chair in a European university. His first teaching position, in Reggio, was followed by posts at the University of Modena and finally at the University of Pavia which he earned thanks to his groundbreaking work debunking the theory of spontaneous generation. Other biological fields in which his experimental skills added to his reputation are blood circulation, digestion, respiration, reproduction, and regeneration.

Of course, Spallanzani did not fail to be intrigued by the phenomenon of living light and he addressed it in three contributions. In the first, a letter to the French biologist Charles Bonnet (Spallanzani, 1784), the first section of which is titled *Luce notturna del mare* ("Nocturnal Light of the Sea"), Spallanzani tried to emulate his predecessors Vianelli and Godeheu de Riville and name specific animals as the source of the sea "phosphorescence." Sampling the water in the Gulf of Spezia (Portovenere), he found no luminous organism remotely resembling the crustacean described by Godeheu, but was able to confirm Vianelli's finding of a "marine glow-worm." To the then prevailing view that marine luminosity was the result of the putrefaction of marine animals – due to oily phosphorescent substances exuding on the surface of the corpses – Spallanzani opposed his own findings and power of reasoning. The oily substances, by their nature, can only float at the surface, yet Spallanzani found a diffuse luminescence at greater depths in the sea. He concluded that the phenomenon was more complex than previously portrayed. He could only offer conjectures, he said, "which I reserve to commit to experiment at the first opportunity presented to me." The opportunity, it seems, never materialized.

Spallanzani's second contribution is a description of jellyfish bioluminescence (Spallanzani, 1794). In this memoir, he first described the anatomy and pulsatile swimming of an (unidentified) jellyfish so as to prepare the reader for his contention that the luminescent flash of the animal was associated with the swimming cycle. To describe the pulsations of the swimming jellyfish he used the terms "systole" and "diastole," and in so doing created an analogy with the heart cycle, an analogy that was resurrected and investigated almost a century later by Romanes and the Cambridge School of Physiology (French, 1970; Geison, 1978). Spallanzani noted that the light emission was restricted to the margin of the bell and the tentacles but also that the light showed off the whole animal in the dark. He observed a brighter luminescence during the systole than the diastole and concluded that luminescence was co-activated in tandem with systolic activity. It is likely that the movement resulting from the systole caused mechanical stimulation, which in turn elicited the light emission. Spallanzani also noted that the luminescence of dead jellyfish liquefied on cloth was revived by freshwater but not by seawater.

In his final contribution Spallanzani addressed the question first raised by Boyle more than a century earlier: what was the relationship of phosphorescence to combustion? He was drawn to it by a controversy initiated by Johann Friedrich August Göttling (1753–1809), a German chemist working at the University of Jena who claimed that nitrogen can revive the light emission of phosphorus. By a series of clever experiments, Spallanzani (1796–67) debunked Göttling's contention. Not only is oxygen the only essential gas for "phosphorescence," he claimed, but the light emission of "shining wood," rotten fish, and live fireflies follows the same rule. Spallanzani curiously suggested that the bioluminescence arises from the slow combustion of hydrogen gas and hydrocarbons present in the organisms.

Spallanzani's brand of scientific investigation was poised to take hold as an enterprise of grander scale in the next century. The investigations of bioluminescence were caught up in this scientific spirit, and the field blossomed in various directions, as we shall see.

3

A Deeper Probing of Nature Aglow

> Although the phenomenon of illumination by some animals garners an extraordinarily great wealth of information, it does not seem ripe yet for decisive understanding.
> –Johann Friedrich Will (1844)

Just as the spirit of the "Siècle des lumières" served as the inescapable springboard for the scientific enterprise of the nineteenth century, so it unfolded for the followers of Réaumur and Spallanzani in investigating bioluminescence. The experimental approach prospered as the tool for understanding the mechanism of light production alongside a proliferation of observations of light emissions from an ever-increasing number of recorded luminescent species worldwide. This chapter examines these currents as they evolved in the first half of the nineteenth century. A new trend emerged early in the century – global circumnavigations with scientific missions that to some extent emulated the pioneering voyage of Captain James Cook.

We have seen in the previous chapter how observations of bioluminescence in the open sea were conducted by gentlemen and amateur naturalists who happened to travel on ships trading the commercial sea routes. In the new century professional naturalists took the stage, some appointed on board vessels commissioned to navigate the globe and make anthropological observations, and take surveys of coastal hydrography and marine and land-living organisms. The first of these expeditions involved two French ships, *Géographe* and *Naturaliste*, under the command of Nicolas Baudin (1754–1803), which left port in October 1800, travelled to Australia, and returned to France in 1804. They arrived home without their captain, who had died of tuberculosis on Mauritius Island, but they still had on board the voyage's naturalist, François Péron, with his reports on luminescent forms observed during this trip.

The historian Ralph Kingston, in his tellingly entitled paper "A not so Pacific voyage" (2007), has used Baudin's command to illustrate the tensions inherent in these early "floating laboratories." Before such science-oriented voyages, it was the surgeon on board who played the self-appointed role of the naturalist, as exemplified by Dr Maturin in Patrick O'Brian's novels and Peter Weir's 2003 movie adaptation, *Master and Commander*. But in the transition represented by Baudin's expedition, the surgeons-cum-naturalists coexisted with professionally trained naturalists. This coexistence created fierce and bitter competition over who would have first right (and consequently credit of discovery) over captured specimens new to science. Péron was one such naturalist embroiled in conflict with a ship's surgeon.

Captain Baudin also had to deal with contentious social as well as professional issues. Some officers and naturalists were displeased "at finding themselves under the leadership of a social inferior – a man who (worse still) had started his career in the merchant navy" (Kingston, 2007). As an amateur naturalist himself, Baudin could hardly portray himself as a credible arbiter of disputes, and the hydrographers spread rumours that Baudin "would rather collect mollusks than chart the Australian coast" (Kingston, 2007). Unsurprisingly, several defections occurred along the way, in addition to deaths from deprivation and illness.

It was in this tense atmosphere, with men rubbing shoulders the wrong way for more than three years, that François Péron (1775–1810) laboured. Born like Baudin to lower-class parents, Péron enlisted in the Revolutionary Army at the age of seventeen and was made prisoner by the Prussians. When he was released three years later, the deprivations of prison had caused him to lose one diseased eye (Girard, 1857). Notables in his hometown, aware of his intellectual potential, sponsored his studies of medicine in Paris. But Péron soon diverted his interests to zoology and anthropology, spending time at the Muséum d'histoire naturelle, where he fell under the spell of Georges Cuvier. When the Baudin Expedition called for naturalists, members of the museum supported his application as an anthropologist on the voyage. He was accepted as a zoologist and gave up his medical studies.

The *Géographe* was sailing in equatorial waters of the Atlantic when it came upon what Péron described as "a large stretch of phosphorescence laying on the waves" (1804a). He soon asserted that this "romantic, imposing and majestic" event "was caused solely by the presence of countless numbers of large animals" that resembled "large cylinders of incandescent iron." Although he described luminescent structures that suggest zooids, he seemed unaware that the observed pyrosome (*Pyrosoma atlanticum*) was a tunicate colony. In Péron's view the dynamics of light emission were associated with a regular contraction cycle; the luminescence was brightest during the contraction phase and gradually disappeared during relaxation. He hypothesized that the duration and intensity of the luminescent phase were enhanced by mechanical stimulation. This account by Péron constituted the first record of pyrosome luminescence. The expedition's long periods of boredom on the high seas were occasionally relieved by entertaining displays of luminosity on the open sea. Péron concluded that such phosphorescent displays are found indiscriminately in all oceans and seas, that the displays are brightest in agitated seas but are still present in calm waters, that the higher temperatures of tropical seas lead to brighter displays there, and finally that more intense displays occur in coastal areas and bays (Girard, 1857).

Péron went on to collect an astounding number of zoological specimens and came back with many anthropological drawings and descriptions of natives, especially in Australia and vicinities. All this was accomplished against a backdrop of hostility from Captain Baudin and hardships that threatened his life more than once. Péron was lucky to be the only zoologist who survived the trip (Girard, 1857). However, his diminished health hastened his death, which occurred only six years after his return.

The Baudin expedition had not yet returned before another circumnavigation with a naturalist on board, which contributed new observations on marine bioluminescence, was underway. The first Russian circumnavigation (1803–06) came about at the instigation of the Russian-American Company, to which the czar had granted the monopoly of the fur trade in their Alaskan colony. Reaching the colony by land was proving difficult, so the primary aim of the voyage was to test the viability of a sea route for trade exchange (Krusenstern, 1813). The scope of the original plan was enlarged, however, to make room for a *bona fide* circumnavigation, to drop the appointed Russian ambassador at his destination in Japan, and to conduct scientific inquiries along the way. Heading the voyage was Adam Johann von Krusenstern, an Estonian career naval officer of German extraction, who captained the vessel *Nadezhda*. The captain of the second participating vessel, the *Neva*, was Yuri Lisiansky, who had previously served under Krusenstern.

One of two physician-naturalists on the expedition was Wilhelm Gottlieb Tilesius von Tilenau (1769–1857). Born in Mühlhausen, Germany, to a family of old nobility, Tilesius was introduced to natural history in his childhood by an uncle (De Bersaques 2011). He studied natural science and medicine at the University of Leipzig, all the while acquiring skills in drawing and music. Frustrated by failures to obtain a university professorship, he accepted a well-paid offer as a naturalist on the *Nadezhda*. Tilesius was boastful about his appointment:

> I did not do it for the money but because of the opportunity; I was really in my element. Even v. Krusenstern had to admit, no one accomplished as much as I did. Despite the many dangers and suffering, I consider these 3 years as the happiest of my life, because in these three I accomplished more than I had in other decades, even though I only collected raw material which will keep me busy for the rest of my life. (Translated in De Bersaques, 2011)

While not as dysfunctional as Baudin's voyage, Krusenstern's had its own tensions, as would be expected when so many inflated egos were crammed into a tight space. As De Bersaques (2011) noted, "Krusenstern mentioned that Tilesius was not the easiest person to get along with and was very sensitive about his status. Tilesius refused to do things outside those mentioned in his contract … His abundant luggage, scientific material, chemicals and other paraphernalia annoyed the other members, as space was very limited on the ships." Tilesius ended up conducting both zoological and anthropological investigations as Péron did, but in addition he drafted his own superb drawings of specimens.

In the course of the voyage Tilesius encountered many manifestations of tropical sea luminescence, using apt metaphors to describe the spectacle (Tilesius, 1819): dull glimmer of light, widespread milk lustre, individual stars, fireballs, light cone, fiery chains, sparks. With microscopes on board, he was able to identify an astounding number of luminous forms: mollusks, crustaceans, polychaetes, salps, jellyfish (*Aurelia* and *Aequoria*), comb-jellies, siphonophores, sea pens, and dinoflagellates; and he gave us the first description of luminous radiolarians. While to him each form produced light in its own way, this phenomenon seemed associated with contractile, respiratory cycles, as Péron had asserted before. To Tilesius it was as if the animals breathed out the light.

After the voyage Tilesius was granted a pension by the Russian government and he remained a few years in St Petersburg, where he was associated with the Imperial Academy of Sciences and published several papers on his findings (De Bersaques, 2011). When he returned to live in Germany (Leipzig and Mühlhausen), however, Tilesius failed again to obtain an academic appointment. He died a bitter man in 1857.

Over a decade after the Russian expedition, between 1817 and 1820, another French circumnavigation was undertaken. One ship, the *Uranie* commanded by Louis de Freycinet (1779–1841), accommodated geographers, hydrologists, and two zoologists/anthropologists – Jean René Constant Quoy (1790–1869) and Joseph Paul Gaimard (1796–1858) – who made observations on marine bioluminescence (Freycinet, 1827). The stated purpose of the voyage was to pursue the quest for knowledge of the earlier circumnavigations, to update information on earth magnetism and meteorology, and to enlarge the number of zoological specimens and the database on human races. Freycinet keenly acknowledged the contributions of Quoy and Gaimard, who brought back a rich zoological collection that included numerous new species, despite the loss of some of the material when the *Uranie* was shipwrecked on the Falkland Islands, then named Îles Malouines.

Quoy was born in a small Vendée town to a family that included several medical professionals. He obtained his medical degree at the University of Montpellier and, even though he suffered terribly from sea sickness, immediately embarked on a career as a naval surgeon (Maher, 1869). He was appointed chief surgeon and naturalist on Freycinet's voyage and chose Gaimard, who had studied naval medicine in Toulon, as his assistant. The pair became particularly known for their anthropological observations of the Papua natives.

In an article dedicated to the "phosphorescence" of various seas during the voyage, Quoy and Gaimard (1825) remarked that some of the luminous organisms appeared to have no control over their light emissions, whereas others did. They noted that in calm waters only luminescence produced by "large mollusks" was visible, whereas in agitated waters the entire surface scintillated. "If at these moments playful dolphins swim around the ship," they wrote, "we see them developing snakelike trails underwater resembling fireworks; when they surface to breathe this illusion is enhanced, and it seems as if one sees and hears the explosion of a rocket." In the waters around Rawak Island near Papua-New Guinea, they observed a swarm of what they called zoophytes but which seem from their description to have been ostracods or copepods:

> By observing them [phosphorescent streaks] we realized they were produced by Zoophytes of tiny size which possessed a phosphorescent principle so subtle and so expandable, that by swimming rapidly and in zigzags they left in their wake dazzling trails, at first one inch wide, but which widened to two or three inches with wave movement ... The fluid produced by these animals was expulsed at will; suddenly one saw a luminous dot shoot out of their surface and expand with stupendous speed.

As a consequence of their impressive accomplishments on the *Uranie*, Quoy and Gaimard were called upon to participate in another circumnavigation, this one led by Jules Dumont d'Urville between 1826 and 1829 (Maher, 1869). However, no observation of luminescence emerged from this voyage. Subsequently Quoy continued to serve in the highest ranks of naval medicine. He used to say, "I like to see patients only on board vessels and in hospitals: docile, submissive, almost always grateful and militarily lined up (1869)." Guimard had a different fate: he was sent to missions in Scandinavia, Iceland, and in the far north, but did not get the recognition showered on Quoy. He died destitute in Paris.

The fourth expedition worthy of mention is the voyage of the *Prinzess Luise* between 1830 and 1832, in which Franz Julius Ferdinand Meyen (1804–1840) participated. Born in Tilsitt in what was then East Prussia, Meyen studied medicine at the Friedrich Wilhelms Institut in Berlin while learning natural history on the side (Querner, 2008). He served as military physician and moonlighted as a naturalist, publishing several scientific articles and the first treatise on plant anatomy in 1830. On the strength of these remarkable achievements at the age of twenty-six, Meyen was recommended by Alexander von Humboldt to serve as the first (and only) naturalist on the *Prinzess*

Luise, one of the Royal Prussian Sea Company's ships. Harry Liebersohn gives some historical details:

> With a small fleet of ships, the company tried to export linen, wool, and other goods to China, South America, Mexico, and the West Indies. The company was never able to compete with the quality or quantity of British goods flooding the world markets; its very existence was a sign of the backwardness of Prussian industry, which needed to turn to the state for the capital and initiative to undertake such a venture … Unlike the great French, British, Russian, and Austrian voyages, with their teams of scientists specially commissioned for scientific research, Meyen went alone on a trip that was supposed to pay for itself. Under the circumstances he did admirably, bringing back large collections for the zoo, the anatomical museum, the botanical garden, the mineralogical museum, and the royal library … Karl von Stein zum Altenstein, the responsible Prussian minister, declared himself very satisfied with Meyen's industry in his report to Friedrich Wilhelm III and arranged for a modest two-year stipend so that Meyen could write up his travel account, which duly appeared in two volumes. (Liebersohn, 2009)

During his trip Meyen also wrote an article of almost a hundred pages devoted entirely to the topic of marine bioluminescence and sightings of luminescent animals (Meyen, 1834). He saw an impressive number of luminescent species, ranging from various radiolarians and hydroids (*Campanularia=Obelia*) to copepods, and wrote the first description of a luminous siphonophore (diphyid). In the latter the light, which according to Meyen seemed to arise in the tentacles, was less bright than in luminous jellyfish. Meyen included illustrations of a crustacean that he named *Carcinium opalinum*, which showed a pair of light organs on the dorsal side of a posterior abdominal segment. The position and cupular shape of the organs suggests a copepod of the current genus *Metridium*. Meyen found that the animal displayed at night a glossy pale green light very similar to that of pyrosomes, but that the light organs from which this light emanated appeared yellowish in the daytime. Meyen also noted, backed by an illustration, that the crustacean's light organs appeared in close association with the nervous system and that the light emission was controlled at will by the animal.

The fifth voyage considered here is doubtless the most famous. Charles Darwin kept a journal of his almost five years (1831–36) sailing around the world with HMS *Beagle*. His journal with accompanying remarks (Darwin, 1839) was published as the third instalment of the *Narrative of the Surveying Voyages of His Majesty's Ships Adventure and Beagle*. In it he recorded no fewer than six episodes of bioluminescent occurrences, according to the compilation of British bioluminescence expert Anthony Campbell (2012). Campbell discovered an entry on 6 January 1832 that was not included in Darwin's book. In it Darwin described a luminous sea near Tenerife which Campbell attributes to dinoflagellates:

> The sea was luminous in specks & in the wake of the vessel, of a uniform, slightly milky colour … When the water was put into a bottle, it gave out sparks for some minutes after having been drawn up … When examined both at night and next morning, it was found full of numerous small (but many bits visible to the naked eye) irregular pieces of (a gelatinous?) matter. The sea next morning was in the same place equally impure. (Quoted from Darwin's zoology notebook held at Cambridge University Library)

In May or June 1832, while visiting Botafogo Bay in Rio de Janeiro, Brazil, Darwin made observations of great accuracy on a species of lampyrid glow-worm, which he documented in discerning detail:

> At these times [after dark] the fireflies are seen flitting about from hedge to hedge. All that I caught belonged to the family of Lampyridae, or glow-worms, and the greater number were *Lampyris occidentalis*. I found that this insect emitted the most brilliant flashes when irritated … The flash was almost co-instantaneous in the two rings [abdominal segments], but it was first just perceptible in the anterior one. The shining matter was fluid and very adhesive: little spots, where the skin had

> been torn, continued bright with a slight scintillation, whilst the uninjured parts were obscured. When the insect was decapitated the rings remained uninterruptedly bright, but not so brilliant as before: local irritation with a needle always increased the vividness of the light … From these facts it would appear probable that the animal has only the power of concealing or extinguishing the light for short intervals, and that at other times the display is involuntary. On the muddy and wet gravel-walks I found the larvae of this lampyris in great numbers: they resembled in general form the female of the English glow-worm. These larvae possessed but feeble luminous powers; very differently from their parents, on the slightest touch they feigned death, and ceased to shine; nor did irritation excite any fresh display.

At about the same period, but in Bahia (Brazil), Darwin also observed a click beetle which he identified as *Pyrophorus luminosus*. In December 1833, as the *Beagle* sailed past La Plata, Argentina, toward Cape Horn, Darwin described in his inspired prose the spectacle of the luminous sea:

> While sailing a little south of the Plata [Argentina] on one very dark night, the sea presented a wonderful and most beautiful spectacle. There was a fresh breeze, and every part of the surface, which during the day is seen as foam, now glowed with a pale light. The vessel drove before her bows two billows of liquid phosphorus, and in her wake she was followed by a milky train. As far as the eye reached, the crest of every wave was bright, and the sky above the horizon, from the reflected glare of these vivid flames, was not so utterly obscure, as over the rest of the heavens.

Darwin spoke of particles "so minute as easily to pass through fine gauze" as the probable agents of this luminous sea. (Only much later was the true nature of the particles of these mysterious "milky seas" discovered, as we shall see in chapter 16.) Darwin also mentioned other luminous organisms observed at about the same period, stating, for instance, that, "having kept a Medusa [jellyfish] of the genus Dianaea, till it was dead, the water in which it was placed became luminous. When the waves scintillate with bright green sparks, I believe it is generally owing to minute crustacea. But there can be no doubt that very many other pelagic animals, when alive, are phosphorescent." On that score Darwin proved prescient.

Finally, Darwin wrote of the luminescence of a hydroid colony, commonly called zoophytes in his days, observed in March 1834 near Tierra del Fuego. He believed that the "zoophyte" was related to the genus *Clytia*. The accuracy of the following description was confirmed over 130 years later in the hydroid *Obelia geniculata* (Morin and Cooke, 1971a):

> Having kept a large tuft of it in a basin of salt water, when it was dark I found that as often as I rubbed any part of a branch, the whole became strongly phosphorescent with a green light: I do not think I ever saw any object more beautifully so. But the remarkable circumstance was, that the flashes of light always proceeded up the branches, from the base towards the extremities.

Darwin mentioned the green colour of the light emission. What's more, in a footnote he noted how remarkable it was "that in all the glow-worms, shining elaters [click beetles], and various marine animals, which I have observed (such as the crustacea, medusa, nereidae, a coralline of the genus Clytia, and Pyrosoma), the light has been a well-marked green colour." He seemed unaware that other colours of bioluminescence do exist. Whether his observed animals all emitted in the green range is uncertain, as no spectrophotometer was available in his days to measure the spectral range of the lights more objectively. Observers like him and many others were left with their own devices: their dark-adapted eyes functioning principally with retinal rod sensitivity, which does not discriminate colours. So, a number of the old reports about the colour of bioluminescence must be taken with a small dose of skepticism.

Darwin's experiences of observing luminous displays left only a small trace in his epochal book, *On the Origin of Species* (1859), in which he wrote of his puzzlement at seeing luminous organs randomly distributed among various phyletic groups. As Campbell (2012) explained, with the benefit of hindsight, "He couldn't see how small change by small change could lead, apparently out of the blue, to a completely new phenomenon such as the electric organs of fishes or the luminous organs of fireflies and jellyfish through his BIG idea of natural selection." It so befuddled him that he never developed the subject further.

~~~

The sixth and last circumnavigation of the half-century that recorded marine bioluminescence was the whaling voyage of the *Tuscan* (1833–36), again with only one naturalist on board. The appointed surgeon for the voyage, Frederick Debell Bennett (1806–1859), was born in Plymouth, Devon, to a wealthy and well-educated family (Bouchier, 1983). Bennett earned his medical diploma in 1829, by which time he could also boast a well-rounded knowledge of natural history and an affiliation with the Zoological Society of London. At twenty-seven he embraced the opportunity of studying whales and other animals by embarking on *Tuscan* as ship's doctor. He was the only naturalist aboard, his companions being three missionaries and their wives posted to the South Seas. The results of his observations were published in a narrative of the voyage (Bennett, 1840) that became a bestseller and is still read today, no doubt thanks to the attraction of its main focus, whales.

Bennett proved to be a trailblazer for his original observations on the bioluminescence of two kinds of fishes. As E. Newton Harvey (1957) pointed out, so far all records of "phosphorescent" fish could be attributed to one of three sources: light reflected from eyes or scales, luminescent bacteria growing on rotten fish, or dinoflagellates sparkling on the fish surface as it swims, the latter remarked on earlier by Quoy and Gaimard. In the first volume of his book Bennett reported unambiguously on the bioluminescence of lanternfish captured in a tow-net:

> They were a species of Scopelus, three inches in length, covered with scales … When handled or swimming, they emitted a vivid phosphorescent light from the scales, or plates, covering the body and head, as well as from the circular depressions on the abdomen and sides … The luminous gleam (which had sometimes an intermittent or twinkling character, and at others shone steadily for several minutes together) entirely *disappeared* after the death of the fish.
~~~

Two years before Bennett's book appeared, Anastasio Cocco (1799–1854), a physician and naturalist born and living in Messina, Sicily, recounted new observations of species of "salmonids" from the straits of Messina. Cocco in fact was the first to alert the scientific community to the unusual confluence of currents in these straits, which brings deep-sea animals to the surface by tidal upwelling, thus creating a bonanza for zoologists (Groeben, 2008). In a letter to his friend Prince Charles Lucien Bonaparte, nephew of the emperor and a noted ornithologist, Cocco described and illustrated the "punta lucidi" on the flank and belly of lanternfishes (*Scopelus*, *Myctophum*) and other so-called salmonoid forms, the stomiatoids *Gonostomus* and *Maurolicus* (Cocco, 1838). Harvey's statement that Cocco was the first on record to see the light organs (photophores) of these fishes appears accurate, but Harvey made no mention of their luminescence. He missed the fact that Bennett was the first to report on their light emission.

Harvey did, by contrast, duly record Bennett's (1840) discovery of the bioluminescence of a shark. Bennett speculated that the shark probably lived in deep water but managed to maintain one alive in a tank. He was dazzled by the display:

> When the larger specimen, taken at night, was removed into a dark apartment, it afforded a very extraordinary spectacle. The entire inferior surface of the body and head emitted a vivid and greenish phosphorescent gleam, imparting to the creature a truly ghastly and terrific appearance. The luminous effect was constant, and not perceptibly increased by agitation or friction … The small size of the fins would appear to denote that this fish is not active in swimming; and since it is highly predaceous, and evidently of nocturnal habits, we may perhaps indulge in the hypothesis, that the phosphorescent power it possesses is of use to attract its prey.

Bennett was not too far off the mark with regard to function, since luring prey is now considered one of the likely roles of shark bioluminescence, as our narrative will later reveal.

For the growing number of less-adventurous naturalists plying their trade in coastal environments or on land, there was no shortage of new luminescent species begging to be discovered. Reflecting the preponderance of aquatic over terrestrial luminescent species, reports of new marine species outnumbered those of land species. Several, from Britain and European countries, are worthy of historical interest.

The earliest report of coastal marine species came from Charles Stewart, of whom nothing apart from his works is known except that he was a Scotsman from Edinburgh and a member of the Linnean and Wernerian societies. He wrote a comprehensive treatise on the fauna of the British Isles (Stewart, 1802), in which the light emission of a hydroid, the sea oak *Sertularia pumila*, is first recorded. However, James Morin (1974) considered this identification "doubtful owing to the confused systematics of that time," and Stewart's *Sertularia*, which is now known not to be luminous, probably was a hodgepodge of species that included the bioluminescent *Obelia geniculata*. Be that as it may, Stewart observed that the luminescent creatures: "give out phosphoric light in the dark. If a leaf of the above Fucus, with the Sertularia upon it, receive a smart stroke with a stick in the dark, the whole Coralline is most beautifully illuminated, every denticle seeming to be on fire." Péron (1804b) was to make similar observations two years later.

Shortly after Stewart's effort, the Italian Domenico Viviani (1772–1840) compiled his own list of observed luminous species, two of which were new additions. Viviani was appointed professor of botany and natural history at the Université impériale de Gênes (Genoa) in 1805, the very year the university was created by decree of Napoléon I (Pellerano, 2013). That same year he published his paper on luminous animals of the Ligurian Sea (Viviani, 1805). Viviani wrote his paper in Latin, which by then had become an anachronistic custom among his contemporaries. In it he listed *Cyclops* (probably a copepod) and various polychaete worms, but the new material included a brittle star to which he gave the name *Asterias noctiluca*, and an amphipod (*Gammarus crassimanus*). Viviani likened the luminescent display of the brittle star to a sparkling star in the water. In contrast, the luminescence of the amphipod was not readily displayed and, to his taste, less than impressive. He gives nothing that approaches a scientific description of the light emissions.

The British naturalists James Macartney (1770–1843), John MacCullough (1773–1835), and John Murray (1778–1820) published within a few years of each other accounts of marine bioluminescence. Altough they worked separately, Macartney and Murray shared the practice of slosh-and-slop experimental research common to learned men lacking the sharp scientific bent of, for example, their contemporary and compatriot Sir Humphry Davy. Macartney, an Irish surgeon who taught comparative anatomy and was instrumental in publishing the first translation of Georges Cuvier's monumental work on the subject (MacAlister, 1900), focused his observations on jellies found at various intervals on British coasts: two jellyfish species and a comb-jelly (Macartney, 1810). He ascribed the luminescence of the comb-jelly, which was enhanced by mechanical stimulation, to its eight "ribs" (comb-plates). He found in jellyfish that "the central part [of the bell] and the spot around the margin are commonly seen to shine on lifting the animal out of the water into the air, presenting the appearance of an illuminated wheel, and when it is exposed to the usual percussion of the water, the transparent parts of the body are alone luminous." He refuted any notion that marine luminescence had a source other than live organisms. Macartney subjected jellyfish to a variety of crude experiments, including exposure to heat, "spirits" (whisky?), vacuum chamber, and electricity. Some of his experiments were flawed, but he concluded sensibly "that heat and electricity increase the exhibition of light, merely by operating like other stimuli upon the vital properties of the animal."

Further, Macartney (1810) must be counted among the first naturalists to cite literature on earthworm bioluminescence. He did so, however, with a good dose of skepticism about the observations of two Frenchmen: "Bruguière [1792] upon one occasion saw, as he supposed, common earth worms in a luminous state; all the hedges were filled with them; he remarked that the light resided principally in the posterior part of the body … Flaugergues [1771] claimed to have seen earth worms luminous in three instances; it was at each time in October; the body shone at every part, but most brilliantly at the genital organs." In her historical survey of oligochaete bioluminescence, Emilia Rota (2009) traced the first account to the Swedish physician and botanist Herman Nicolas Grimm (1641–1711) who sighted luminescent earthworms in the East Indies (Grimm, 1683). Rota rightly blamed Macartney's skepticism on the fact that at the time earthworms phylogenetically distant from the common species *Lumbricus terrestris* were confused with the latter, which in everyone's experience was not found to be luminous.

John Murray, an eminent Scottish chemist and lecturer of the early nineteenth century, is unfortunately best remembered as the opponent of Sir Humphry Davy's theory that chlorine is a gas of its own, and not the result of a mixture of oxygen and anhydrous muriatic acid, as Murray and others believed. Davy proved him wrong by dint of elegant experiments. For our purposes, Murray did little more than enlarge on Macartney's descriptions of jellyfish and comb-jelly luminescent displays (Murray, 1821).

John MacCullough, a physician born on the Island of Guernsey, actually made his fame as a geologist who produced on commission the first geological map of Scotland (Hull, 2007). While surveying the Orkney and Shetland islands of Scotland he observed luminescent animals and produced a seminal paper of great importance (MacCullough, 1819). Contrary to many of his predecessors, MacCullough dismissed any link between locomotion and light emission, believing instead that mechanical disturbance was the key trigger or enhancer. The greater service contributed by MacCullough to the community of naturalists was his sound analysis of the pitfalls they faced in realistically assessing the population statistics of marine bioluminescent fauna: a calm sea makes one underestimate the quantity of luminescent animals; a stormy sea or the disturbance by the hull of a vessel betrays the real luminescent potential; bright displays irradiating light beyond the size of the emitting animal induce one to assume the population density at the sea surface is greater than it actually is; misidentification of the luminous source can take place even under a microscope because luminous animals move so swiftly that a non-luminous species appear in the field when the amber light is turned on; and so on.

Guarding against these potential pitfalls, MacCullough identified as bioluminescent about twenty species of jellyfish in addition to a few crustaceans, polychaete worms, radiolarians, and dinoflagellates. He mentioned one fish (eel larva: leptocephalus) as luminous, and then entered into a prescient discussion of how fishes living in deep waters – some as deep as 250 fathoms (457 meters) such as the common ling caught by Shetland fishermen – can see their food at depths where solar light does not reach. He came to the conclusion that food finding by the fish "can only be effected by the luminous property, either of his prey, or of the animals which abound in the sea, or else by that elicited from his own body." In subsequent chapters I show that MacCullough hit here upon an important contribution of bioluminescence to deep-sea ecosystems. However, this remarkable man's life was unfortunately cut short. At sixty-one, he married for the first time but soon met a pitfall of his own – during his honeymoon in Cornwall he died in a carriage accident (Hull, 2007).

William Baird (1803–1872) was a Scottish physician who worked for the East India Company. Returning to his work base in India in 1832, he happened on luminescent displays in equatorial waters. He was also a self-trained zoologist who specialized in crustaceans, so in his journal of the trip (Baird 1843) he was able to associate small luminous dots on the water surface with "Cyclops":

> Drawing a bucket-ful of water up, about 8, P.M., I allowed it to remain quiet for some time, when upon looking into it in a dark place, the animals could be distinctly seen emitting a bright speck of light. Sometimes this was like a sudden flash, at others appearing like an oblong or round luminous point, which continued bright for a short time, like a lamp lit beneath the water, and moving through it, still possessing its definite shape, and then suddenly disappearing … They evidently appeared to have it under their own will, giving out their light frequently at various depths in the water, without any agitation being given to the bucket. At times might be seen minute but pretty bright specks of light dart across a piece of water, and then vanish, the motion of the light being exactly that of the Cyclops through the water.

Baird's description and illustration of these crustaceans leave no doubt that he, like Meyen before him, had witnessed the light of a copepod, probably *Metridia*. Harvey (1957) seems to ascribe to Baird the first irrefutable observation of copepod luminescence, but he apparently failed to notice Meyen's contribution in this regard, published nine years prior to Baird's. By the time Baird published his paper, he had left the East India Company and taken the position of assistant in the Zoology Department of the British Museum, where he remained for the rest of his life.

Figure 3.1 Jean Louis Armand de Quatrefages de Bréau.
Source unknown in Wikipedia.

That same year (1843), across the Channel, appeared the first of two articles on marine bioluminescence by Jean Louis Armand de Quatrefages de Bréau (1810–1892). Born in the Cévennes, the precocious Quatrefages earned a doctorate in mathematics at the age of twenty, a medical degree at twenty-two and, after a few years of medical practice in Toulouse, a doctorate in zoology in Paris as well (Hamy, 1894). Under the influence of Henri Milne Edwards (1800–1885), who had recently been appointed professor at the Muséum d'histoire naturelle, Quatrefages did research on marine invertebrates at a number of seaside locations. His publications on luminous organisms resulted from such forays. In the first paper Quatrefages (1843) deplored the material conditions under which naturalists had to work, which to him explained the neglect suffered by bioluminescence as an object of study:

> This neglect is undoubtedly caused by the precise difficulty of the subject: indeed, the largest number of animals which display this remarkable property with particular intensity dwell generally in the sea; and the paucity of research centers located near the ocean or the Mediterranean, the near impossibility of transporting on location the precision instruments needed for the serious examination of these phenomena, have so far prevented the treatment of this question with the care it deserves.

Quatrefages went on to present his understanding of the source of the "phosphorescence" of the annelid worms (*Syllis*, *Polynoe*) and brittle stars observed on the coast of Normandy. With the help of a microscope, he noticed that luminescence was associated mainly with the articulation of appendages, and that muscle contraction at these articulations was linked with light emission. Today the light production would be interpreted merely as the result of the innervation of both muscle and light organs – or luminous cells – being stimulated. He wrongly identified muscles as the source of light emission; epidermal cells were later found to be responsible in polynoid worms. In the brittle star, Quatrefages described the light source as streaks running along the arms, but again he attributed the substrate of this luminescence to muscle cords. He developed the analogy of this muscle-derived luminescence with the electric organs of some fishes. His description of the dynamics of light emission provided the first evidence of the phenomena of flash facilitation and fatigue, which a future French biologist, Jean-Marie Bassot, would eventually elucidate.

In the second, longer, paper Quatrefages reviewed the literature and made further observations on marine luminous animals. This paper invites analysis in greater depth here because of its historical importance and its discussion of Christian Gottfried Ehrenberg's 1835 monograph on the luminescence of the seas. Harvey (1957) called Ehrenberg's contribution a monumental work that provided the most thorough and critical review of the literature on the subject up to that point. Ehrenberg (1795–1876) was born in a town near Leipzig, Germany, and was awarded his medical degree at the University of Berlin in 1818 (Siesser, 1981). Alexander von Humboldt soon took him under his wing and had him participate in a scientific expedition to Egypt and the Middle East (1820–25), which was decisive in orienting Ehrenberg's career toward zoology. He joined the zoology staff at the University of Berlin in 1827 and shortly afterward started research work on "infusoria" (unicellular organisms or protists), on which he became the greatest authority in the academic world. It is precisely because he suspected infusoria (especially peridinians) of being responsible for the luminescence of the Red Sea during a trip there, that he undertook the study that led to his monograph on bioluminescence (Siesser, 1981).

Ehrenberg's extensive review of the literature on luminescence and his own observations on many luminous species led to a synthesis aptly summarized by Quatrefages (1850). Ehrenberg came to eight important conclusions on the sources and modes of "phosphorescence": (1) the phosphorescence of the sea is solely accounted for by living organisms; (2) a large number of organic and inorganic bodies shine in and out of water in different ways; (3) the light emission of terrestrial animals is also a vital act; (4) active organic light emissions appear as spontaneous or elicited single or consecutive flashes which recall electric sparks; (5) a luminous slime oozes from some of the luminescent sources and may be absorbed by the ovaries which shine it back; (6) there is a clear relationship between light emission and sexual functions in luminous beetles, whereas luminescence in marine animals serves primarily a role of defense or protection; (7) among worms, only in annelids can one find phosphorescent organs, such as the hyperdeveloped median cirri [elytra] of *Photocaris* [probably a scale-worm], with a cellular structure; and (8) light production is a vital act similar to the development of electricity, susceptible to fatigue, which sometimes shows a direct connection with the nervous system.

On the whole, Ehrenberg's statements have stood the test of time, apart from his obsessions with ovarian functions and heavy electrical metaphors, traits he shared with several of his contemporaries. In particular, he had hit on the true arrangement of the luminescent system of polynoid worms, doing away with Quatrefages' far-fetched notion of muscles as seat of light production. As Quatrefages himself explained Ehrenberg's discovery: "The light starts from two enlarged and fleshy cirri belonging to the dorsal branch of the feet. The author saw sparks, isolated at first, spread over the cirrus until the latter was fully luminous; after which the phosphorescence reached the entire dorsal aspect of the body, and the animal resembled a fiery sulfurous wick."

Figure 3.2 Luminous dinoflagellates (*Noctiluca*, two upper left images), a fragment of *Noctiluca* (upper right image) and an enlargement of the luminescent zone of *Noctiluca* showing the constellation of sparks (lower image). From Quatrefages (1850).

Quatrefages (1850) distinguished two modes of light production: "by the secretion of a specific substance exuding either from the entire body or from a specific organ," or "by a vital act, from which issues the production of a pure light free of any material secretion." This dichotomy was to endure and later appear in a slightly different cloak: "by teasing extrinsic (luminous glands) from intrinsic (light emitted from inside cells) luminescent systems." It is symptomatic of Quatrefages not having kept pace with the cell theory expounded by Theodor Schwann and Mathias Schleiden eleven years earlier that he chose to speak quaintly of "vital act" and "pure light" rather than cell-based light emission.

The main source of cell-based luminescence, according to Quatrefages, was what he referred to as the rhizopod *Noctiluca* (a dinoflagellate). The French naturalist made new and important discoveries on the luminescence of this protist. For one, he found that *Noctiluc*a displayed a dual mode of light emission: short-duration flashes of a pale blue colour and long-duration glows of a whitish hue. As he increased the magnification of his microscope, he saw stable, punctate sources of light associated with the glows, but increased magnification further showed him that those sources were in fact clouds of micro sparks that lit up intermittently. This description, as we shall see later, endured the test of time. But Quatrefages also engaged in

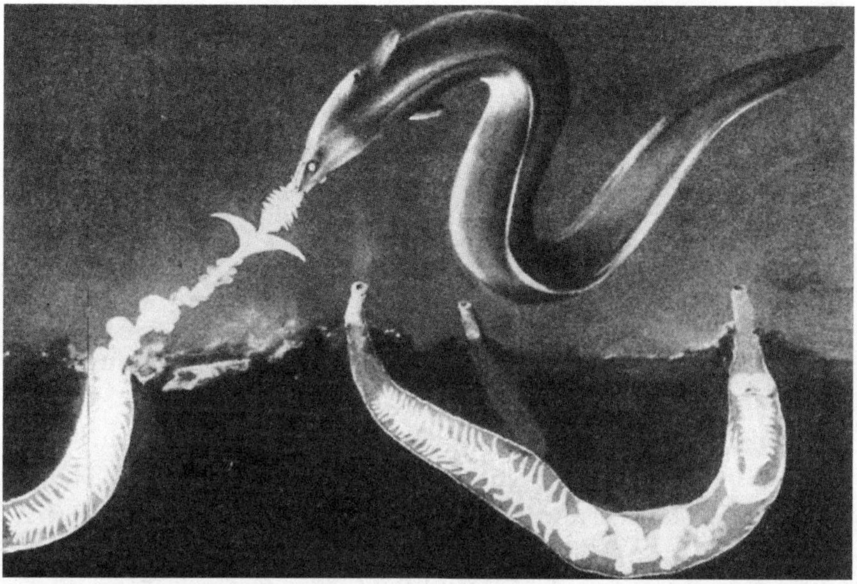

Figure 3.3 Representation of tube-worms (*Chaetopterus*) secreting luminescent clouds as an eel attempts to pull the worm out of its tube buried in the sandy bottom. Drawing by Bruce Horsfall in Dahlgren (1916e).

slapdash experiments (exposure to oxygen-poor atmospheres, electricity, acids, and so forth) that led to shaky interpretations.

It was the German naturalist Johann Friedrich Will (1815–1868) who in 1844 put forward the first description of the unusual luminescent display of the tube-worm *Chaetopterus variopedatus*. Will studied under Rudolf Wagner and Karl Theodor von Siebold in Erlangen, replacing the latter in the medical faculty as assistant professor in 1845 and full professor in 1848 (Nyhart, 1995). On a scholarship trip to Trieste in 1842–43, Will studied some luminous animals of the Adriatic coast: ctenophores, the clam *Pholas*, and an ascidian, in addition to the tube-worm (Will, 1844). This polychaete worm builds its own U-shaped parchment tube, which lies submerged in the bottom sediments with the exception of its two distal openings. When threatened, the worm secretes a copious, luminous slime from numerous epithelial glands on its specialized appendages, and the luminous mucous mass bolts out of one tube opening as a decoy to distract the attacker. Will observed that touching any part of the epithelial glands elicits the secretion of mucus that displays a bright green luminescence. The luminescence appears gradually and also vanishes gradually. Scrutinizing the mucus under the microscope revealed the presence in it of a multitude of tiny luminescent dots. Will's experiments allowed him to conclude correctly that the phenomenon was under nervous control.

Crossing back to the United Kingdom, we encounter the figure of Edward Forbes (1815–1854), who made observations on the luminescence of various coelenterates. The son of a banker of Scottish descent living on the Isle of Man, Forbes developed a passion for natural history in childhood and never wavered (Huxley, 1854). His prodigious industry and keen intelligence propelled him to the Regius Chair of Natural History at the University of Edinburgh at the age of thirty-nine. Unfortunately, that same year, he died after a short illness. Leaving our discussion of his famous Azoic hypothesis for chapter 4, I focus here on two noteworthy contributions on coelenterate animals. In the first, Forbes's research on luminous hydroids and sea pens is reported through correspondence with the Scottish naturalist George Johnston (1797–1855) and recorded in the latter's monograph, *A History of the British Zoophytes* (1847). As strange as it may seem as a way of disseminating knowledge, correspondence was accepted, if not current practice at the time. Here Forbes mentions the light emission of a hydroid polyp that he takes to be *Sertularia*, a non-luminescent genus, but which must have included a luminescent form such as *Obelia geniculata*. He discusses at greater length the sea pen *Pennatula phosphorea*, in which "the phosphorescence appears at the place touched, whether it be the stalk or the polypiferous part, and proceeds from there in an undulating wave to the extremity of the polypiferous portion, and never in the other direction." Louis Agassiz, the famous Swiss-Amercan naturalist, also observed the luminescence of another pennatulid, the sea pansy *Renilla reniformis*, from the South Carolina and Georgia coasts: "It shines at night with a golden green light of a most wonderful softness. When excited, it flashes up more intensely, and when suddenly immersed into alcohol, throws out the most brilliant light" (Agassiz, 1850). Agassiz did not witness Forbes's luminescent waves, which would later define all the luminous soft corals.

The following year Forbes published his own *Monograph on the British Naked-Eyed Medusae* (1848), in which he recorded his observations made in 1845 on the light emission of several true jellyfishes (hydromedusae). "In no case," he reports, "have I seen it [luminescence] continuous or constant in any one species. In every instance, the light has been given out only under circumstances of irritation, and not always even then." Immersing the jellyfish in freshwater or spirits elicited a vivid light emission that gradually vanished until death. The light originated from the tentacle bulbs or bell margin, but it also appeared "to radiate from the reproductive gland" as a greenish light. As others before him, Forbes entertained a misconception about reproductive structures serving as light organs.

The first scientific description of a luminescent squid was reported by the French amateur naturalist Jean-Baptiste Vérany (1800–1865). The son of a pharmacist in Nice, Vérany obtained his own pharmacy diploma at the University of Turin in 1819 (Fredj and Meinardi, 2007). His passion for mollusks, and especially squids, soon turned him into an amateur naturalist who acquired specimens from the Nice and Genoa fishermen, described and illustrated them, and sold them to collectors. Vérany committed all his findings about cephalopods to a lavishly illustrated monograph that appeared in 1851. In it he described the luminescence of a single specimen of the squid, *Histioteuthis bonelliana*, which he suspected came from deep waters. He placed the specimen in a tank and watched – and wondered.

> At that moment I enjoyed the astonishing spectacle of brilliant spots which adorn the skin of this cephalopod whose body shape is extraordinary in itself; at times the spots looked like the shine of sapphire which dazzled me; at others it was like the opalescence of topaz which made them even more remarkable; at others still the two rich hues blended with each other to produce a magnificent radiance. During the night the opalescent spots projected a phosphorescent light, which distinguishes this mollusk as one of the most brilliant productions of nature. (Vérany, 1851)

The specimen unfortunately did not survive for long, so Vérany said nothing about the dynamics of the luminescent display. About the time the monograph was published, Vérany resigned from his civil service post in Genoa and returned to Nice, where he dedicated the rest of his life to his studies of natural history. Another mollusk, the nudibranch *Tethys fimbria*, was recorded as "phosphorescent" by the German zoologist Adolf Grube

(1812–1880) in the waters around Trieste, along with the luminescence of a polychaete worm (Grube, 1861).

~~~

Observations of terrestrial bioluminescence also fared relatively well among landlubber naturalists of the early nineteenth century. Even though the land species are more within reach, too few are luminous to expand the list in the way coastal roamers and sea-going naturalists were able to do. The first record of this era was the description of the railroad worm two and a half centuries after Oviedo's tentative account (see chapter 1). The railroad worm, actually the caterpillar-like female of the beetle *Phrixothrix*, must be counted as one of the most spectacular bioluminescent animals, what with her adornment of lights on the head and along the body which recall an illuminated train at night (Harvey, 1957).

Felix de Azara (1746–1821), responsible for this first record, was a Spanish military engineer commissioned "to take part in delimiting the boundaries between the Spanish and Portuguese possessions in South America" (Beddall, 1975). His commission lasted throughout the last twenty years of the eighteenth century and left him plenty of time to take stock of the natural history, anthropology, and geography of the area – the confluence of Brazil, Argentina, and Paraguay. Although an amateur naturalist who tended to belittle his accomplishments (Beddall, 1975), Azara managed to publish his discoveries in France after the Spanish Crown's show of disinterest. Charles Darwin is known to have consulted Azara's book attentively during the voyage of the *Beagle*.

In his book, Azara (1807) gives as accurate a description of the luminescence of the railroad worm as can be expected of an eyewitness account: "In Paraguay I saw a large worm, about two inches in length, whose head resembled at night a blazing red coal, and who possesses in addition, on each side and all along the body, a row of round, eye-like pits from which emanate a weaker and yellowish light."

Turning to the firefly, the beetle most familiar to everyday people, it can be said that the first half of the nineteenth century afforded significant advancement of knowledge, thanks largely to the experimental spirit. The three prominent figures who made it happen are Jean-François Macaire (1796–
~~~

1869), Wilhelm Karl Hartwig Peters (1815–1883), and Carlo Matteucci (1811–1868). Macaire, a Swiss-born professor of medical chemistry at the Académie de Genève, published numerous papers on topics of interest for the pharmaceutical enterprise, focusing on botany and chemistry, according to entry 7319-47 of the *Livre du Recteur de l'Académie de Genève: 1559–1878* (Librairie Droz, 1972). So, studying fireflies was a foray beyond his professional activities, but, as he so haughtily asserts, "the most learned men have not disdained to share with the vulgar the admiration called for by the sight of phosphorescent insects" (Macaire, 1821).

At the outset Macaire made an interesting observation; namely, that the female firefly twists its tail at the same time the light organ flashes. The coordination of the two activities led him to suggest that nervous control was involved. He also found that oxygen was necessary for the light emission to occur; but even more interesting was his observation that the gas nitric oxide stimulates flashing. The key role of nitric oxide in firefly flashing was only discovered recently (Trimmer et al., 2001). Sending galvanic current to the light organ through platinum wire electrodes also stimulates light emission. After various chemical treatments of dissected light organ tissues, Macaire came to this erroneous, but unsurprising conclusion given the state of chemical knowledge at the time: "According to these properties, I am inclined to view the luminous substance as basically, if not totally, composed of albumin in a semi-transparent state; and the cause of the termination of the light seems to be the coagulation of this albumin and its conversion to the opaque state."

The contribution of Wilhelm Peters is of a different order, but just as significant as Macaire's. A student of the great pioneer of German physiology, Johannes Müller, Peters was his assistant when he made a study of the firefly *Lampyris italica* during a sojourn in Nice (Peters, 1841). Peters's record of his encounter with this firefly species is worth citing as it fairly epitomizes the common human response when confronted with this spectacle:

> From the middle of May to the middle of July, when taking a walk around Nice at dusk, one is surprised by the curious spectacle of thousands of minute sparkling lights which crawl here and there, at times illuminating the tip of a rock and at others lighting up a deep cavity, still at others producing suddenly as by a magician's wand on the black

> stumps of olive trees a brilliant light show which in its mobile and shifting scenery affords the greatest interest. This phenomenon repeats itself every evening, but it seemed to me more brilliant when the atmosphere is saturated with humidity.

Peters put paid to one of the myths about firefly bioluminescence which had enjoyed currency for centuries; namely, that fireflies store sunlight during the day and offload it through the outlet of the light organ at night. He was struck by the high frequency of flashing that this firefly was capable of, producing between eighty and a hundred flashes per minute. More important, Peters noticed that severing the head from the trunk extinguished the flashing, thus confirming that the brain exercised the central control. But Peters's key contribution was certainly his discovery of the basic functional units composing the firefly light organ, the lantern rosette, although this name was coined later. Thanks to his deft use of the microscope, Peters found that "the whole organ consists in a regular array of small globules in which penetrate the fine tracheal branches, distributed there in the most elegant manner and forming, so to speak, its basic scaffold." These globules, which are known today as the rosettes, are formed of respiratory tracheoles delivering oxygen to the light organs and surrounded in a rosette fashion by the light-emitting cells, the photocytes.

It appears that Harvey (1957) not only confused Wilhelm Peters with a namesake (Amos W. Peters, an American physiologist), but that he also missed this Peters's discovery of the firefly rosette pattern as the source of flashing activity. Shortly after his analysis of firefly luminescence, Peters, yet another young naturalist sponsored by Alexander von Humboldt, went on an extensive excursion of Mozambique and returned with an impressive collection of specimens, the basis of a book that made his fame among naturalists (Hilgendorf, 1887). He became a curator of the Berlin Zoological Museum in 1858.

Carlo Matteucci, considered one of the great pioneers of the study of bioelectricity as applied to the electric organ of the torpedo fish and muscle contraction (Bresadola, 2011), published a paper two years after Peters on the same firefly species (Matteucci, 1843). In his contribution, the Italian naturalist shows that he was aware of Macaire's work but not Peters's, which led him to unnecessarily repeat experiments already conducted by the German. But Matteucci also designed a series of clever experiments in which gases and temperature were manipulated. He also compared the luminescent responses of intact fireflies and isolated light organs. While he confirmed results by Macaire and others, he was the first to clearly state that "this phosphorescent substance, when it emits light in oxygen or in air, consumes a portion of oxygen which is replaced by an equivalent portion of carbonic acid." This observation led to the conclusion that "light production in this insect is entirely bound to the combination of oxygen with carbon, which is one of the elements of the phosphorescent substance." This was an astute and far-reaching biochemical statement for his time.

~

In addition to studies of luminous insects, more records of earthworm bioluminescence accrued in the first half of the nineteenth century, as Rota (2003, 2009) has shown. This time it was the French physicians and naturalists Antoine Louis Dugès (1797–1838) and Victor Audouin (1797–1841) who dominated the scene. Dugès who, like Audouin, published on earthworm luminescence shortly before his untimely death, recorded his observations in a paper describing numerous annelid species. One of them attracted his attention "by the luminous liquor which it excretes from the body surface, and which undoubtedly is analogous to the coloured liquor expelled by the dorsal pores in many other earthworms" (Dugès, 1837). Audouin, in his contribution, reported on an earthworm that gave out a whitish light in sexually mature specimens, hinting at a role for its light emission in reproduction (Audouin, 1840).

Also of interest, another Frenchman, the famous entomologist Jean-Henri Fabre (1823–1915), made a detailed investigation of another terrestrial luminous organism: not an earthworm as Rota (2009) mistakenly believed, but the mushroom *Agaricus olearius*, which, as its scientific name implies, thrives around olive trees. The soft glow of the agaric's luminescence is spontaneous and continuous day and night, according to Fabre (1855). It is not inhibited by light and is impervious to heat within limits. It disappears in oxygen-free or oxygen-poor environments. The mushroom produces more carbon dioxide while luminescent than when the glow is extinguished by removing the oxygen; Fabre concluded cautiously, after critical experiments, that a causal relationship may exist between the oxidation process and light emission.

For all the efforts on the part of some to conduct as rigorous experiments as the scientific standards and instruments of the period allowed, these experiments still stood out for their crudeness or absence of unimpeachable controls. Students of bioluminescence, far from systematically addressing the origin and mechanism of the phenomenon, contented themselves with accruing basic observations and expanding the list of luminous species. Even then, those who produced essays on "phosphorescence" and reviews of the literature, with the possible exception of Ehrenberg, failed to include the few fishes already recorded as truly luminous, or engaged in fantasies with regard to the mechanism of bioluminescence (Bernoulli, 1803; Phipson, 1862).

For the majority of these naturalists, watching luminous animals was like a fleeting fascination for a beautiful object. Mesmerizing, yes; enlightening, no. It is as if the little time spent with the subject – none made a career of studying luminous organisms – could only be indulged in fantasies about the object of fascination. There were moments of lucidity that dispelled some misconceptions, but the object still retained its aura of mystery. And observers were not yet ready to shake off their misleading infatuations: fireflies called glow-worms (in fact the larva-like females of some firefly species), crustaceans called marine insects, the sparkling sea an electrical phenomenon akin to the aurora borealis. When scientific data are lacking and reasoning ability falters, substitute metaphors sometimes creep in. But from the mid-nineteenth century forward, reliance on metaphor gradually yielded to the scientific urge to name things as they are, to see them as they are, and to interpret them on a sounder basis, within the limits of contemporary conceptual frameworks.

As we shall see in the next two sections, it was time to break the spell – time to take a cold look at the cold light.

PART TWO

THE LIGHTS BENEATH THE SURFACE

4

The Birth of Scientific Ocean Exploration

The handful of scientific men who reawakened interest in marine exploration about 1870 could not have imagined the consequence of their efforts.
–Harold L. Burstyn (1972)

We can make little sense of what happened next in the field of bioluminescence without an account of the great scientific enterprises of worldwide ocean exploration that occurred from around 1870 to the first decade of the twentieth century. None of these expeditions explicitly stated that the search for new marine luminous species was one of their goals, but the discovery of quite a few of these exotic and often grotesque-looking creatures certainly proved to be a collateral harvest of the projected work.

What we are speaking about is nothing less than the birth of oceanography as a scientific discipline in the modern sense. It all started with the British *Challenger* expedition of 1872–76. The history of this epochal voyage has been recounted many times from different perspectives. On the occasion of the centenary of the voyage, historical essays by Maurice Yonge, Daniel Merriman, and Harold L. Burstyn were published in the *Proceedings of the Royal Society of Edinburgh*, and the novelist and biographer Eric Linklater published a coffee-table book with excellent illustrations (Linklater, 1972). The following account owes much to these secondary sources, but it also relies substantially on the original narratives of the voyage.

It was more than mere chance that the Royal Society of Edinburgh hosted the celebration of the hundreth year of the launch of the *Challenger* voyage. Indeed, evidence points to Edinburgh as "the birth place and the early home of modern oceanography, and Edinburgh men and Edinburgh ideas played a leading part during the nineteenth century in establishing this comprehensive science of the sea" (Herdman, 1921). To bolster his case, Herdman two years later highlighted the roles of the Scotsman Charles Wyville Thomson (1830–1882) as the spearheading thrust behind the *Challenger* venture and of Canadian-born but Scottish-descended Sir John Murray – not to be confused with his namesake in the previous chapter – as the pioneer of modern oceanography (Herdman, 1923). Let us begin with the story of Charles Wyville Thomson.

As explained by Herdman (1923) and Yonge (1972), it was an obscure animal form – the crinoids, or feather-stars, which are relatives of sea urchins and sea cucumbers – that triggered the chain of events leading to the *Challenger* expedition. Owing to a shared interest in stalked crinoids, and particularly the fossil forms, Wyville Thomson became a colleague and friend of William Benjamin Carpenter (1813–1885). Carpenter, a native of Exeter, in the west of England, studied medicine in Bristol and, after seeking in vain the medical physiology chair at the University of Edinburgh, settled as a lecturer at the London Hospital and eventually reached top administrative posts at University College and the University of London. A scholar of eclectic interests, Carpenter published profusely on topics related to medical physiology and also natural history and invertebrates. Following a trek on the Scottish Isle of Arran, he dredged specimens of larval crinoids resembling some stalked crinoids in the fossil record, and this "culminated in a paper on their structure, physiology and development in the Philosophical Transactions in 1866" (Yonge, 1972).

Wyville Thomson studied at the University of Edinburgh but never graduated. His appointments in zoology, botany, and geology, culminating in his post at the University of Belfast in 1860, reflect a range of interests at least as wide as Carpenter's. As Yonge (1972) further explains, "it seems to have been palaeontology that concentrated his marine biological activities on the crinoids, and it was their common interests in these animals which brought visits from Carpenter in 1863 and 1864." Then they learned from Michael Sars, the Norwegian theologian and natural historian, that his son (George Ossian Sars) had dredged up a living adult stalked crinoid from a depth of three hundred fathoms (550 meters) in the Lofoten fjords. The British duo treated the finding as a living fossil, and it made them wonder what else in deep water could be collected that bore a resemblance to fossil forms already known to them.

Figure 4.1 Early pioneers of oceanography: Charles Wyville Thomson (left) and William B. Carpenter (right). From Linklater (1972).

They had never asked themselves this question before because Wyville Thomson's teacher in Edinburgh, Edward Forbes, had advanced the hypothesis that life did not exist in marine depths greater than 300 fathoms (~ 550 meters) – the so-called Azoic zone. (This was the same Forbes with whom we became acquainted in the previous chapter, who gave accounts of jellyfish luminescence.) In 1841 Forbes was appointed naturalist on board a British surveying ship commissioned to engage in hydrographical work in the Aegean Sea of the Eastern Mediterranean (Anderson and Rice, 2006). He successfully dredged to depths of over 200 fathoms (365 meters), bringing up shells and other "curios," as the ship's crew called them. In a report submitted to the British Association for the Advancement of Science in 1843, Forbes described the vertical distribution of animals in the Aegean Sea and "conjectured that the zero of animal life would probably be found somewhere about 300 fathoms." As no animal life could in his view exist at greater depths, he called anything deeper the Azoic zone.

Wyville Thomson became engrossed in this conundrum of the deep sea, especially as anecdotal reports kept coming in of submarine telegraph cables being hauled up for repair with abyssal animals clinging to them. In 1868 he struck on the idea that the British government might be cajoled into sponsoring a scientific expedition to the deep waters wedged between the Shetlands and the Färoe Islands. The Royal Society was approached to apply to the government, and the application was accepted. The Admiralty made available a barely sea-worthy steamship, the *Lightning* – "the old tub," as Carpenter, appointed expedition director, called her. Despite weather that made work extremely difficult if not impossible – and made even the officers sick – they managed to haul in catches from over 600 feet deep (100 fathoms). Although this was hardly a breakthrough in depth reckoning, Carpenter, apparently desperate to convince his sponsors that the expedition was not a waste, reported "to the Royal Society that results fully confirmed the opinion ... that life did occur in previously considered azoic depths" (Yonge, 1972). But despite their progress, a feeling of disappointment, fuelled by the poor weather and the dreadful vessel, impelled the investigators to appeal to the Admiralty once more: in order to cast the collecting equipment in a wider area and at greater depths, they needed a better ship and a better season for their work.

Their wish was granted with the appointment of the 400-ton survey vessel *Porcupine*. Its three cruises, all between 18 May and 7 September 1869, spanned the Atlantic Ocean from the Färoe Channel in the north, to the west of Ireland and further south to the Bay of Biscay. The first survey covered the Irish waters and was supervised by John Gwyn Jeffreys (1809–1885), a wealthy mollusk specialist who frequently used his own yacht for dredging excursions. The second cruise, led by Wyville Thomson, searched the Bay of Biscay; and the third, led by Carpenter, explored the seabed between Scotland and the Färoe Islands. In their report on the expedition (Carpenter et al., 1870) the authors set great store by the methods and equipment used to collect physico-chemical data and specimens. We get an accurate assessment here of the technical developments of oceanographic research to that point: derricks especially designed to reduce strain on cables during deep-sea sounding and dredging under vessel pitch conditions, the use of multiple heavy sinkers for sounding at great depths, construction of thermometers tested to withstand the high pressures at great depths, the addition of "hempen brushes" to the bottom dredges so that bottom organisms could be swept more effectively inside the dredge, to list but a few.

Thanks to these new developments, the *Porcupine* expeditions set new records of dredging depths, down to 2,435 fathoms (4.5 kilometers). They caught numerous and diverse invertebrates, but few or no fish. At shallower depths, between 300 and 600 fathoms, they caught many luminous invertebrates:

> Many of the animals were most brillantly phosphorescent; and we were afterwards even most struck by this phenomenon in our Northern Cruise. In some places nearly everything brought up seemed to emit light, and the mud itself was perfectly full of luminous specks. The Alcyonarians, the brittle stars and some annelids were the most brilliant. The *Pennatulae*, the *Virgulariae*, and the *Gorgoniae* shone with a lambent white light, so bright that it showed quite distinctly the hour on a watch. The light from *Pavonaria quadrangularis* was pale lilac, like the flame of cyanogen; while that from *Ophiocantha spinulosa* was of a brilliant green, coruscating from the centre of the disk, now along one arm, now along another, and sometimes illuminating the whole outline of the starfish [brittle star].

The members of the expedition were confronted by this apparent paradox: there was obviously light at these depths, as many of the crustaceans and other higher animals possessed unusually large eyes presumably designed for optimal sensitivity to light, and yet their measuring instruments, imperfect as they were, suggested that no sunlight penetrated beyond 200 fathoms. The only solution to this paradox, they reasoned, was to envisage that the luminescence of deep-sea animals served the purpose that sunlight does near or at the surface: making the prey visible to their predators or attracting a prey by the luminescence of a lure. (We may recall from the previous chapter that John MacCullough had entertained a similar theory half a century earlier.)

Hardly two years had passed, however, before the Scottish physician and marine zoologist William Carmichael M'Intosh (1838–1931) published an essay refuting what he called the "abyssal theory of light" proposed by the *Porcupine* investigators (M'Intosh, 1872). By arguing, among other points, that luminous and non-luminous congener species shared similar feeding or prey-evading habits, he exposed the flaws of the theory of Carpenter, Wyville Thomson, and Jeffreys with seemingly unassailable logic.

Wyville Thomson reaped considerable rewards from these endeavours. For one, he made a name for himself, a name that circulated and landed him admission to the Royal Society in 1869 and to the prestigious Regius Chair of Natural History at the University of Edinburgh in 1870. As a side note, his predecessor to the chair was George James Allman (1812–1898), an Irish-born graduate of Trinity College, Dublin, and of Oxford University. As a zoologist, Allman was an expert of lower invertebrates, particularly focusing on bryozoans and hydrozoans (Calder, 2015), but I mention him in this context because he published short notes on the luminescence of an insect, the springtail *Anurophorus* (Allman, 1851), and of the comb-jelly *Beröe* (Allman 1862). He was the first to observe that the eggs and embryos of the comb-jelly were also luminous, and that the luminescence is inhibited by light. As it happened, his successor, Wyville Thomson, published a book on the deep sea from his experiences on the *Lightning* and *Porcupine* (Wyville Thomson, 1873), in which bioluminescence was also discussed.

One of the benefits accruing from Wyville Thomson's fame was that he could now move mountains to promote oceanographic research on a grander scale. His friend Carpenter still worked in the shadows to make this happen, to the point of lobbying the First Lord of the Admiralty for a vessel adequately equipped with oceanographic instruments and a scientific staff for a circumnavigation (Yonge, 1972). The Hydrographer of the Navy, Admiral George Richards, greatly impressed by Carpenter's and Wyville Thomson's achievements to date, gave a favourable recommendation to the First Lord. As a result, Her Majesty's Government worked hand in hand with the Royal Society to move the project along. As Yonge explains:

> Matters proceeded with unusual celerity because both parties, Society and Admiralty, were in complete accord. Moreover, the time was ripe; scientific and public opinion alike had been aroused by the prospect of revealing the inhabitants of the deep oceans, the importance of cur-

rents and other ocean movements had been demonstrated by [US Navy Admiral] Maury while the needs of cable laying demanded the fullest information about the contours of the ocean floor.

The Navy recruited the corvette HMS *Challenger* for the expedition. According to Linklater (1972), she "was a three-masted, square-rigged, wooden ship of 2,200 tons [over two million kilograms] displacement and some 200 feet [60 meters] over all." For the purpose of the voyage the steam engine was used largely for dredging tasks and harbour manoeuvres, while sails served for cruising between working stations. The requirements of the scientific work necessitated drastic alterations to the ship. "All but two of the 18 main-deck 68-pounders were removed," wrote Daniel Merriman (1972), "and this space, along with that normally housing stores and ammunition, was turned to scientific purpose." On departure day, 21 December 1872, 243 men crammed the ship's nooks and crannies, among them a scientific staff of six.

The Royal Society did not waver long in appointing the scientific leader of the expedition; Wyville Thomson held the winning ticket, thanks to his experience and newly earned prestige, and on account of Carpenter's age – seventeen years older than his friend and now competitor (Merriman, 1972). The happy few scientists to accompany Wyville Thomson were John Murray (1841–1914), John Young Buchanan (1844–1925), Henry Nottidge Moseley (1844–1891), Rudolf von Willemoes-Suhm (1847–1875) and John James Wild (1824–1900). These were the scientists whom a young officer aboard, navigating sub-lieutenant Herbert Swire, derisively called "the philosophers"; and "poked fun at their clothes both afloat and ashore, [so that] even Wyville Thomson – rotund as befitted his years and inclined to a vanity that his eminence excused – did not escape his mockery" (Linklater, 1972). The disrespect heaped by crewmen on the scientists – who after all were the *raison d'être* of the expeditions – has a long history in oceanographic enterprises, beginning, as we have seen, with the *Challenger*. But let us now look at some snapshots of the "philosophers" from the various, often divergent optics of peers and outsiders.

The most important figure after Wyville Thomson himself turned out to be John (later Sir John) Murray. Born in Cobourg, Ontario, to Scottish settlers, he left Canada at the age of seventeen to attend Stirling High School in Scotland in preparation for entrance at Edinburgh University (Agassiz, 1917). Agassiz's portrayal of Murray's independent attitude as a youth in Edinburgh foreshadows many later writings about the man:

> [Murray's] career at the University appears to have been stamped by some of the qualities that distinguished him in after life. Impatient of dogmatic authority, he was somewhat scornful of inherited tradition, and treated his prescribed studies with a cheerful sans gêne. For even in those days he desired to find out things for himself, and delve for knowledge independently. The capacity of clear and original thought, with a genius of disentangling the heart of a subject from its enveloping details, was as characteristic of the youth as of the man. From the small circle of scientific men who then made Edinburgh famous, he gathered, during his student days, what was most worth having, and went his way. That one of the facets of his personality drew him into a friendship with [Robert] Louis Stevenson, offers a suggestive glimpse into a byway of his character.

Like Wyville Thomson, Murray never completed a university degree. In 1868 he travelled on a whaler bound for the Spitzbergen to study the Arctic Sea (Agassiz, 1917). This initial foray into what was developing as the science of oceanography should have induced the Royal Society *Challenger* committee to shortlist him for the scientific team, but the fact is that they failed at first to consider him. It took the withdrawal of a recruited candidate for the committee to assign him as a replacement. He was thirty-one when *Challenger* left port. As we shall see, his contribution to the enterprise was immense, to the point that this young Canadian is said to have later coined the word "oceanography" itself.

John Buchanan, with whom Murray worked closely, was twenty-nine years old when the ship sailed. He had a solid rooting in chemistry, having studied the discipline at various German universities and in Paris. He was a laboratory assistant in Edinburgh when he was chosen for the *Challenger* expedition, on the strength of "his reputation as a practical chemist, his interest in natural phenomena, and his technical ability in devising and making apparatus" (Merriman, 1972). As a trail-blazing chemical oceanographer, Buchanan made analyses of water and sediment samples that proved critical to the success of the voyage. His labmate Henry Moseley was recruited as zoologist on the team. Born near London, he had an unremarkable education, but he was quickly developing into a serious amateur naturalist. Sent to Oxford by his clergyman father, who fancied him a mathematician or classics scholar, Moseley felt utterly miserable. But, as Richard Corfield (2004) explains, there was a way out for the young man:

> It was in this state that he was discovered one day by an old – and rather more liberal – friend of his father's. This friend, also a clergyman, realizing that Moseley was wasting his life in Oxford, interceded on his behalf with George Rolliston, professor of anatomy. Moseley was enrolled in the recently established Honor School of Natural Science at Oxford, where he immediately blossomed. He won a first-class degree in natural sciences in 1868 and, after a four-year dalliance with a career in medicine, was chosen for the *Challenger* expedition.

Moseley worked with incredible energy and motivation during the voyage. However, his centre of interest being terrestrial natural history, he grew exasperated with the sounding and dredging stations that kept him away from the ports of call. Not so with the precocious Rudolf von Willemoes-Suhm, who felt comfortable working on marine animals. He was born a German national in Schlewig-Holstein when it was still under Danish rule. He studied zoology in Munich and earned his doctorate at the University of Göttingen when he was only twenty-three. He started collecting specimens of "lower animals" in the Bay of Kiel for his Habilitation (post-doctoral studies). After military service he began teaching at the University of Munich. In 1872 he joined the staff of the Danish ship *Phoenix* to study invertebrates around the Färoe Islands. It was during a coaling stop of his ship in Scotland that Willemoes-Suhm's life turned around and he became a member of the *Challenger* crew. This is how he described the excitement of the moment in letters to his mother:

> This morning [10 October] I visited Professor Wyville Thomson, delightful man, who will leave on the 1st of November as the leader of a three-year expedition around the world on a warship that is especially rigged for the purpose. While we were talking in his lovely drawing-room,

> he became serious and asked how much I had been to sea and then if I "would like to go around the world." Of course I answered that this would be a magnificent opportunity. – Three years on an English ship that goes to all parts of the world would be wonderful! October 14: I was invited to six o'clock dinner by Professor Thomson. Mrs Thomson was a gracious hostess and we enjoyed a sumptuous repast. I talked alone with the professor after dinner about the future … he showed me some fascinating things he had fished up from the depths of the sea. It was left that I should talk with Professor [Thomas] Huxley about joining the expedition; then Professor Thomson would see what to do. Consequently, I left [for London] on the following morning … On Sunday morning I met Professor Huxley in the South Kensington Museum. He received me with great warmth and gave it as his opinion that I should take advantage of the opportunity to participate in the expedition. He promised to do everything for me on the following Thursday at the meeting of the "Royal Society Committee" … At three o'clock Professor Huxley took me to the Zoo, which was open only to "Fellows" today …. At six o'clock I had dinner at Professor Huxley's … [and the next night he rejoined the *Phoenix*]. October 20: Hotel Phoenix, Copenhagen. This night I got a telegram informing me that the English Admiralty has appointed me as a zoologist on board the *Challenger* … For the next three years I shall have the advantage of living in a house that moves slowly but surely through all five parts of the world. (Willemoes-Suhm, *Challenger Briefe*, translated from German by Merriman, 1972)

Willemoes-Suhm was the youngest scientist, but also the only scientist who came on board with his own servant, so officers derided him with the title of Baron, to add to that of philosopher (Corfield, 2004).

The last member of the scientific team did not attract attention to himself, so he was spared by the crew. He was born Jean-Jacques Wild in Zurich, Switzerland, and made his life in Great Britain after a stint teaching languages in Belfast, where he met his future wife and anglicized his first name to John James. He became the secretary of Wyville Thomson, and it was in this capacity that he joined the scientific team, occupying part of his boss's cabin. He worked as the drawing artist of the expedition, thereby contributing critically to the narratives of the voyage.

The stated aims of the *Challenger* expedition, which emphasized the investigation of the deeper layers of the oceans and the bottom, were clearly laid down in a report of the Royal Society committee dated 30 November 1871. In a nutshell, they were itemized as follows (Murray, 1885):

1 To investigate the Physical Conditions of the Deep Sea, in the great Ocean-basins.
2 To determine the Chemical Composition of Sea Water, not merely at the surface and bottom, but at various intermediate depths.
3 To ascertain the Physical and Chemical characters of the Deposits everywhere in progress on the Sea bottom.
4 To examine the Distribution of Organic Life throughout the areas traversed, especially in the deep Ocean-bottoms and at different depths.

To accomplish these lofty goals, unprecedented in scale at that time, they were granted three and a half years to follow a route from Portsmouth to Gibraltar and Tenerife; across to the West Indies and up to Halifax, Nova Scotia; back to the North African coast; down the South Atlantic to Brazil; crossing back to the Cape of South Africa and to the Antarctic Great Ice Barrier; up to southeast Australia, Papua New Guinea; northwest to Cape York and the Dutch colonies (Indonesia); north to Hong Kong and back, then north to Japan; crossing the West Pacific to the Sandwich Islands (Hawaii); then due south and east to Chile and round the cape to the Atlantic, with the final lap from Argentina to the port of departure, Portsmouth. Along the way they conducted numerous depth soundings and bottom sediment analyses, temperature recordings, and water analyses, in addition to dredging and trawling for faunal specimens. No wonder that the crew, used to straightforward cruises from port to port, joined Moseley in decrying these interuptions in mid-ocean for "drudging," as they called it (Linklater, 1972).

But tedious as it may have seemed, the work stations yielded an incredible number of exotic species new to science, many of which displayed light emissions or are now known to be bioluminescent. In fact, this revelation came early to the scientists on board. From the very first dredges, as *Challenger* was making her way towards Madeira, strange-looking gorgonians and hydroid colonies were hauled up on deck with their brilliant displays of "phosphorescence" for everyone to see (of this more in the next chapter). Fishes, which the *Lightning* and *Porcupine* had failed to capture at depth, now showed up in number, thanks to an equipment change from the classical dredge to the large-framed trawls from which fishes found it more difficult to escape as they were hauled up.

One of the first catches was the hatchet-fish, a monstrous little fish with upwardly directed tubular eyes. The hatchet-fish is bioluminescent, but the *Challenger* scientists could not tell because the specimens arrived on deck dead, due to decompression. Decompression occurs when animals with gas spaces are exposed to lower pressures than normal as they are dragged to the surface faster than they can accommodate, causing pent-up gases (such as those of fish swimbladders) to reach out, with fatal consequences. On one occasion Willemoes-Suhm (1875) saw a hatchetfish shine "like a luminous star hanging in the trawl net." Many new fishes were discovered at unheard-of depths, such as *Ipnops murrayi* at 1900 fathoms (nearly 3.5 kilometers) off Brazil and angler-fishes at 2450 fathoms (nearly 4.5 kilometers) between the Carolina Islands and the Marianas in the Pacific (Murray, 1885).

The deep-sea was not the exclusive domain of bioluminescence during the voyage. Like the occasional naturalist of past circumnavigations, the *Challenger* naturalists also delighted in experiences of surface "phosphorescence." One such phenomenon, seen off the west coast of Africa, was vividly recorded by sub-Lieutenant Lord George Campbell in the language of a visual artist:

> On the night of the 14th [August 1873] the sea was most gloriously phosphorescent to a degree unequalled in our experience. A fresh breeze was blowing and every wave and wavelet as far as one could see from the ship on all sides to the distant horizon flashed brightly as they broke, while above the horizon hung a faint but visible white light. Astern of the ship, deep down where the keel cut the water, glowed a broad band of blue, emerald-green light, from which came streaming up or floated on the surface, myriads of yellow sparks which glittered

> and sparkled against the brilliant cloud-light below, until both mingled and died out astern far away in our wake. Ahead of the ship, where the old bluff bows of the *Challenger* went ploughing and churning through the sea, there was light enough to read the smallest print with ease. It was as if the Milky Way … had dropped down on the ocean, and we were sailing through it. (Quoted in Corfield, 2004)

On *Challenger*'s return to Spithead on 24 May 1876, only five scientists disembarked, and their future lives bore the imprint of the difficulties of the voyage. Willemoes-Suhm, who had been looking forward to "the advantage of living in a house that moves slowly but surely through all five parts of the world," had died of a severe skin infection (erysipelas) in 1875 as the ship was sailing to Tahiti. Moseley, deeply affected by this death, worked so hard on the zoological results of the expedition and on the writing of his narrative (Moseley, 1879) that he soon burnt himself out and died at forty-seven. Wyville Thomson, whose health had not been good to start with, suffered from seasickness in addition to the long hours of work on the ship. He only had two years to produce a two-volume preliminary account of the voyage (Wyville Thomson, 1878) before exhaustion hastened his death at fifty-two. From then on, the Herculean task of recruiting specialists to study the collections of the *Challenger* and produce the fifty thick volumes of scientific results – a task completed in 1895 – fell on the shoulders of John Murray. In that way Murray, who after all was always on deck to help sort out the bottom samples and trawled fauna, became the great figurehead of the pioneer expedition. His reputation as a deep-sea oceanographer – the "father of oceanography" – was sealed with his book *The Depths of the Ocean* (Murray and Hjort, 1912); his fate was to have been killed two years after its publication in a car accident.

~

The voyage of the *Challenger* had far-reaching implications. "The expedition did more than fire the imagination of the British people," wrote Sir Maurice Yonge (1972); "it stimulated corresponding activities in all the major maritime nations." The ensuing expeditions that are of special interest for us here are of course those that brought results for bioluminescence research.

In chronological order, these were expeditions led by Prince Albert I of Monaco (1848–1922), an American expedition led by Alexander Agassiz (1835–1910), the *Valdivia* expedition, led by the German Carl Chun (1852–1914), and the Dutch *Siboga* expedition in Indonesia, led by Max Weber (1852–1937).

It is remarkable that in the wake of the *Challenger* expedition the tiny Principality of Monaco should move ahead of the major maritime nations as the indisputable powerhouse of oceanography. This peculiar fact owes much to the personality and driving ambition of one man, Albert Honoré Charles Grimaldi, born in Paris, the son of Prince Charles III of Monaco. Much of the information gathered here we owe to Monaco historian Jacqueline Carpine-Lancre (1998). The awakening of Albert's passion for the sea occurred through his readings and interactions with Monaco fishermen during his teenage years. After completing his secondary education, he chose the career of navy officer, training first in the French Imperial Navy, and then in the Spanish Navy for two years, during which time he became acquainted with Cuba, Puerto Rico, and the United States. To gain experience as a navigator he would for many years acquire vessels or yachts and sail far and wide. One such vessel was a schooner purchased in England in 1873, which he rebaptized *Hirondelle* (Albert de Monaco, 1889). During these travels the Monaco heir sought contacts with scientists and read up on scientific subjects, and he gradually came up with a plan for fostering oceanographic research. Curiously, it was France's tepid attempts at oceanographic research by the *Talisman* and the *Travailleur* in the early 1880s that decisively inspired the prince to try his own vision of the science. Between 1885 and 1887 he put the *Hirondelle* to use in a study of the near-surface water currents in the North Atlantic by launching almost 1,700 special floaters (Carpine-Lancre, 1998). In parallel, variously designed fishing gear were grafted onto the *Hirondelle* to harvest animals from the surface to depths approaching 3,000 meters, particularly in the Azores.

By 1888 the *Hirondelle*'s useful life as an oceanographic vessel was over. Shortly after the prince succeeded to the throne as Prince Albert I in 1889, he commissioned a British shipyard to build a three-masted vessel with an auxiliary steam engine, designed and fitted specifically for oceanographic work. The *Princesse-Alice*, named after his second wife, was only the second dedicated research vessel built (after the US steamer *Albatross* – of which

Figure 4.2 Prince Albert I of Monaco, photographed by French artist Jean-Baptiste Édouard Detaille. From the Photo Library of the National Oceanic and Atmospheric Administration (NOAA).

more to come). The *Princesse-Alice* conducted seven long-range cruises in the Mediterranean and the Atlantic between 1891 and 1897. To enable work in deeper waters, a second, larger and more powerful *Princesse-Alice* was built in 1897, to be succeeded in 1911 by *Hirondelle II*, whose activities shut down with the advent of the First World War. In all, the prince conducted twenty-eight expeditions between 1885 and 1915, a feat unparalleled in oceanographic enterprises of the era.

The prince captained his ships on every cruise, and Jules Richard (1863–1945), an outstanding and inventive French oceanographer, supervised the scientific tasks of all the cruises from 1887 on, as well as the publication of the scientific reports (*Résultats des campagnes scientifiques*). Strangely, Richard's only direct contribution to bioluminescence was an article published in his youth, not on a marine species, but on a terrestrial animal, the myriapod *Scolioplanes crassipes* (Richard, 1885).

Surveying the Atlantic from tropical zones to the Arctic, the Monaco cruises dredged animals from record depths (over 6,000 meters); their record was broken only in 1947. The prince's team built a watertight bathymetric recorder that accompanied the trawling net and helped determine the precise depth at which animals were captured as the net meandered its course to the surface (Albert de Monaco, 1921). They also devised an on-deck mechanism enabling them to control the closure of the net's mouth at a fixed depth, so as to avoid contamination by animals from upper layers as the net was hauled up (Albert de Monaco, 1889). These clever contrivances, which cast doubt on the accuracy of the *Challenger*'s measurements of pelagic depths, contributed to the discovery of the daily vertical migration of midwater animals and, more pertinent to our topic, to the important discovery that nearly all the vertically migrating animals possess luminous organs (Albert de Monaco, 1921).

Paul Regnard, an old schoolmate of the prince at the Collège Stanislas in Paris and a professor of physiology at the Sorbonne, participated in some of the early Monaco cruises. He described an experience of the luminescence apparently associated with vertical migrations (Regnard, 1891): "Offshore, prince Albert of Monaco, witnessing a phosphorescent sea, moved to catch swarms of small luminous crustaceans which covered the surface. As morning approached, these myriad organisms vanished and sank in deeper waters. But their luminescence had not ceased as they sank, and those who had illuminated the water surface now shed a similar light on deeper layers." So, we know that the prince and his team, like the *Challenger* crew before them, became acquainted with and intrigued by bioluminescent phenomena early in their cruises.

The collections of deep-sea animals reaped by the Monegasque ships revealed a diversity of abyssal life unsuspected even by the *Challenger* expedition (see the following chapter for greater detail). To a greater degree than invertebrates, the strange-looking deep-sea fish fauna started to arouse the imagination of the public – and even of writers like Jules Verne. Other nations took note of the growing impetus for ocean exploration and began to act on it. Around 1880 the US Fish Commission, which regulated coastal fisheries, moved to expand its surveys to the open ocean. To this end they obtained appropriations from Congress for a custom-built ocean-going research vessel. The *Albatross* was launched in March 1882 from the shipyard of Pusey and Jones in Wilmington, Delaware (Hedgpeth, 1946). Run more along the lines of the *Challenger* than the Monaco vessels, as explained by Joel Hedgpeth (1946), the *Albatross* "was a 'bastard' ship, manned by naval personnel but controlled by a civilian department, and assignment to her was not always considered advantageous to advancement in the service. For scientists, however, a cruise on the *Albatross* was an almost essential requirement of their career, and the roster of scientists who have sailed on the *Albatross* reads like the *Who's Who* of American zoölogy."

One of those zoologists was Alexander Agassiz (1835–1910). He was born in Neuchâtel, Switzerland, where his father, Louis Agassiz, was director of the local Museum of Natural History (G.R. Agassiz, 1913; Lurie, 1960). When his parents separated, Alexander stayed behind with his mother in Germany, while his father moved to Boston. The year after his mother died of tuberculosis, the thirteen-year-old Alexander was reunited in Boston with his father, who had by then been appointed professor of zoology and geology at Harvard College. As Alexander grew to adulthood he witnessed the fulfillment of his father's great scientific accomplishments, such as the creation of seaside laboratories and of the Museum of Comparative Zoology at Harvard. Alexander graduated in engineering and later in natural history at Harvard.

After his marriage Alexander started working at his father's museum as agent and curator of Radiata (coelenterates and echinoderms), but even in the land of opportunity, dedication to natural history brought few financial rewards; his salary could not support his young family. However, he and some relatives had bought shares in a major copper-mining operation in Northern Michigan and by 1866, with Calumet and Hecla Mines facing difficulties, Alexander was sent to Michigan to manage the company, drawing on his engineering training. As he explained to a colleague: "I want to make money; it is impossible to be a productive naturalist in this country without money. I am going to get some money if I can and then I will be a naturalist. If I succeed, I can then get my own papers and drawings printed and help my father at the Museum" (G.R. Agassiz, 1913). His expert management of the mines ensured their success and by 1871 he became president of what became a very profitable mining company that employed up to fifteen thousand men and made him a very wealthy man for the rest of his life.

Alexander was now able to carry on with his studies of sea urchins happily and serenely at the Museum of Comparative Zoology until two deaths crushed him utterly. In December 1873 his wife and his famous father died within eight days of each other. These sudden events – his wife's demise in particular – changed him into a permanently set, gloomier personality. In 1874 he took charge of the museum but without an academic appointment at Harvard. He continued to work on sea urchins and, as a means of distracting himself from the memories of his personal losses, travelled extensively to add specimens to his museum's collections. Interestingly, it was Alexander, who knew both Wyville Thomson and Sir John Murray personally, who in 1881 wrote the report on the sea urchins that had been collected by the *Challenger*.

Alexander Agassiz's own contribution to oceanography was bound up with his first *Albatross* expedition of 1891. The ship had until then basically done the US Fish Commission's bidding – surveying ocean basins with practical fisheries in mind – but paid only lip service to oceanographic science. Alexander would change all that. The results of his collecting trips in the Caribbean over the years had raised this simple question in his mind: what happened to the deep-sea fauna on the Pacific side when the formation of the Panama isthmus isolated it from the Caribbean side? To what extent had the Pacific fauna diverged from the Caribbean's over time? Ever since 1879 he had wanted to launch an expedition to address this inquiry, but was frustrated at every turn. Finally, in 1890, he was approached by the US Fish Commissioner to take charge of a deep-sea expedition off Panama the following year in the *Albatross* – under certain conditions: "The conditions under which Agassiz was offered the ship included his supplying the coal, assisting in thoroughly re-equipping the boat, and paying part of the running expenses. In return, he was to get a first set of the collections which especially interested him" (G.R. Agassiz, 1913). In other words, his wealth was well known to the commissioner and it afforded a bargain for the government. But Agassiz, who was reaching the grand age of fifty-six, had yearned so long for this expedition that he swallowed the pill and accepted the offer.

Agassiz joined the *Albatross* in February 1891 in Panama, just as the canal was being built by the French, whose progress in the work, he reported, left him unimpressed. For over two months the *Albatross* surveyed the Eastern Pacific off Baja California, Central America, and around Cocos Island and the Galapagos Archipelago, conducting numerous dredges and soundings along the way. To his question on the nature of the deep-sea fauna in the

Eastern Pacific, Agassiz found evidence confirming his suspicion about a former connection between the Pacific and the Atlantic, as expressed in a letter dated 14 March 1891 to the Fish Commissioner (G.R. Agassiz, 1913):

> I have found, in the first place, a great many of my old West Indian friends. In nearly all the groups of marine forms among the Fishes, Crustacea, Worms, Mollusks, Echinoderms, and Polyps, we have found familiar West Indian types or East Coast forms, and have also found quite a number of forms whose wide geographical distribution was already known, and is now extended to the Eastern Pacific.

The *Albatross* here was covering territory overlooked by the *Challenger* expedition; the British ship had sailed from Hawaii straight to the Juan Fernández Islands off Chile, thus circumventing the equatorial and subequatorial zones of the Eastern Pacific. Agassiz made the most of this circumstance, with the aid of the excellent facilities offered by the ship's equipment and roomy laboratory. Like his predecessors he was particularly interested in studying the vertical distribution of animals, but to his taste the gears designed until that point to open and close at determined depths were unsatisfactory, if not downright untrustworthy. So, with his skipper, Zera Luther Tanner, who had served on the *Albatross* from her launch and who excelled also as hydrographer and oceanographer, he designed the Tanner net. The purpose of this device was to prevent "anything from getting into the net on the way [down]. It was then towed for a time at any desired depth, and before being hauled to the surface a messenger was sent down that released two weights which tightly closed the lower part of the net, leaving the upper part open, to catch specimens on the way to the surface" (G.R. Agassiz, 1913).

Thus equipped, the *Albatross*'s trawling suggested that intermediate depths – roughly equivalent to today's mesopelagic zone – are poorly represented in animal life. This finding ran counter to those of his predecessors on the *Challenger* or the Monaco cruises, and to the results of Carl Chun in the Gulf of Naples (Chun, 1888), but Agassiz was adamant that the others' techniques had been at fault, and he persisted in his belief (Agassiz, 1892). As it turned out, he was right: the Eastern Pacific is a local anomaly due to the oxygen-minimum layer found at mid-water depths there, where only organisms physiologically adapted to such anoxic conditions can thrive (Mills, 1980). Of course, Agassiz could not have known or guessed this peculiarity at the time.

In this and a subsequent expedition to the Eastern Pacific (Agassiz, 1906), Agassiz's reports never made any mention of live displays of bioluminescence. Nor did he draw readers' attention to the inescapable fact that many of the enumerated dredged animals must have carried light organs. In contrast, he rhapsodized on the striking diversity of animal colours at depth. Why was it so? Perhaps no animal hauled from significant depth turned up alive; many descriptions of the catches suggest the towing of the nets caused serious damage to the specimens. Mention of luminous animals in these zones only appeared in some scientific reports by experts in the ensuing years.

Carl Chun was different. Even before his stewardship of the *Valdivia* expedition, bioluminescence and light organs were integral to his vision of what the oceanic environment was all about. Chun was born and raised in a suburb of Frankfurt am Main, and his partiality to the natural world was nurtured by visits to the local Senckenberg Museum. He studied zoology at the University of Göttingen and earned his doctorate in medicine at the University of Leipzig, where the great zoologist Rudolf Leuckart acted as his graduate supervisor (Mertens, 1957). As assistant to Leuckart in the ensuing years, he spent considerable time at the Naples Zoological Station, where he studied comb-jellies. The publication of his monograph on these transparent, iridescent pelagic animals in the *Fauna and Flora of the Gulf of Naples* series in 1880 earned him wide recognition as a zoologist and facilitated his appointment at the University of Königsberg in East Prussia – today the city of Kaliningrad in Russia.

Chun used his vacations in the Mediterranean to tinker with fishing nets in his quest for perfect sampling techniques to capture pelagic animals at various depths. This research resulted in an important monograph on the vertical distribution of pelagic fauna which was published in 1888. In this monograph Chun asserted that pelagic fauna were significantly present through the entire water column, but varied in species composition at different depths (Chun,

Figure 4.3 Carl Chun. From Winter (1914).

1888). We may recall that Alexander Agassiz criticized this monograph precisely because of Chun's sampling methods, which he considered to be flawed. But Agassiz was a single discordant voice drowned by other appraisals, and Chun became recognized as a pioneer of plankton research. His reputation preceded him to Breslau (now Wroclaw, Poland) where he was appointed professor of zoology in 1891 – considered a promotion, given the rating of German universities in that period. From Breslau he kept active in plankton work, concentrating on comb-jellies and siphonophores.

It was during his tenure in Breslau that Chun's plankton research led him to his interest in bioluminescence. Indeed, an important component of the bathypelagic fauna captured by his nets were the euphausiids – known to fishermen as "krill," key to the diet of many whales. In his first paper on the topic, Chun asked if the structure of the light organs and the stalked compound eyes of bathypelagic crustaceans (those who lived 1,000–3,000 meters deep) revealed special adaptations to low light levels compared with epipelagic (200 meters or less) species (Chun, 1893). His findings suggested that the light sensitivity of euphausiid eyes increased with depth through

Figure 4.4 The trawling net of the *Valdivia*, ready for lowering in deep water. From Chun (1905).

structural reorganizations such as loss of retinal pigment, to allow as much of the remaining ambient light as possible to reach the photoreceptors. Turning to light organs, Chun found that species living at great depths possessed fewer light organs than shallower species, while those living at depths where no solar light reaches had no light organs at all. He gave a more detailed description of the light organs three years later (Chun, 1896) – of which more in the following pages.

For Chun it was a stain on Imperial Germany that it had not joined the ranks of nations engaged in global oceanographic surveys. He took it upon himself to present such a project to his colleagues at the 1897 meeting of German natural historians and physicians held in Braunschweig (Klewitz, 2013). They endorsed his plan and facilitated financial authorizations by Kaiser Wilhelm II, the Bundesrat (Federal Council), and the Reichstag (Parliament) in January 1898. In the meantime, his old mentor in Leipzig, Rudolf Leuckart, had died and Chun was named to replace him as chair of zoology in July 1898. A mail steamer, the *Valdivia*, was put at his disposal and it was retrofitted for oceanographic and dredging tasks in an astonishingly short time, so that it was ready to sail on 1 August 1898, leaving Chun little time to enjoy his academic promotion.

For assistance Chun could count on zoologists Carl Apstein and Ernst Vanhöffen from Kiel, Fritz Braem from Breslau, and August Brauer from Marburg. The botanist Wilhelm Schemper (Basel), the oceanographer Gerhard Schott (Hamburg), the chemist Paul Schmidt (Leipzig), and the physician and bacteriologist Martin Bachmann (Breslau) completed the scientific team (Klewitz, 2013). Friedrich Wilhelm Winter, a scientific draftsman and photographer from Chun's hometown (Frankfurt am Main), proved an invaluable addition to the team. Chun engineered a publicity coup by having Sir John Murray as guest of honour on board for the leg of the journey between Hamburg and Edinburgh. Thus began their 32,000-nautical-mile journey, which took them to the South Atlantic, near the Antarctic mainland, across the Indian Ocean to Sumatra, northwest to the Bay of Bengal and Ceylon (Sri Lanka), off to the East African coast, and up to the Gulf of Aden and the Red Sea. It was then a simple matter to cross the Suez Canal, head to Gibraltar, and sail the North Atlantic back to Hamburg on 1 May 1899.

Three years after the *Valdivia*'s return, Chun sought to arouse interest in the deep sea from a wider circle than that of scientists by writing a popular book, *Aus den Tiefen des Weltmeeres*. The following passage vividly describes the rainbow of emotions elicited by their first encounter with deep pelagic fauna as they sailed west on the Atlantic:

> After the relatively sparse deep-sea bottom fauna, we were compensated on leaving Cape Verde by the absolutely amazing trophies caught in our vertical nets at greater depths. For the first time the magic of the pelagic deep-sea fauna caught up with us, what with a true abundance of new and remarkable animal forms. We will describe them in a different context later, but let us say for now that for the first time those black deep-sea fishes appeared in our nets, which thanks to their sets of light organs and to their bizarre habits always greatly excited the interest of researchers. Also collected were large, blood-red crustaceans, giant forms of ostracods the size of hazelnuts, transparent squids, arrow

> worms with red intestines, violet colored jellyfish, fragrant, uncommonly delicate sea cucumbers, deepwater forms of comb jellies of unprecedented behaviour, and an overabundance of radiolarians with exotic pebble skeletons. The scientific crew was in constant excitement about this unexpected glory brought up by the nets; all hands on deck were fully at work to draw and preserve specimens, and often they were enthralled and astonished by the mixture of colors, the transparency and bizarre shape of some forms. When our artist saw and tried to draw the bizarre deep-sea fish Melanocetus for the first time, this utterance escaped him: "It is as though God had hidden in the deep sea all the stupidities he ever committed." (Chun, 1905)

The *Valdivia*'s greatest sounding depth read nearly 6,000 meters. The most interesting, if bizarre catch occurred in the middle of the Indian Ocean, where they hauled up a humpback frogfish, a fish larva with eyes stuck at the distal end of long stalks, and squids also with stalked eyes. As it happened, the richness and diversity of deep-sea squids and fishes dominated the collection of animal species. The prevalence of bioluminescent species among these did not, to be sure, escape Chun's notice, partial as he already was to the importance of bioluminescence in the deep ocean. But more important for Chun, the expedition confirmed him in his deep-seated belief that the fauna of the "intermediate depths" of the ocean, where temperatures are low, are uniformly rich from the Arctic to the Antarctic and in all oceans, in contrast to shallower waters, where higher but also more variable temperatures due to the whims of prevalent currents produce a patchier distribution of animal assemblages (Chun, 1905). The twenty-four published volumes of the scientific results of the expedition documented these findings very effectively.

~~~~

The final expedition considered in this chapter boasted less ambitious goals than the *Valdivia*. The voyage of the *Siboga* confined itself to waters in and around the Indonesian archipelago (Van Aken, 2005). Calls from Dutch academics for an oceanographic expedition to the Indonesian Seas dated back to 1888. After all, was it not time the home country took seriously the
~~~~

Figure 4.5 Max Wilhelm Carl Weber. Courtesy of Artis Library, Special Collections, University of Amsterdam.

survey of its own colonial waters? But general pleas for action were not sufficient. It took the determined initiative of a single individual, in the spirit of Carl Chun's drive, to set things in motion. It is the contention of Florence Pieters and Jaap de Visser "that the pinnacle of Max Weber's scientific career was the organization and leadership of the Siboga Expedition to the former Netherlands East Indies (now Indonesia) in the years 1899–1900" (Pieters and de Visser, 1993).

Max Wilhelm Carl Weber (1852–1937) was born in Bonn of a German father and a Dutch mother. His father died before he reached his second birthday, so his privileged contact with his mother ensured that he spoke German and Dutch fluently (Pieters and De Visser, 1993). He studied medicine and natural history at the University of Bonn, where he fell under the influence of the leading zoologist and comparative anatomist Franz Leydig, whose assistant he became. After securing a PhD in zoology in 1878, he was appointed lecturer in anatomy at the University of Utrecht, a post he held until 1883, when he was promoted, at the unusually precocious age of thirty, to the prestigious chair of zoology and comparative anatomy at the University of Amsterdam. In the same year his attachment to the Netherlands was welded by his submission of his naturalization papers and by his marriage to a young widow, Anne Antoinette van Bosse, who became an authority on marine algae (Pieters and De Visser, 1993).

Before the *Siboga* adventure Weber was already no stranger to sea voyages or to excursions in the Dutch East Indies. He had travelled on ships operated by the Dutch Navy, collecting animals, acting as medical officer and generally impressing Dutch navy officers, all of which stood him in good stead for the future. The catch he prized most was whales and, in addition to his many other eclectic zoological interests, he became a pioneer expert on their comparative anatomy. In 1888 he had travelled to Sumatra, Java, and the Celebes, where he studied the geographical distribution of East Indies animals, especially the freshwater fauna. Later on, he travelled to South Africa with his wife. All these experiences had shaped him into a well-rounded zoologist.

Weber seized the opportunity of the 1896 annual meeting of the Dutch Commission for the Advancement of Scientific Research in the Dutch Colonies, during which a vague proposal for a modest expedition was being discussed, to submit a far more ambitious plan "for an oceanographic expedition to explore the marine fauna in the deep basins of the Indonesian archipelago," and "he declared his availability to lead that expedition" (Van Aken, 2005). The commission accepted Weber's plan and it was agreed that the Dutch colonial government should supply a research vessel and seek financial support for the voyage. "In May 1898," writes Hendrik van Aken, "the Governor General in Batavia [today's Jakarta] decided that the newly built gunboat *Siboga* of the colonial navy should be made available to Professor Weber." All the chips fell into place, financially and logistically, and Weber was officially appointed the leader of the expedition.

To fit a 170-foot-long steam-powered gunner boat for oceanographic work was no simple matter. Guns had to be removed to make space for the electrical and steam winches for sounding and dredging work. For all his trawling and other equipment, Weber sought advice from Sir John Murray of *Challenger* fame and from Carl Chun, visiting the *Valdivia* in Hamburg shortly before her departure (Weber, 1902). Weber's good standing with the Dutch navy ensured the navy's willing cooperation in readying the ship for her mission. As recorded in Weber's own account (1902) of the expedition, inexplicably written in French, the *Siboga* finally sailed from Amsterdam on

16 December 1898 and arrived in Batavia on 7 February 1899. From there she sailed to the East Java port of Surabaya, where the final preparations were completed.

Weber's team included: his first assistant, Jan Versluys; second assistant, Hugo Frederik Nierstrasz; draughtsman J.W. Huysman; and his wife, Anna van Bosse-Weber, the first woman ever to participate in an oceanographic expedition, in charge of the marine flora. The hydrographic surveys were put in the care of the captain, Gustaaf Tydeman, a highly skilled mariner who got along well with the likeable Weber, leader of the expedition (Van Aken, 2005). The *Siboga* departed from Surabaya on 7 March 1899 and returned to the port on 27 February 1900 after nearly a year of circuitous sailing through the region, during which over two hundred deep soundings were performed and numerous deep-sea animals were collected as per the model of their predecessors. The soundings allowed the expedition to determine that the Wallace line – named after Alfred Russell Wallace, the zoologist and co-proponent of the theory of natural selection with Darwin – supposedly representing an impassable barrier between the fauna of South Asia and Australia, is in fact blurred by a broad transition zone of mixed faunal composition between the two sides.

This expedition sounded a note of disagreement with the previous expeditions, however, in that serious attention was again paid to surveys of shallower and coastal waters. This came about at the urging of Sir John Murray, who regretted the *Challenger*'s neglect in this regard. Heeding this advice paid off handsomely, as the *Siboga* collections showed that the coastal fauna was the richest of its kind in the world. This decision also brought dividends for bioluminescence. Among the captured coastal fishes Weber's curiosity was drawn to two unusual luminous species, as narrated at length in his introduction to the voyage (Weber, 1902). Weber noted that the fishermen of the Banda Islands used the light organs of these fish as lures:

> The luminous organs used by the fishermen of Banda are truly extraordinary because the fish that possesses them can of its own volition expose or completely hide them: in the first case they broadcast their light; in the second no trace of light can be gleaned … The indigenous people of Banda have long known that the "Ikan leweri laut" and "Ikan leweri batu," as they call *Anomalops* and *Photoblepharon*, have under

their eyes a mobile disk which emits an intense light. This disk can easily be excised and its luminous power remains intact for several hours. That is why the fishermen fasten it to the hook and use it as a lure to catch larger fish. (My translation from French)

Unbeknownst to Weber at the time of the voyage, a Dutch physician and naturalist living in Indonesia, Adolphe Guillaume Vorderman (1844–1902), had observed the blinking light of these organs in 1897 (Vorderman, 1900). But Vorderman's short paper was written in Dutch and therefore failed to attract the readership that Weber's account enjoyed two years later. The exact mechanisms by which these light organs switch on and off and the role of these "living flashlights" in the life of these fish became known only later (see chapters 9 and 16).

When the *Siboga* expedition ended, the ship returned to its naval duties as a gunner. But for science it was just the beginning. The pattern of assigning experts to study the collections seen in the previous expeditions repeated itself. Between 1901 and 1982, over 130 monographs of results from the *Siboga* expedition were published (Van Aken, 2005). Obviously, this follow-through involved several generations of scientists. As the font of knowledge from this and the previous oceanographic odysseys started to accumulate – not only in scientific publications but also in the consciousness of the general public – the inhabitants of the deep, and especially bioluminescence, gained unprecedented exposure.

5

The Mystery of a Lit Underworld

> Luminous phenomena in the dark of night presented an erotic spectacle of sparkling arcs and serpentine fires pouring into the obscure womb of the fertile depths.
> –Natasha Adamowsky (2015)

The scientific reports of the pioneer oceanographic expeditions swelled to such a volume in the latter part of the nineteenth century and the early years of the twentieth that their production can truly be regarded as a cottage industry on a grand scale. They served not only to disseminate the newly acquired knowledge of the deep seas among the experts but also to arouse the general public's awareness of the strange inhabitants of the dark underworld. After all, what was left to be explored now that distant lands had largely been charted, if not the deep cosmos – the purview of the astronomers – and the deeper recesses of the oceans?

Just a few years before the *Challenger* expedition, Jules Verne's novel *Twenty Thousand Leagues Under the Sea* had articulated questions much on the minds of his readers: "The great depths of the ocean are totally unknown to us. Soundings have failed to reach them. What is going on in these recessed abysses? What creatures dwell or may dwell twelve or fifteen miles below the surface? What is the make-up of these animals? We can hardly conjecture" (my translation). But what Verne conjectured in his novel is what many feared: that giant menacing monsters made their abode in those deeps.

These questions found no clear answers in the scientific results of the ensuing years. Why not? Early marine scientists could not embrace in their conceptual field the three-dimensional living world of the deep seas the way we can embrace our terrestrial environment and watch a firefly flash her signal to the opposite sex and make sense of it without resorting to extrapolation. The only way these scientists could sample the deep-sea world was to

Figure 5.1 Composite image of Sir John Murray with the laboratory of the ship *Challenger* in the background. Still photo from the documentary film *The Man Who Challenged the Deep*, by Andy Crabb (SAMSCOMFILMS). Courtesy of Andy Crabb.

haul to the surface animals from a minuscule portion of the sea and stare at them out of their environmental context. Sir John Murray, who had in person sorted catches hundreds of times on the deck of the *Challenger*, put it this way:

> It has often been said that studying the depths of the sea is like hovering in a balloon high above an unknown land which is hidden by clouds, for it is a peculiarity of oceanic research that direct observations of the abyss are impracticable. Instead of the complete picture which vision gives, we have to rely upon a patiently put together mosaic representation of the discoveries made from time to time by sinking instruments and appliances into the deep, and bringing to the surface material for examination and study. (Murray and Hjort, 1912)

So, they were left with specimens pickled in formaldehyde or spirit, and with snapshots of animal physiognomies, drawn or photographed by the resident artists of the research vessels. Even these results often could not fail to disappoint. An early pessimist was the *Challenger* naturalist Henry Moseley. In true Darwinian spirit the *Challenger* naturalists had gone out fishing for the ancestral deep-sea forms of fossils found in geological deposits. This quest went hand in hand with the long-held view that life had begun in the sea; that the primaeval womb of life lurked in the depths. According to Moseley their efforts went unrewarded:

> In many respects, the zoological results of the deep-sea dredging were rather disappointing. Most enthusiastic expectations were held by many naturalists, and such were especially put forward by the late Prof. [Louis] Agassiz, who had hopes of finding almost all important fossil forms existing in life and vigour at great depths … Large numbers of interesting new genera and species of well-known families of animals were obtained by the dredge, but very few which were widely different in their essential anatomical structure from hitherto known forms, and thus of first rate zoological importance. We picked up no missing links to fill up the gaps in the great zoological family tree. The results of the "Challenger's" voyage have gone to prove that the missing links are to be sought out rather by more careful investigation of the structure of animals already partially known, than by hunting for entirely new ones in the deep sea. (Moseley, 1879)

But to many, Moseley had missed the point. The deep seas swarmed with creatures; that was the message the oceanographers brought back to the landlubbers. What these creatures were up to down there was a mystery. Add to that the finding that many of the creatures were luminous, and the mystery was heightened. Natasha Adamowsky explained the cultural process that raised consciousness of bioluminescence to a new level in her insightful essay:

> For centuries, marine luminescence, like other phenomena, had likely not been perceived as anything mysterious by coastal inhabitants. Living in a world filled with willful forces and dark powers, they would have considered it more important to know the spawning grounds of herring than the reasons why the sea glowed. Bioluminescence was a matter one simply observed – a riddle, perhaps, but not a mystery.

> Only when natural science developed an interest in marine phenomena in the course of the eighteenth century did it demand explanation. Mysteries emerge at the height of knowledge, as if to spite it. In other words: Knowing too little always also means not knowing that there is a mystery. However, as soon as the glow of the sea was recognized as something mysterious, it became possible to view it in its complex, theatrical dimension – which constitutes the essence of mystery. (Adamowsky, 2015)

Initially, to be sure, bioluminescence did not feature among the phenomena of the deep sea to be anticipated in the course of the voyage of the *Challenger*. And yet its crew experienced several sightings of luminous displays by a variety of deep-sea animals and could not fail to notice organs that they suspected were the source of the light emissions. But Moseley found nothing extraordinary about these findings and, as the following quotation tends to show, he seems to have gone out of his way to downplay them:

> All the Alcyonarians [octocorals] dredged by the "Challenger" in deep water, were found to be brilliantly phosphorescent when brought to the surface, and their phosphorescence was found to agree in its manner of exhibition with that observed in the case of shallow-water forms. There seems no reason why these animals should not emit light when living in deep water, just as do their shallow-water relatives.
>
> The light emitted by phosphorescent animals is quite possibly in some instances to be regarded only as an accidental product, and of no use to the animal producing it, although of course, in some cases, it has been turned to account for sexual purposes, and may have other uses occasionally. There is no reason why a constant emission of light should be more beneficial than a constant emission of heat, such as takes place in the case of our own bodies, and it is quite conceivable that animals might exist to which obscure heat-rays might be visible, and to which men and Mammals generally, would appear constantly luminous.
>
> However, whether be the light beneficial to them or not, it seems certain that the deep sea must be brighted here and there by greater or smaller patches of these luminous Alcyonarians, with wide intervals,

> probably, of total darkness intervening; very possibly the animals with eyes congregate round these sources of light. (Moseley, 1879: 590–1)

Moseley's teammate on the *Challenger*, Sir John Murray, held a contrasting view:

> Phosphorescent light plays a most important role in the deep sea, and is correlated with the prevailing red and brown colours of deep-sea organisms. Phosphorescent organs appear sometimes to act as a bull's-eye lantern to enable particles of food to be picked up, and at other times as a lure or a warning. All these peculiar adaptations indicate that the struggle for life may be not much less severe in the deep-sea than in the shallower waters of the ocean. (Murray, 1899)

One fauna that contradicted Moseley most persuasively was deep-sea fishes. We have seen in previous chapters how little naturalists knew about luminous fishes, even doubting their existence. The fish hauls of the early oceanographic expeditions changed all that, to the point that luminous fishes took centre stage with cephalopods. It was Albert Günther (1830–1914), the zoologist whom Wyville Thomson and Sir John Murray put in charge of the *Challenger* fish collection, who was largely responsible for this change. Albert Karl Ludwig Gotthilf Günther was born in Southern Germany and initially studied for the ministry in the Lutheran Church as family tradition demanded (M'Intosh, 1915). But under the influence of the great physiologist and zoologist Johannes Müller in Berlin, he soon switched his interests to zoology. However, he followed the example of the majority of naturalists of that period in pursuing a medical degree on the side. In 1857, after early ichthyological work on German freshwater fish, Günther moved to England, where his mother then lived, to arrange, at the invitation of curator Richard Owen, the fish, amphibian, and reptilian collections of the British Museum (M'Intosh, 1915). His reputation as a leading systematic zoologist rose with the years, and in 1875 he was appointed the Keeper of Zoology at the British Museum.

Günther's expertise in ichthyology was such that his assignment to study the *Challenger* fish collection was no surprise. But, due to the heavy demands of his duties at the British Museum, it was ten years before he saw his labour

Figure 5.2 Albert Karl Ludwig Gotthilf Günther. From Smithsonian Institution Archives, image #85-4441, with permission.

in print. The result was worth the delay. Günther's bread and butter as a museum employee consisted in naming, describing, and cataloguing collection specimens – which he did for all 177 deep-sea fish species collected by the *Challenger*, many of which were new to science. But his monograph (Günther, 1887) offered so much more. He provided a description and interpretation of the features that distinguish deep-sea fish from their shallower counterparts, thereby helping to lift some, but not all, of the mystery around these creatures.

Günther noticed, for instance, their extremely weak, frail-looking skeletons made of thin, cavernous bones that would not support the body at the water surface. This and their correspondingly thin, spare muscles he interpreted as adaptations to the tremendous pressures under which these fish swim. A definite curiosity to him was the colour of the walls of the internal organs. "An extremely common, almost general characteristic of deep-sea fishes," Günther remarked, "is the black coloration of some of the body-cavities; this is limited to the pharynx in many of the fishes which live about the 100-fathoms [near 200-meter] limit, but the colour is more intense, and spread all over the oral, branchial, and peritoneal cavities in strictly typical deep-sea forms." It is now believed that this phenomenon has a connotation for bioluminescence. Donald McAllister (1934–2001), an ichthyologist then working at the National Museum of Canada, proposed that black peritoneum and stomachs serve to conceal the bioluminescence of recently ingested prey so as not to advertise themselves to their own predators (McAllister, 1961). That role is still accepted today.

The features that loom largest in Günther's discussions are surely the light organs, whose description stretches over several pages. His interest in the subject dated back many years. In the 1887 *Challenger* report he wrote:

> In 1864, when engaged in the systematic arrangement of the fishes of the families Sternoptychidae, Scopelidae, and Stomiatidae, I ascribed phosphorescent properties to all these organs, whilst from their histological examinations Leuckart and Ussow declared them, or at least part of them, to be accessory eyes. Leydig holds the opinion that they are "pseudo-electric" organs, which sometimes may have the function of emitting light; Emery adopts the view of their phosphorescent nature in Scopelus. As these organs occur, not only in deep-sea fishes, but also in nocturnal pelagic forms, their function might have been expected to be readily ascertained by actual observation; however, so far as I am aware, this has been done twice only, viz., by myself, when I happened to notice distinct flashes of light to issue from a dying specimen of Scopelus floating on the surface in the British Channel; and by Dr. Guppy, who examined some freshly caught specimens of the same genus.

This passage is very instructive. Günther ascribed to organs a luminescent role for which he had no direct evidence save observations by him and by Dr Guppy (1881), each on a single *Scopelus* specimen, and a dying one at that. No wonder his German colleagues, Franz Leydig (Max Weber's mentor), Rudolf Leuckart (Carl Chun's mentor), and Mikhail Ussow in Russia, refrained from committing themselves to the existence of such a role. But their interpretation of what these structures stood for played into the hands of those all too keen to view these fishes as monstrosities. Both Leuckart and Ussow assumed that since these skin organs, distributed at various locations over the body, had all the structural trappings of eyes, they therefore should be construed as accessory or modified eyes (Leuckart, 1865; Ussow, 1879). Leydig, on the other hand, in a full monograph dedicated to the subject (1881), concluded that these structures were "pseudo-electric" organs which in some cases could also assume a light-emitting function. Altogether, these claims, based on superficial observations were, even by the standards of the day, runaway fantasies. But pondering a fish capable of "skin vision" or electric discharge added a new exotic touch to the beast.

Günther decided that the matter could only be settled by undertaking detailed studies of the histology of the light organs (Günther, 1887). He had intended to assign this task to Moseley, but the *Challenger* naturalist cited other duties to excuse himself. He then offered the job to Robert von Lendenfeld, who accepted. The latter's work, to be discussed in the next chapter, reassured Günther that he was dealing with *bona fide* light organs and he felt justified in discussing them as such. An example of his conclusions follows:

> Light-producing organs are very generally distributed in the abyssal fauna, and those parts of the depths of the ocean in which phosphorescent animals are abundant must be sufficiently illuminated to enable such of them as are provided with well-developed eyes to perceive objects with as much distinctness as do the pelagic forms which sport at the surface at night, and are dependent on the light of the moon and stars and the general phosphorescent light around them. There is no doubt that fishes contribute a considerable amount of this luminosity of the abyssal depths; but the various degrees of differentiation of the luminous organs, as well as their location on very different parts of the fish, prove that the production of light is dependent on a variety of circumstances and subserves different purposes.

Günther went on to enumerate these purposes of the luminous organs. Of course, the first purpose that comes to the mind of those used to carrying a torch or an oil lamp in the night is "to enable its possessor to see." Just as humans control when to light a torch or turn on a lamp, light emission must be controlled by the fish, for "if the production of light were constant, or

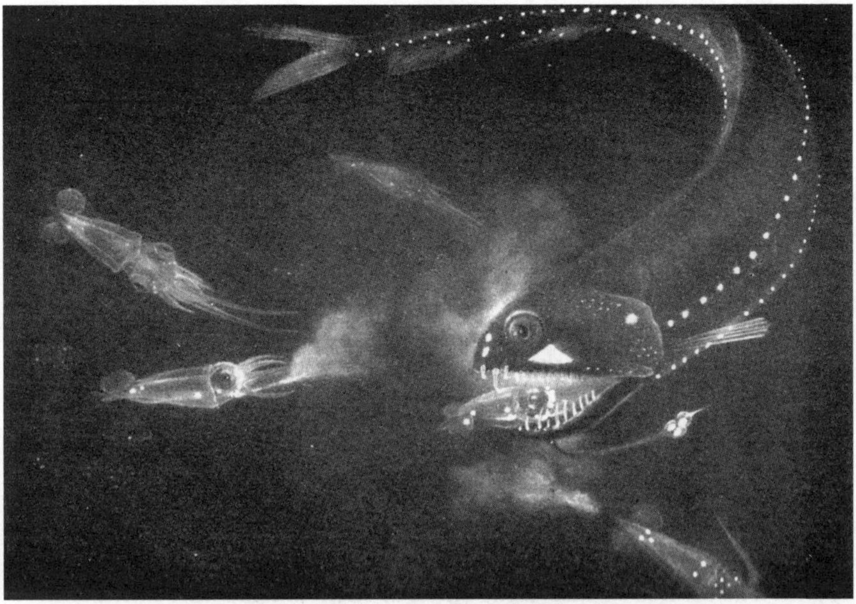

Figure 5.3 Dragonfish with subocular organ and body photophores lit up, preying on luminous squids. Drawing of Bruce Horsfall in Harvey (1940).

could not be suppressed instantaneously, the fish would be a most conspicuous object and fall a ready prey to its enemies." The large light organ under each eye of dragon fishes (stomiids), he suggests, is ideally placed to illuminate the field of vision where their prey may be spotted. Although the intensity of light emissions may vary between species and they therefore may be more or less effective as visual aids, Günther was adamant that "the light which issues from the large pearly organs of the Scopelidae (lanternfishes), the infraorbital organs of the Stomiatidae, and from the lenticular organs of the Halosaurids (sea lizards), must be intense and penetrate to a considerable distance."

By analogy with the fisherman's traditional use of light as a lure, Günther assumed that any light organ located on the barbels, filamentous fin-rays, or tentacles of a fish functions as a lure for the attraction of other animals. These fish are caught at depths where solar light does not penetrate and cannot therefore be used for vision, he argued; as confirmation, he noted that their eyes were degenerated like those of cave-fish. He theorized that these appendages would instead "prove most efficient lures in the abyssal darkness, when, with one or several bright phosphorescent spots at the end of the tentacle, they are played about by the fish." Likewise, he assigned this function to the light organs on the trunk tail of lanternfishes and hatchetfishes, for "we can understand that these posterior organs are of great assistance to the fish in picking up any creatures which, attracted by the gleam of light issuing from its tail, are lured into too near a proximity." Posterity has not favoured this theory; the caudal luminous organs, which differ between the sexes, are probably used for sexual attraction, not prey attraction.

The ichthyologist who wrote the monograph on the fish harvested by the *Albatross* expedition also mused at length on the functions of deep-sea fish bioluminescence. Samuel Walton Garman (1843–1927) was born in Pennsylvania into a Quaker family. Garman impressed Louis Agassiz during a trip west, with the result that Agassiz hired him to work on the fish collections at Harvard's Museum of Comparative Zoology (Summers and Koob, 1997). After Louis Agassiz's death his son, Alexander, took Garman under his wing and appointed him curator of fishes, amphibians, and reptiles at the museum, even though Garman held no doctoral diploma. It was in the order of things, then, that the deep-water fish collections of the *Albatross* fell on Garman's lap. As for character, Adam Summers and Thomas Koob had this to say: "Garman has fascinated modern ichthyologists by virtue of the rumors and facts of his eccentricity. In many ways he epitomizes our notion of a 19th century museum-based systematist: anti-social, obsessed with trivia, slightly tyrannical and extremely productive, if not imaginative, in his pursuit of scientific truth."

In the general discussion preceding his systematic descriptions of the fish, Garman confused theory with fact when he stated: "many of them [are] provided with lanterns, flash lights, or other luminous organs to prevent mates or individuals of a school from losing one another, or with light organs to lure the prey" (1899). He made a distinction between "abyssal" fishes that possess diffuse luminous tissue (occultative luminosity) such as sharks and macrurids, and those harbouring specialized light organs (demonstrative luminosity) as found in lanternfishes and stomiid fish. The latter organs are said to be controlled by the bearer and to serve mainly "as means of recognition by kindred and on others to decoy and bring the prey within reach."

Garman had this interesting observation about fish that used lures (luminous barbels at the tips of a stalk):

> Creatures living in the ooze in many cases are of an intense black in which no luminosity can positively be asserted to exist; on such of these as are possessed of lures the latter are most often directed upward as if to capture a prey swimming above them, for instance *Dolopichthys* … Many of the Stomiatoids [stomiids] and the Muraenoids also are of the same deep black; it would appear as if the former inhabited the lower edge of the dark or azoic zone, their lures are extended downward as if to secure prey approaching from below.

The deep-sea fish harvest of the *Valdivia* expedition at least equalled, if not exceeded in importance those of the *Challenger* or the *Albatross*. Of 90 genera and 206 species collected, 14 genera and 56 species were new to science (Brauer, 1906). The *Valdivia* expedition leader, Carl Chun, an invertebrate specialist, entrusted the care of the deep-sea fish collection to August Bernhard Brauer (1863–1917). Born in Oldenburg, Germany, Brauer trained in zoology at various universities (Bonn, Berlin, and Freiburg) and earned his PhD in 1885 with a thesis on ciliates (Vanhöffen, 1918). His interests drifted from unicellulars to *Hydra*, whose development he studied, and to zoogeography. After a few years of professional stagnation, Brauer in 1892 finally completed his habilitation at the University of Marburg, where he was hired as lecturer. As others in our narrative have done, he prioritized his *Wanderlust* over academic advancement. He travelled to the Seychelles in the Indian Ocean in 1894–95 to conduct zoogeographical research sponsored by the Akademie der Wissenschaften in Berlin (Vanhöffen, 1918). He was still publishing results from this trip when he was called to join the *Valdivia* expedition.

Chun chose Brauer on the basis of his expertise in zoogeography, certainly an asset for a global research enterprise, and for his familiarity with the Indian Ocean, where the *Valdivia* was to spend considerable time. The advantage of assigning the fish collection to Brauer was that, contrary to Günther,

Figure 5.4 August Brauer. From Vanhöffen (1918).

who only had access to specimens after the fact, he had been right there on the deck of *Valdivia* when the catches were hauled in, and had valuable contextual information about the fish he was studying. What's more, he also had the incredible luck of having an exceptional draftsman, Friedrich Wilhelm Winter, as his assistant. Brauer knew his good fortune and acknowledged his helper in his monograph on the systematics part of the work (Brauer, 1906):

> It was no small advantage that so many exquisitely coloured illustrations of already known and new forms were produced to accompany this work. The merit goes to my friend and travelling companion Fritz Winter, not only on board the ship where he rendered in masterly fashion the fresh colours of the fish immediately after their arrival, but also out of pure interest in science he has been willing to take on the job of producing the definitive version of the sketches. That the portraits of the deep-sea fishes in this book are so much lifelike can

only be attributed to the circumstance that he is not only an artist but also a zoologist.

Without minimizing any of Winter's merits, it should be emphasized that the quality of the specimens captured by the *Valdivia* expedition overall proved superior to that of all the other expeditions, with the possible exception of the Monaco cruises. Winter could work on better-preserved material because, as Brauer explained, they used vertical trawl nets equipped with a glass bucket instead of the usual cod-end. The smoother surface of the glass helped minimize the "laceration and blemishes" incurred when catches suffer great drops of pressure as they are hauled up.

Today, thanks to the trawling techniques of the *Valdivia* and the artwork of Winter, it is the illustrations of "his dear monsters," as Brauer called them, not his systematic descriptions and his insightful discussions of their vertical distribution, that fascinate readers. If Winter was the illustrator of the otherworldly physiognomies of these fish – and also particularly Chun's squids – Chun played the artistic director of the page layout for the *Valdivia* monographs. His artistic flair was such that, as Adamowsky (2015) remarked: "Fishes swim out at the viewer with murderous jaws or rush 'busily' over the pages and the centre fold, as if the book were a paper aquarium." Although Adamowsky failed to discern that neither Chun nor Brauer created the illustrations – or at best gave zoological guidance to Winter – she strikes the right note when she suggests the following: "Many of Chun's contemporaries shared his view of the wonder-inducing, marvellous and incommensurable as an important component of research work; however, he united aesthetic praxis and scientific insight at a level that others seem to have failed to attain even to this day."

Light organs are well represented in the elegant illustrations of Brauer's monograph. This degree of representation is no surprise, since it reflected Brauer's view that, ever since the *Challenger* expedition, bioluminescence was revealed as one of the most characteristic phenomena of the deep sea (Brauer, 1908). As of 1908 Brauer was the first to provide a census of bioluminescent fishes. By far the larger bulk consisted of deep-sea species, including: ten species of sharks, and among the bony fishes, the families Stomiidae (dragonfishes, eighteen genera); Sternoptychidae (hatchetfishes and bristlemouths, sixteen genera); Scopelidae (lanternfishes and blackchins, four

Figure 5.5 An assortment of anglerfishes with their luminous fishing lure. Modified from Colour Plate 18 in Brauer (1906).

genera); Saccopharyngidae (gulper eels, one genus); Ceratiidae (anglerfishes, eleven genera); Gigantactinidae (whipnose anglers, one genus); Antennariidae (frogfishes, one genus); and Malthidae (batfishes, eight genera). Shallow-water fishes were underrepresented, including the families Anomalopidae (flashlight fishes, two genera) and Batrachidae (toadfishes, one genus). Genera may include more than one species, so the number of deep-sea species likely to be bioluminescent in Brauer's era was probably a small multiple of the seventy genera listed by Brauer – maybe around one hundred and fifty species with light organs.

Given this number and the relatively gentle treatment the specimens enjoyed as they were hauled up, it seems surprising that so few sightings of bioluminescence were recorded by the *Valdivia* scientists. Brauer compiled the sightings in his second monograph, which covered the anatomy of the light organs and eyes of deep-sea fishes (Brauer, 1908). They saw the greenish glow of the head and the abdominal light organs of the hatchetfish *Sternoptyx*; a similar observation had previously been made on the *Challenger* by Rudolf von Willemoes-Suhm (1875). Similarly, Ernst Vanhöffen, a member of the *Valdivia* expedition and a lifelong friend of Brauer, observed green "phosphorescence" from the ventral side of the dragonfish *Astronesthes*. What really caught their attention was the large light organ just below each eye (suborbital) of several dragonfishes, reminiscent of the similarly situated light organ of the shallow-water flashlight fishes. The most spectacular was the suborbital organ of the stoplight loosejaw *Malacosteus* (Stomiidae), which "shone like a ruby"; its unusual red bioluminescence was rediscovered only many decades later (Denton et al., 1970). The suborbital organ of *Melanostomias* (Stomiidae), in contrast, "flashes like a pearl on velvet black" and a "splendid bluish phosphorescence showed forth." And Brauer finally mentioned the blue light emitted by the postorbital light organ of the black dragonfish *Idiacanthus*.

The *Valdivia* experience put Brauer at the vanguard of German zoology and was certainly instrumental in landing him the prestigious directorship of the Berlin Zoological Museum in 1906, the same year his monumental monograph on the *Valdivia* deep-sea fishes was published. Because of his new administrative duties, the publication of the anatomical section on deep-sea fishes had to be postponed until 1908, and even then, in view of the immensity of the task, Brauer confined himself to the histology of the light organs and eyes (see chapter 6 below).

Meanwhile, in Leipzig, Carl Chun experienced even further delay than Brauer in bringing out his monograph on cephalopods. Chun had valid reasons, not the least of which was his supervision of the publication of the expedition's results, as monumental a task as it had been for Wyville Thomson's and Sir John Murray's *Challenger*'s results. And of course Chun gave priority to publicizing the major achievements of the expedition to a larger readership, a concession to Kaiser Wilhelm II, who had funded the expedition and at the turn of the century was anxious for any source of prestige that could accrue to his empire (Klewitz, 2013). The popularizing book was *Aus den Tiefen der Weltmeeres* [Out of the Depths of the World's Oceans] (Chun, 1905). Chun had published an article on the light organs and eyes of cephalopods (1903) and a preliminary diagnostic note on the cephalopods of the *Valdivia* (1908), but the final monographs came out between 1910 and 1915. As Chun died in 1914, Brauer was appointed the new supervising editor and became responsible for ensuring the appearance of Chun's last instalments. One may speculate whether the inhuman demands of

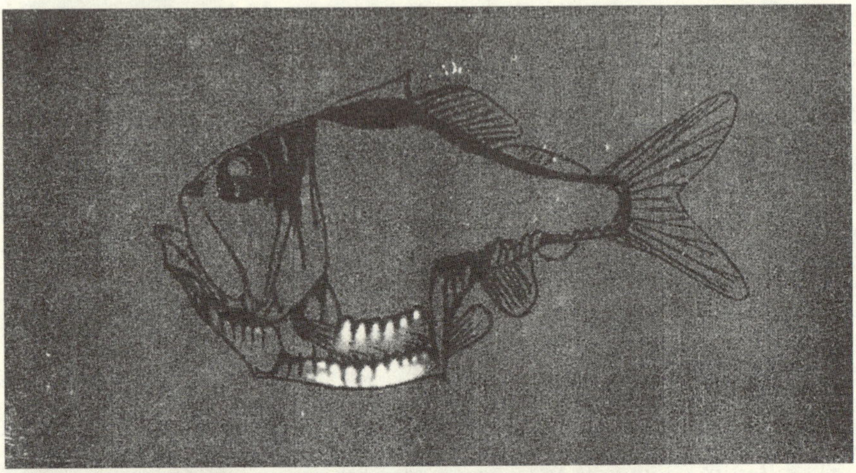

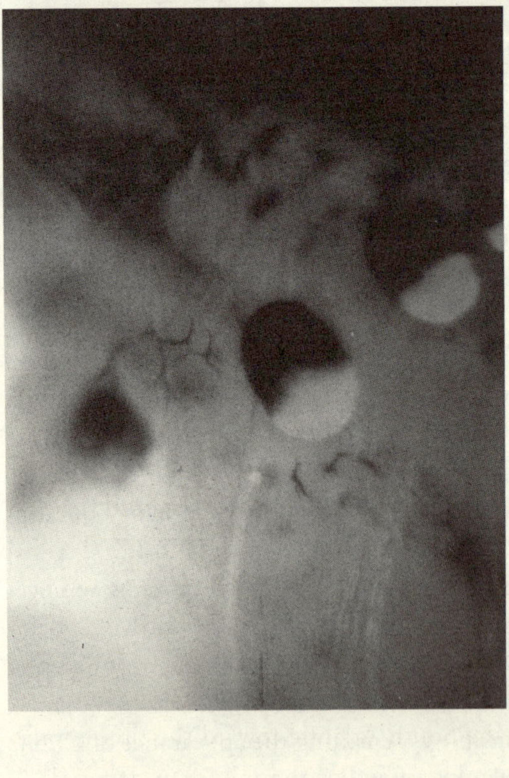

Figure 5.6a *Above*
The hatchetfish *Argyropelecus olfersi* photographed by its own light. The outline of the fish was ghost drawn afterwards. Note that light from the photophores is directed downward. From Harvey (1952) after Bertelsen and Grøntveld (1949).

Figure 5.6b *Left*
The fluorescence of viperfish *Chauliodus photophores* shows the dorsal pigmented cap of the organs and the downward (ventral) direction of light. Modified from an original colour photograph by Fernand Baguet, in which the fluorescence appears red.

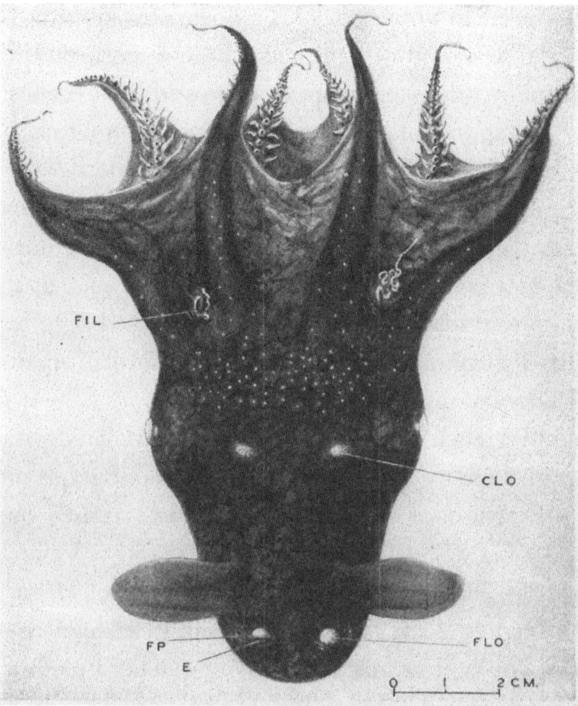

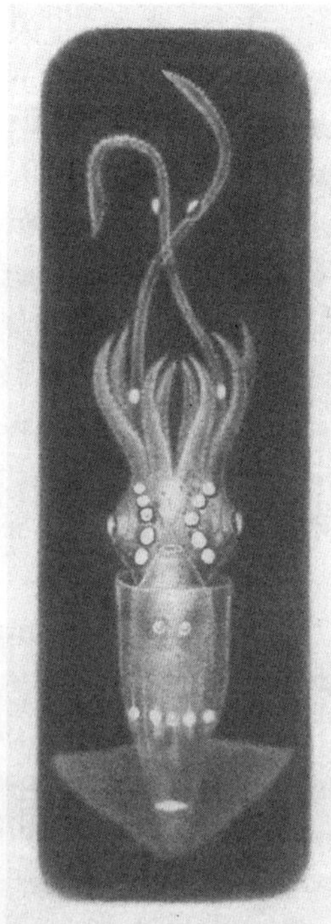

Figure 5.7a *Left* The vampire squid *Vampyroteuthis infernalis*, showing the location of its light organs: fin light organs (FLO), composite light organs (CLO) and above the latter minute organs spreading to the arms. From Harvey (1952) after Pickford (1949). Figure 5.7b *Right* The squid *Thaumatolampas diadema* with its luminous organs. From Plate 2 in Chun (1910).]

these mega-expeditions hastened the deaths of Wyville Thomson at fifty-one, Chun at sixty-one, and Brauer at fifty-four. One might say that these great oceanographers and zoologists fell victim to the overwhelming success of their adventures.

In the minds of many today, what Chun is most famous for is the discovery of the "vampire squid from hell," *Vampyroteuthis infernalis*, so named by Chun himself. Three specimens of this squid, which epitomized the kind of abyssal monster that had most captivated the public's imagination, were caught as the *Valdivia* crossed the South Atlantic toward the tip of South

America. Chun failed, however, to notice that his emblematic squid possessed light organs, a feature that was only discovered decades later (Pickford, 1949). But he found that many other cephalopods displayed light organs. In fact, he identified forty-two species of cephalopods known by him to harbour light organs, twenty-four of which were caught in the nets of the *Valdivia* (Chun, 1910). And yet, he recorded only one instance of light emission by a cephalopod; in *Valdivia*'s darkroom Chun "observed slight phosphorescence in a specimen of the new genus *Thaumatolampas.*" But in spite of so little observed evidence, he was adamant that "only exaggerated skeptics would reject the possibility that phosphorescence occurs in similar organs of other families of Cephalopoda, too" (Chun, 1910).

Louis Joubin (1861–1935), a French zoologist who was then affiliated with the University of Rennes in Brittany and was preparing a monograph on the cephalopods of the early Monaco cruises (Joubin, 1895), studied the putative light organs of the squid whose luminescence Jean-Baptiste Vérany had witnessed (see chapter 3). His description of these organs of *Histioteuthis ruppelli* (Joubin, 1893a) could suggest superficially to his contemporaries the trappings of eyes or light organs: reflector, pigment cells, lenses, a deep-seated mass of innervated cells. But to Joubin these organs, which were seen to luminesce, were organized in such a way as to preclude a visual role. Chun relied on Joubin's account and on his own correlation between the similar organization of the corresponding organs in *Thaumatolampas* and the light emission of these organs to draw a generalization that all such organs are "phosphorescent."

In his comparisons Chun (1914) remarked that light organs tended to cover the mantle surface and the arms, with variations in distribution between species:

> [In some forms], a certain preference for the ventral surface is evident. Thus the organs may be restricted to the ventral arms (*Chiroteuthis*) or irregularly scattered in straight or oblique rows on the ventral side of the mantle, the funnel, the ventral side of the head and the ventral arms ... A marked deviation from the preference for ventral surfaces is found in *Benthoteuthis*; they have 6 luminous organs on the base of the dorsal pairs of arms but none on the ventral arms. In addition to these

> organs, many genera have characteristic eye organs, situated almost without exception on the ventral side of the eyeball.

The *Valdivia* material allowed Chun to discover two new types of these organs.

> One of them [3rd type], the organs on the tentacles, occurs in *Thaumatolampas* in small numbers (2 on each tentacle) embedded in the middle of the stalk. The 4th type of luminous organs are those which I name 'ventral organs.' That they have been overlooked until now is mainly due to the fact that in the live animal they are visible only through the transparent abdominal wall, while in preserved specimens they are invisible from the outside.

It is perplexing that Chun, a specialist of the pelagic environment, only stated the fact of the ventral distribution of light organs and showed no curiosity about the significance of this phenomenon. He was not alone; many deep-sea fishes also displayed a predominantly ventral location for the light organs, and yet the likes of Günther and Brauer never made a fuss about it. Only much later did a British marine biologist, James H. Fraser (1909–1990), take the question seriously and come up with a plausible role for ventral bioluminescence; that is, camouflage of the silhouette (Fraser, 1962):

> Most photophores [specialized light organs] point downwards … and the reason is not really understood. Photophores, as distinct from a general all-over phosphorescence, seem most common amongst those organisms that live in the middle depths and migrate towards the surface after dark … One plausible reason is that one creature below another would see it as a black silhouette against the slightly illuminated blue background above. Bluish photophores under the body would tend to neutralize this effect, but if photophores were on the back, then the animal would show up against the black background of the depths. This is, in effect, the same type of adaptation that makes a fish silvery underneath but dark on top.

The written reports on the different components of the fauna collected by the Monaco cruises are difficult to assess, as they cover decades of surveys and the expeditions of several yachts (as explained in chapter 4). Joubin, for instance, wrote four monographs on cephalopods, the most important being his 1895 report on collections from the North Atlantic's cruises of the *Princesse-Alice*, and another in 1924 from specimens collected in the same region by the *Hirondelle II*. Although Joubin studied the histology of squid light organs, there is little of interest in his Monaco monographs about bioluminescence or light organs. A possible exception is his last opus (Joubin, 1924), in which he describes a new family and species of squid and its light organs in detail. The squid, *Cycloteuthis sirventi*, he wrote, possesses two types of light organs: the "visceral photophore" which corresponds to Chun's "ventral organ," and the light organs along the ventral border of the eyelids. The visceral photophore is a large organ associated with the ink sac and digestive tract, and its luminescence is probably derived from luminous bacteria cultured inside the organ (McFall-Ngai and Montgomery, 1990). A more complex visceral photophore apparatus, over many viscera and covered by lenses, was observed by Joubin in another squid, *Megalocranchia abyssicola*. "Owing to the great transparency of the mantle and the almost complete absence of pigment cells," remarked Joubin, "the photogenic organs must produce the same luminous effect as if they were located on the external surface of the mantle."

Similarly, fishes collected by the Monaco cruises were covered in several monographs. The first two, authored by Robert Collett, a Norwegian zoologist, appeared in 1895 and 1896 and provided next to nothing on light organs. In contrast, the contribution of Erich Zugmayer (1879–1938) deserves special attention. Born in Vienna, the son of an industrialist and amateur geologist, Zugmayer was destined to enter the coppersmith firm of his father (Mliner, 2010). He prepared by attending the Vienna Commercial Academy and by completing an apprenticeship in England. But he decided against the business model and turned instead to natural history, studying zoology, geology, and geography at the University of Heidelberg. Thanks to his father, who was a research associate and fundraiser at the Zoologische Staatssammlung München, the Bavarian zoological museum, Erich was appointed curator there. Starting in 1899, Zugmayer travelled extensively in Scandinavia, Iceland, Persia (Iran), Russia, Tibet, and British India, where he collected many freshwater fishes, some new to science (Neumann, 2006). He was also a skilled explorer, recording many aspects of the lands he visited. Given his lack of credentials regarding marine fishes, why and how he was approached to study the deep-sea fish collections of the *Princesse-Alice* is unknown, but he certainly turned out to be an excellent choice.

Zugmayer's contribution is remarkable in several ways. The task of dealing with the fish collection from the *Princesse-Alice* cruises extending from 1901 to 1910, which forms the basis of his monograph (Zugmayer, 1911b), was in itself a mammoth undertaking: twenty-three species were new to science: twelve new genera and eleven new species belonging to extant genera (Zugmayer, 1911a). But Zugmayer also pioneered the diagnostic use of external morphology, number, and distribution of light organs for taxonomic assignments; he consciously made this point in the monograph. And last but not least, the artwork that went into the illustrations was every bit the equal of those in Brauer's own *Valdivia* monograph.

Beyond the exacting descriptions of the arrangement of the light organs, Zugmayer offered a few interesting observations. Lanternfish present the greatest variation in the distribution and development of their light organs, probably because of the slow maturation of the organs during development. In contrast, the light organs of hatchetfishes seem the least variable in their deployment, being already fixed in early developmental stages. The caudal luminous organs of lanternfishes present a dimorphic character, with only one ventral organ in the female and both a dorsal and ventral organ in the male. In the male the arrangement of the large caudal organs can appear ornamental, to the point that the scales that bear the organs sit in some cases on top of the caudal peduncle like a saddle. This Zugmayer compared to a *parure nuptiale* (bridal dress), implying a key role of the caudal luminous organs in attracting females. And some of the dragonfishes are literally showered with tiny light organs, up to 1,500 in *Photostomias guernei* and even more in *Eustomias braueri*. In another dragonfish, *Malacosteus niger*, Zugmayer made the surprising observation that one of the two subocular light organs "is rotary and allows the fish to hide the emitted light."

One of the exotic deep-sea fishes first discovered by Zugmayer is what is called today Zugmayer's pearleye, *Benthalbella infans*. Zugmayer overlooked four relatively large organs along the ventral surface of the trunk, which were identified as bioluminescent only much later by Merrett et al. (1973).

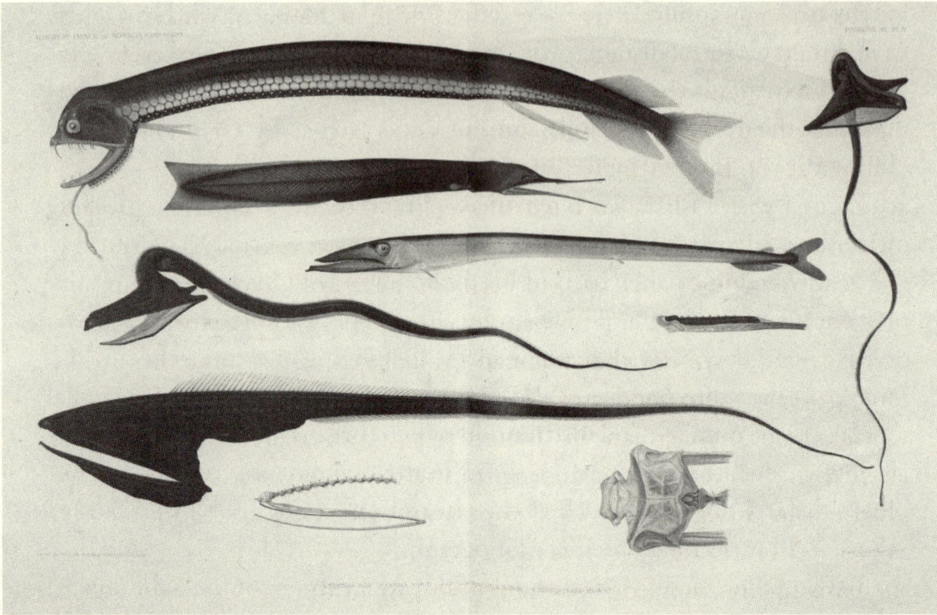

Figure 5.8 Dragonfish *Stomias boa* with a fishing lure projecting from the lower jaw (upper left image), and assorted monstrous looking fishes, particularly gulpers. Modified from Colour Plate 4 in Zugmayer (1912).

A surprising finding by Merrett's team was that the light-emitting cells in the organs are actually modified muscle cells. This is the only organ known to be derived from muscle besides the electric organs of marine and freshwater fishes (torpedo fish, electric eel, and weakly electric fishes).

The Prince of Monaco, highly satisfied with Zugmayer's work, entrusted him with the bathypelagic fish collection of the 1911 and 1912 cruises of the *Hirondelle II*. Again, Zugmayer surpassed himself by discovering a new genus, the dragonfish *Aristostomias*, and five new species, also dragonfishes (Zugmayer, 1913). From the same collection he found, besides the dragonfishes, six new species of other kinds of fishes, which he wrote of in a paper published the following year (Zugmayer, 1914). All in all, Zugmayer left an indelible mark on deep-sea biology.

His last contribution to deep-sea ichthyology, when Zugmayer was thirty-five, coincided with the outbreak of the First World War. The war put on ice the oceanographic enterprises of the Prince of Monaco, who struggled to maintain a neutral distance from the belligerents. Not so Zugmayer, who felt strongly about the Austro-Hungarian Empire; after all, it had helped make his family's coppersmith business the largest in Central Europe (Mliner, 2010). But Zugmayer was now based in Munich, and there became acquainted with Oskar von Niedermayer (1885–1948), a Bavarian artillery officer and Islamic scholar on the side (McMeekin, 2010). Niedermayer had made reconnaissance trips to Persia on behalf of German military intelligence between 1912 and 1914, under the cover of geological and anthropological studies. Now that war had broken out, it was time to put his intelligence work to good use.

The aim of the Germans in "Islamic Asia" (Persia, Afghanistan, and adjacent British India) was to undermine the British position by stirring up jihad revolts in segments of local populations (McMeekin, 2010). As early as October 1914, Zugmayer was recruited by Niedermayer to assist him, on the basis of his knowledge of the region gained during his ichthyologic explorations. Zugmayer was actively and successfully creating trouble in Istafan (Persia, now Iran), and in 1915 "had mustered up a djihadi army of some 300 gendarmes and Kashgai tribesmen, which assaulted the British Consulate in mid-September, killing fifteen to twenty British officers" (McMeekin, 2010). The Germans had the British on the run throughout Persia, but their luck soon turned and the British gradually restored their hegemony, so that by April 1916 the Germans had fled Persia. Zugmayer himself was captured by the British in Baluchistan province. All trace of Zugmayer was lost after that; he was presumably freed after the war, but his academic career never resumed. He finished his days in Vienna and died in his sleep in 1938, only fifty-eight. Zugmayer contributed with many others to dispel some of the mysteries of the deep, but he himself remains an enigma to this day.

6

Inside the Light-Producing Organs

> I was particularly anxious that as much as possible of the materials of the *Challenger* should be utilized for a thorough histological investigation of these organs by zoologists thoroughly versed in the method of histological research.
> –Albert Günther (1887)

One question that hounded the zoologists sorting through the deep-sea specimens harvested by the various oceanographic expeditions was: if the skin organs and glands are indeed light organs – and they are inferred to be by virtue of the fact that some of them are seen emitting light – then should their internal tissue composition and organization tell us something about how the light is produced? Several zoologists skilled in histological techniques were called upon to undertake the ambitious anatomical task in attempts to answer this question.

Knowledge of the histology of light organs was almost non-existent at the time the *Challenger* started on its voyage. Microscopy, although an ancient art and technique, had only grown into maturity by the mid-nineteenth century. Aided by improved histological processing methods, the field of microscopic anatomy was ripe for the studies of the small structures typical of many light organs. Albert Günther, the great ichthyologist who classified all 177 deep-sea fish species collected by the *Challenger*, knew that, and accordingly asked Henry Moseley, who had travelled with the *Challenger* and was reputed to be skilled in histology, to take on the study of the structure of light organs of the fish collection in his care. But Moseley declined, citing many conflicting demands on his time. So Günther turned to Robert von Lendenfeld (1858–1913), who accepted.

Robert Lendlmayer von Lendenfeld was born in Graz, Austria, where he received his university training in natural history and earned his doctorate (Menzel, 2004). In 1881, right after completing his doctorate, he departed for Australia with his young and wealthy wife. An accomplished mountaineer, he explored Australia and New Zealand, and had several mountains named by him or after him. Over a span of five years "down under," he studied the local coelenterates and sponges on the side. In 1886 he returned to Europe and started working under the zoologist E. Ray Lankester at the British Museum. So Günther, himself on the staff of the museum, had not far to go to call on Lendenfeld. The Austrian had no research experience with fishes, but Günther's paramount criterion for the job was histological, not ichthyologic expertise.

Lendenfeld's opus appeared as an appendix in Günther's monograph (Lendenfeld, 1887). He selected the best-preserved specimens that Günther could supply, although the preservatives used by the *Challenger* scientists for these fishes was not specified. It came down to eleven species: five dragonfishes, two hatchetfishes, one lanternfish, two halosaurs (Halosauridae), and one slickhead (Alepocephalidae). Lendenfeld was rather spare in describing his histological method, mentioning only that the "phosphorescent organs were studied principally by means of series of sections, cut in the usual manner after embedding in paraffin; they were stained with Kleinenberg's haematoxylin, borax-carmine, alum-carmine, or picrocarmine. The last named produced the best results." From his examination Lendenfeld classified the organs in two broad categories: regular ocellar organs (today's photophores) and irregular glandular organs. Lendenfeld's zeal for classification drove him to break down the two types of light organs into numerous subcategories according to their level of structural complexity: presence or not of pigment layers, reflectors, and lenses as accessories to the suspected light-emitting cell mass.

The core of the ocellar organs, recognizable regardless of the level of complexity achieved by specific subtypes, is a mass of secretory cells arranged in tubes – "acini" in the jargon of endocrinologists. This is the photogenic tissue; that is, the tissue generating the light. These granular cells can be polarized or unpolarized. When polarized, the elongated cell's granules develop away from the cell body where the nucleus is located, until they reach maturity and are released at the tip of the cell into a canal. However, Lendenfeld was uncertain about the fate of the secretion; he seemed inclined to believe that for some organs the granular secretions remain in the canal of the organ, whereas for others the luminous secretion would exit to the skin surface and mix with the slime already produced by other skin glands and covering the skin surface. Adjacent to the glandular tissue Lendenfeld noticed a small mass of semi-transparent cells devoid of granules and which he (mistakenly) identified as ganglion cells. He found that all the light organs are innervated by branches of spinal nerves.

The more complex organs add optical accessories. "The pigment coat prevents the light from issuing in any other direction than that indicated by the axis of the organ ... The membrane which reflects the light [reflector] produced within the organ is, in consequence of its shape, very well adapted for the purpose of concentrating the light in one cone (Lendenfeld, 1887)." In other words, the reflector bundles the light like an optical fibre. Lendenfeld also mentioned a structure near the skin that he interpreted ambiguously as either a cornea or a lens.

In composite ocellar organs the core part is considerably more elaborate. It consists of a superficial spherical part separated from a cup-shaped part by a constriction looking like a diaphragm (disc). Heavily granulated cells organized in acini dominate as usual, but other cells more transparent and devoid of granules are also present. Nerves penetrate at the level of the constriction (disc) and nerve fibres irradiate into the glandular tissue. From these arrangements Lendenfeld reached the following functional interpretation:

> The fish can at its option incite the organ, which under ordinary circumstances is non-luminous, to phosphorescence. The voluntary impulse is transmitted by the thick nerve of Leydig to the disc of ganglion cells, which excite the phosphorescent cells in the cup to action by means of the radial nerve-fibres.
>
> The phosphorescence of the latter requires, however, the secretion of the gland in the sphere as fuel, in a manner similar to that which has been described in the simple organs. The secretion passes through the disc and the radial fibres to its destination, where it is consumed to produce light.

In the lanternfish *Scopelus* (now *Hygophum benoiti*), Lendenfeld noticed large light organs behind the dorsal fin. He called them "stern-chasers" because their projected light beam is directed backward. He classified them as a variety of composite ocellar organs:

> It seems not improbable that the glands in the sac-shaped proximate portion of these organs produce a secretion which is poured out into the cup-shaped distal part, and there a mutual chemical reaction between this slime, in which also cells and nuclei are found, and the typical phosphorescent clavate cells may take place at the will of the fish, and a certain amount of light may be produced, which is reflected by the parabolic spicule-layer [reflector], and thrown out as a strong flash.
>
> The ventral organs can illuminate the dark water below the fish for any purpose, whilst the dorsal stern-chasers or the solitary dorsal stern-chaser, which are invariably directed backwards, probably serve for purposes of defence, inasmuch as a strong ray of light shot forth from the stern-chaser may dazzle or frighten an enemy which is in hot pursuit of a *Scopelus*.

Today we know that these organs, called caudal luminous organs, are not glandular in nature (Anctil and Case, 1977). But Lendenfeld was right in deducing that they produce strong flashes. And more significantly, this passage hinted for the first time at an important notion that was destined to develop much further a few years later; namely, that the luminescent reaction involves the mixture of two ingredients that are held separately until joined to effect the light emission.

Among the irregular glandular organs are the light organs at the tip of "the extraordinary barbels" of dragonfishes, "which are either attached to the mouth or developed from the first ray of the pectoral fin." Lendenfeld described the structure as composed of gland-tubes reminiscent of the glandular arrangement of the photogenic tissue of the body photophores. Its function, Lendenfeld suggested, "would be to act as lures to attract other animals." Another striking feature of dragonfishes is the suborbital light organs, which in *Pachystomias microdon* "appear as two very conspicuous white masses below the eye." The orbital cavity of the eye is especially enlarged downward and backward to accommodate the large organs. The organs lie loosely inside a pouch, and are formed of a deep cup emerging to an extended shelf superficially. The tissue inside, as for the other light organs, is made up of gland-tubes. The entire organs are richly innervated by the trigeminal (fifth cranial) nerve and supplied with the usual pigment coat and reflector. Lendenfeld concluded that the "light, which is presumably emitted from these organs, is thrown in such a direction, that it illuminates the field of vision of the fish; they thus enable the fish to see, and must be considered as aggressive organs."

In his general conclusion, Lendenfeld proposed that the light organs of fishes are modified glands that have evolved partly from mucous glands of the skin and partly from the lateral-line canals. The connection to lateral-line canals was not supported by Brauer (1908) in stomiatoid fishes or Gatti (1904) in lanternfishes, although Brauer agreed that the light organs are modified glands (see below). Lendenfeld also suggested that the accessory reflectors and sphincters originate in the skin around the light organs. He firmly believed that gland cells are shared in one form or another by all luminous animals, whether they are integrated into organs as in fishes or disseminated over the skin as in mollusks (*Phyllirhoë*), ascidians (pyrosomes), and jellyfish.

Lendenfeld's work on the light organs of the *Challenger* fishes so impressed Alexander Agassiz that he enlisted the Austrian to do a similar job with the deep-sea fishes collected by the *Albatross*. By then Lendenfeld had moved on, fulfilling appointments at various universities in continental Europe, until he secured a full professorship at Charles University in Prague, where students were taught in the German language. He was working there when he received the collection of *Albatross* fishes, all but one new species discovered by Samuel Garman. However, Lendenfeld was underwhelmed by Garman's depictions in his monograph (Garman, 1899) of the distribution of light organs over the body of the fishes. "For this reason," Lendenfeld defiantly asserted, "the entire fishes are here figured again" (Lendenfeld, 1905). Among the fishes examined were such classics of the deep as the dragonfishes *Chauliodus* and *Stomias*, the bristlemouth *Cyclothone*, and the lanternfish *Myctophum*.

In his work on the *Albatross* specimens, Lendenfeld departed from his *Challenger* study in several respects. First, the tissues were embedded in celloidin as well as paraffin, and this time it was Van Gieson's stain (haematoxylin-picric acid-fuchsin) that produced the best results. Second, Lendenfeld strangely shifted his terminology for naming light organs, now using the phrase "radiating organs" instead of "phosphorescent organs" as all his colleagues were wont to do. And third, his classification of the organs differs somewhat from the preceding study; this time he discriminated between ocellar organs, disc organs, and tubular organs. In retrospect, the classification appears meaningless, as almost half of the species examined are now deemed to be non-luminescent: *Bassozetus*, *Halosaurus*, *Ipnops*, *Leuciocorus*, *Lychnopodes*, *Malthopsis*, and *Mixonus*.

Prior to Lendenfeld's *Albatross* monograph, two German investigators had studied the histology of hatchetfish light organs. The first was Gustav Brandes (1862–1941), an assistant professor at the University of Halle at the time of his contribution (Brandes, 1898), and later director of the Zoological Garden of Dresden (Kleinschmidt, 1955). Brandes provided no supporting illustrations; nor did he add to Lendenfeld's description, except for his stronger focus on the lens apparatus of the organs and his contention that the luminous barbel of *Chauliodus* might be used to startle predators or prey. Nothing is known of the second contributor, Kurt Handrick (1873–?), save that his detailed study of the nervous system and light organs of the hatchetfish *Argyropelecus* (Handrick, 1901) was the published form of his inaugural dissertation at the University of Leipzig. His thesis supervisor was likely Carl Chun, who provided him with material from his collections dating back to short voyages in the Mediterranean before the *Valdivia* expedition.

The hatchetfish body is extremely flattened laterally, so perhaps Handrick wondered whether the brain morphology showed adaptations to this constraint. Whatever his motivation, it led to a detailed mapping of the innervation of the hatchetfish's various types of light organs. The preorbital organ is innervated by the cranial trigeminal nerve, and the postorbital, opercular, and jaw organs by various branches of the facial nerve. All the postopercular organs and those of the abdominal flank, belly keel, anal region, and tail are innervated by branches of the spinal nerves. Although Handrick gave less priority to the histology of the light organs themselves, his study revealed the extraordinary development of the paired light organs of the belly keel, where the two members of the pair coalesce to form a massive organ with a great quantity of gland-tubes. Although Lendenfeld (1887) mentioned this phenomenon, neither his text nor his illustrations did it justice, and the reader misses the importance of these coalesced organs, which generate intense light downward.

~

If the few hatchetfish specimens that Chun provided to Kurt Handrick generated a thesis, what he gave of the *Valdivia* fish collection to August Brauer turned to gold. After his systematic treatment of the deep-sea fishes, as chronicled in chapter 5 above, Brauer (1908) published his monumental monograph on the light organs and eyes of these fishes. Amalgamating light organs and eyes in the same monograph was not merely a matter of chance. The eyes of many of these deep-sea fishes are in their own right otherworldly in appearance and puzzled spectators on deck as the nets were brought up. But, as we discuss at intervals in this book, the strange forms of these eyes in fact suggest adaptations to the dimly lit environment of the deep sea and to the broadcast of light emissions by animals around them. The awakening to the interrelationship between vision and bioluminescence in the adaptive sense can thus be construed as an unintended consequence of the first global oceanographic surveys, and especially of the visionary minds of Carl Chun and August Brauer.

Brauer's monograph stands head and shoulders above Lendenfeld's two contributions in several ways. For one, the sample of species examined is larger (twenty-five) and their status as to luminescence capability is unassailable. Brauer organized his material in a systematic fashion, assembling it not by category of light organs – a type of assignment potentially marred by subjectivity – but by family of species as recognized in his days: Sternoptychidae (hatchetfishes), Stomiidae (dragonfishes), Scopelidae (lanternfishes), Ceratiidae (deep-sea anglerfishes), Gigantactinidae (whipnose anglers), Antennariidae (frogfishes), and Malthidae (batfishes). The illustrations of the histological sections, keeping pace with those of the fishes themselves, are so exceptional that even microphotography, had it been available, would have been hard pressed to compete. Granted, the material handed to Brauer was superiorly preserved, thanks to the technical refinement of the trawling nets of the *Valdivia* and care in the choice of preserving fluids; Lendenfeld had no say in this, and he lacked the advantage that direct participation in the expedition gave Brauer. As Brauer acknowledged no draftsman in the monograph – and, as we have seen, he took care to thank Fritz Winter in the 1906 monograph – the likelihood is that Brauer executed his illustrations himself, which makes the comparison even more damning for Lendenfeld.

It is worth taking the time to look closely at Brauer's considerable contributions to the study of the light organs of deap-sea luminescent fishes. Brauer discerned two general shapes of light organs: tubular (*schlauchförmige*) organs and cup-shaped (*becherförmige*) organs. However, all the organs, especially among the stomiid fishes, shared similar structural designs along the lines of those previously delineated by Lendenfeld. In contrast to the latter, Brauer was able to provide more detailed descriptions of the minutiae of these organs. But beyond the detailed accounts of the microscopical anatomy of the light organs, Brauer also offered many timely and incisive reflections from the comparative studies he performed: on the conclusions to draw from the comparative analysis, on the susprisingly few direct observations of light emissions from these fishes, on the physiological proccesses that can be inferred from the dynamics of light emission and histological organization, and on the biological significance of the light organs.

Brauer's literature review brought him to the consensus at the time of his study that all light organs were glandular, with the possible exception of the lanternfishes of the supergenus *Myctophum*. For Brauer, the organs that met the criteria for typical glandular status were those of the stomiatoids *Gonostoma elongatum*, *Diplophos tacnia*, *Neoscopelus*, and the anglerfishes; these are gland cells producing secretions released in a lumen and opening to the exterior through a duct – in other words, what are known today as exocrine glands, as opposed to endocrine glands, which empty their secretions into the blood stream. The light organs of the other luminous deep-sea fishes only have, in Brauer's own words, "closed glands" – gland cells that have no duct to dispatch the secretion. Brauer seems to have believed that all these light organs were originally typical glands and that they evolved to varying degrees toward the closed type. For this reason he referred to all these light organs as "modified glands."

That the source of bioluminescence should be gland cells was, to Brauer, quite evident: their position in the organ as the ideal location to have their light manipulated by accessory structures around them (pigment coat, reflector, lens) and to fashion a specific light beam to the outside; and the

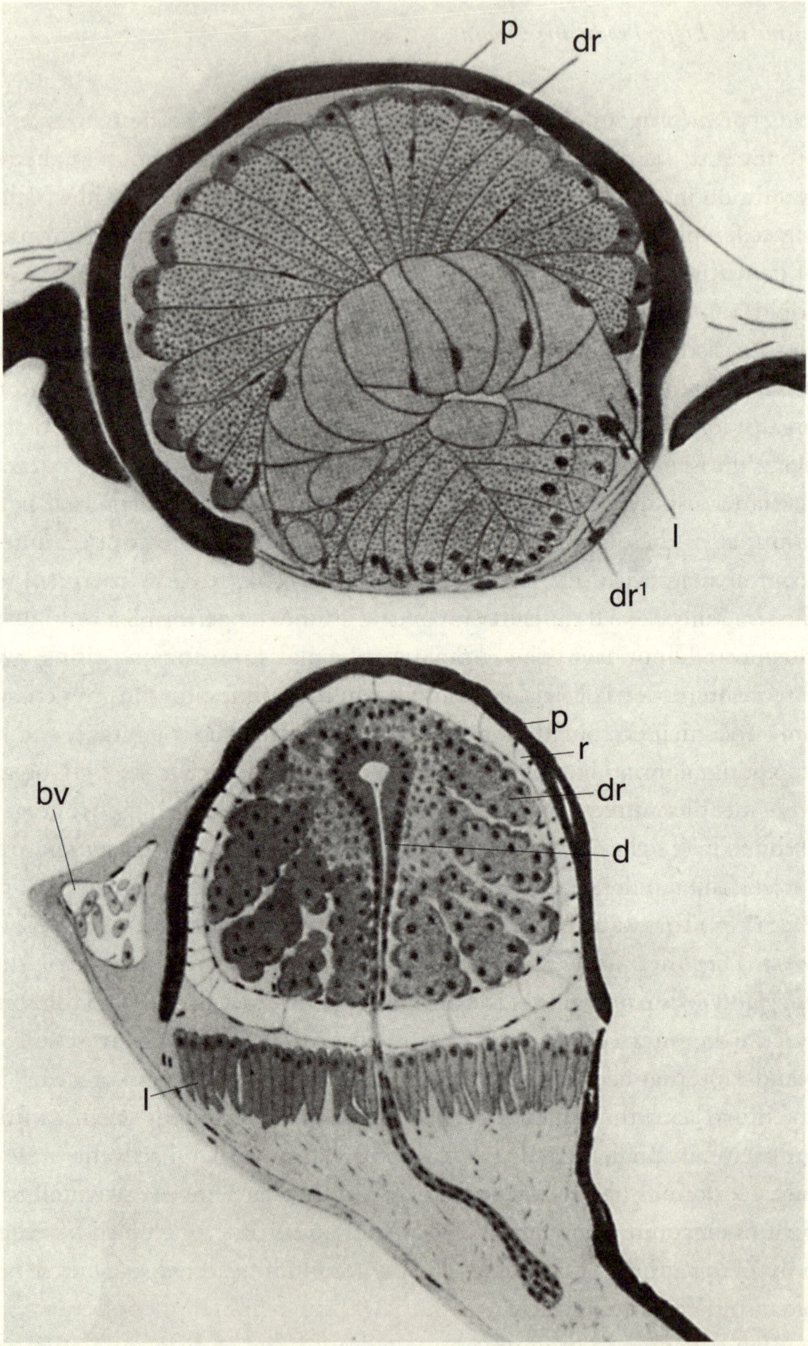

Figure 6.1 Structure of stomiatoid fish light organs. *Top*, cup-like organ of dragonfish Stomias; *bottom*, light organ of bristlemouth Cyclothone, with (d), secretory duct; (bv), blood vessel; (dr), luminous gland cells; (l), lens cells, (p), pigment sheath, (r), reflector. From Plates 29 (top) and 20 (bottom) in Brauer (1908), as reproduced in Harvey (1952).

fact that even in simple organs devoid of optical accessories the gland cells are the only cells of significance. But how did Brauer recognize them as gland cells in the first place? It was undoubtedly the way they stain, displaying eosinophilic secretion granules and the bluish non-granular cytoplasm obtained with haematoxylin. In the few cases where developmental stages could be followed, Brauer found that light organs, and especially gland cells, are of ectodermal origin; this was important to him because it drove the final nail into the coffin of Leuckart's notion that these organs were electric or pseudo-electric organs, which were by then already known to be of mesodermal origin.

Brauer also regarded the light organ associated with the barbel of anglerfishes – the lure organ or fishing rod – as glandular. He felt impelled to do so by analogy with the barbel of dragonfishes, although he could see that the structure showed differences. Given the limited power of his microscope, Brauer could not have guessed that what he took for secretion granules were actually luminous bacteria. That fishes could harbour luminous bacteria was only documented four years after Brauer's anatomical monograph.

A short note by Portuguese ichthyologist Balthazar Osorio (1855–1926), who was then director of the Zoological Section of the Museo Bocage in Lisbon (Beolens et al., 2011), told of the Portuguese fishermen who press the abdomen of a fish, the softhead grenadier *Malacocephalus laevis*, to collect a cloudy liquid that exudes a blue luminescence in the dark (Osorio, 1912). They then smear the expelled liquid onto the underlying muscle layer of a piece of shark skin. As the luminescence lasts for hours, the fishermen can use the smeared pieces as bait to attract fishes to the light. It was Osorio who found that bacteria were the source of the light emission. A quarter of a century later, the organization of the bacteria-filled light organ abutting the rectum was elucidated (Haneda, 1938). Many other discoveries of symbiotic luminous bacteria were to follow in the course of the twentieth century, as I recount below in this book. In Brauer's time, anglerfishes were the only major group of deep-sea fishes whose luminescence had never been seen, so their luminescence capability was still a matter of conjecture.

Brauer wanted to promote physiological studies of deep-sea bioluminescence as a prerequisite to understanding how the light is produced, but he was all too aware of the challenge this represented. Even when the odd fish lands on deck still alive it often displays various states of body damage and a fish in the best condition survives at most for only an hour or two. For this reason, Brauer exhorted future researchers to visit the Straits of Messina, because the unusual currents there brings to the surface hatchetfishes and lanternfishes in good living conditions such as to make them the best experimental models for deep-sea physiology. And these fishes can be transferred in short order to the Sicilian coast so that experiments can be conducted in a laboratory, however improvised, in better conditions than on a pitching vessel in the middle of the ocean. His advice was heeded only seventy years later (Baguet and Maréchal, 1976). Failing such an approach, Brauer could only rely on his histological studies to offer speculations on mechanisms of light production.

By interpreting the light organs as glands, Brauer realized that he faced a paradox. If glands have a duct to drain the secretion into the water, why then do some of them have accessory optics as if the organ was designed to produce its light internally. He was aware that the glands of copepods (crustaceans), for example, eject a cloudy secretion that only becomes luminous outside the body (Giesbrecht, 1895), but the light organs of these copepods have no optical accessories. And what's more, some of Brauer's fishes possessed light organs that looked glandular but had no duct to drain the secretion. So Brauer was forced to conclude that light emission was intracellular; and he agreed with previous researchers that the oxygen needed (after all, the luminescent reaction was believed to be an oxidation) was provided by the rich blood supply of these organs. However, in cases where light organs have poor blood supply, there were exceptions; and in these cases, Brauer suggested, secretions in contact with water at the entrance of the duct are necessary, to provide exposure to sufficient oxygen and to allow light emission.

Brauer found light organ innervation only in lanternfishes and in the postorbital (rotatable) organ of stomiids. In all other light organs he could find no innervation. This refuted Lendenfeld's observations of innervation in all light organs. Brauer thought Lendenfeld must have been fooled by nerves coming close to the light organ but passing it by, or by collapsed blood vessels that looked like nerves. Brauer's observation of lack of innervation concurred in his mind with observations that luminescence was primarily seen as continuous; that is, not under the control of the fish. But because endocrinology – the study of hormones – was only in its infancy at the time of Brauer's writing (Wass, 2015), he could not have envisioned a hormonal alternative to nervous control in the case of long-lasting episodes of light emission.

Brauer struggled also with the biological significance of light organs and bioluminescence in the deep sea. For a start, he had trouble defining the boundary between depth zones in the open ocean. He settled for the view that the majority of luminous fishes are not really deep-sea fishes; that is, according to his controversial criterion that anything from 400 meters or less does not belong to the deep sea. Most luminous fishes span between the twilight zone (300–400 meters) and the surface, but some dwell as deep as 700 meters. Many luminous fishes, especially lanternfishes, undergo daily vertical migration to the surface at night. He believed that no fishes living on the ocean bottom were luminous.

Depending on the type and position of light organs over the body, their roles may differ. Brauer found light organs in unlikely places, on unlikely "perches": on fin rays, on the gill cover and other parts of the gill apparatus, and on the tongue, for instance. The luminous barbels of stomiids are believed to serve as lures to attract prey. Some of these perches, Brauer found, lack muscle, so are not movable at will; but he also found that many have strong muscles that enable the tentacle to be moved in any direction, just as a fisherman can move his fishing line in different directions. The fish, of course, once it has attracted a prey to the luminous bulb, needs to move the perch to bring the bulb close to the mouth. That these fishes are powerful predators is readily evidenced by their big jaws and intimidating fang teeth. For fishes that actively pursue their prey, Brauer suggested that other uses must be looked at, but he was at a loss to identify them.

Brauer further remarked that, in fish with huge numbers of light organs, the more dorsally placed tend to be less numerous and smaller – sometimes even degenerated – whereas those in the lower part of the body (ventrolateral) are larger and more numerous. Also, he paid attention to the orientation of the light beam as suggested by the layout of the organ's optical apparatus – beaming toward the front, back, laterally, or downward – and what the meaning of these orientations might be. He noted, for instance, that the tubular eyes of the hatchetfish *Argyropelecus* are directed upward, and yet their light organs are mostly directed downward. So Brauer found it difficult to pinpoint roles when the use of luminescence as bait did not apply and other roles were not self-evident.

One hypothetical role of luminescence is for recognition among species or conspecifics according to patterns of light organ distribution, intensity of light emission, or the colour of the light. Brauer dismissed the latter because his study of the retina of these fishes, in which rods but not cones are present as photoreceptors, left little doubt that no colour discrimination is possible in deep-sea fishes. A role for recognition appears unlikely, too, when many small organs crowd the body surface. Brauer compared the light show of the dragonfish *Dactylostomias*, with its "positively extravagant wealth of light organs" on a bronze skin, to the iridescence of butterflies. But otherwise he was inclined to believe that it was the pattern of distribution of light organs that served as the main recognition mechanism. He particularly called attention to sex difference in the distribution of caudal luminous organs in lanternfishes – dorsal for males, ventral for females – and the fact that these organs develop later than the other light organs as evidence of their role in sexual attraction. Brauer admitted that flashing of caudal organs for reproductive purposes exposes the lanternfish to visually guided predators, but so do the nuptial colours of land animals, Brauer added, and the benefits seem to offset the risks.

Brauer saw no value in having light organs located on the gill cover, as occurs in many dragonfishes, unless the organ had been of value to ancestors that had their eyes on the tips of long movable stalks (periscope eyes) like the bizarre *Stylophthalmus*, thus allowing the fish to see the luminescence of their own light organs behind them. He implied that the elaborate infolding seen around the eyes of modern dragonfishes represents a vestige of the past existence of such periscope eyes. *Stylophthalmus* was one of the fantastic discoveries of the *Valdivia* expedition that caught the attention of the public, and Brauer identified it as a new genus of deep-sea fish (1906). However, it turned out that *Stylophthalmus* is in fact the larval and post-larval form of the dragonfish *Idiacanthus fasciola* (Beebe, 1933). But in spite of Brauer's error, which is understandable given the extensive sampling required to obtain series of developmental stages of deep-sea fish, this finding actually reinforced Brauer's stance, as it suggested that the periscope eye phase in the life cycle of the dragonfish could represent a recapitulation of an ancestral form that possessed this peculiar feature.

Brauer's detailed histological study of deep-sea fish eyes led him to conclude that their eyes were adapted for night vision. But even so, the scope of visual capability would be limited. The distance at which decent visual discrimination is possible seemed to him no farther than the distance the fish's shadow would be cast if in sunlight. Beyond that, the luminescence of the ensemble of light organs from another animal would just look like a luminous blur. Brauer noted the downward direction of the light emitted by many of these fishes and the upwardly directed telescope eyes but, as many researchers did in the second half of the nineteenth century, he failed to connect the dots; that is, to consider that the eyes were designed to look at the ventral silhouette of a potential prey above, and that downwardly directed luminescence was designed in turn to break up the silhouette as a camouflage tactic. It is crucial here that the potential predator not see the individual light-emitting organs, in which case camouflage would not work; but if, according to Brauer's estimate of visual capability, the scene for the predator is a luminous blur matching in intensity that of the ambient downwelling light, then the camouflage ploy can be effective and predation avoided.

The last vision-related issue Brauer addressed was the existence of subocular or postorbital light organs. "What all these orbital organs have in common," he remarked, "is, first, the location on the eye and, secondly, the direction of the light. All are located so that their light must fall into the anterior chamber of the eye, reaching the lens and thus penetrating into the eye" (1908). Brauer was genuinely puzzled by this feature. It seemed to him that letting light inside the eye at such an awkward angle must reduce image quality. On the other hand, he saw the possibility that it "might allow the animal using these orbital organs to have a more secure feeling of its own light so as to make discrimination of conspecifics versus other alien species easier in terms of light intensity … such that these organs are another means to ensure collaboration between sexes particularly during the breeding season, or even to help discriminate between friend, enemy or prey."

Brauer mused in the end about his expectation that in such an environment as the deep sea, as monotonous as the desert on land, little diversity in animal morphology would be present. So he found himself all the more surprised when he surveyed the bewildering diversity of organizations of light organs and eyes. How, he asked, could such a seemingly unvarying environment provide selective pressures for the amazing adaptations he observed? Brauer proposed that, as some coastal and demersal fishes moved gradually into the pelagic zones of the open ocean and descended to greater depths, adaptations to the reduced yet varied light conditions of mesopelagic and then bathypelagic environments must have occurred variously according to the different body designs initially in place. The design of light organs and ocular adaptations fit that trend. But then other pressures intruded, not least of which would have been the new trophic chain resulting from the need to face new feeding and reproductive challenges in the dark. This need to adapt would have led to extraordinary reshaping of fish bodies, culminating in such forms as the gulpers, and to deployments of light organs designed to make the most of the new struggle-for-life conditions. Such was the impressive intellectual journey that Brauer had embarked on.

~

The person who played a role comparable to Brauer's for the Prince of Monaco could hardly have been expected to improve on the German's achievements. From biographical information culled from the published memoirs of Josef Nusbaum-Hilarowicz (1859–1917), we learn that he was born in Warsaw to a merchant of Jewish descent and, as a consequence of the third partition of Poland, which had gone on for close to a century, was educated at the Russian Imperial University of Warsaw (Nusbaum-Hilarowicz, 1916). Benefiting from the benevolent sponsorship of two famous Russian biologists – Aleksandr Kowalewski and Nobel Prize winner Ilya Mechnikov – he studied for a master's degree in zoology at the University of Odessa and returned to Warsaw, where he obtained his doctorate in 1888. He received his habilitation in 1891 at the University of Lwow – or Lviv, then the capital of Galicia under Austro-Hungarian rule and now part of western Ukraine – and the next year was appointed assistant professor there. In 1907 Nusbaum-Hilarowicz converted to Catholicism in order to be eligible for the Chair of Zoology and Comparative Anatomy at the University of Lwow. His stewardship of the department put the backwater town on the zoological map and led to the nurturing of many talented researchers, who later made their mark in Polish zoology and medicine. Nusbaum-Hilarowicz proved also to be a dedicated evolutionist who translated Darwin's *On the Origin of Species* into Polish.

Nusbaum-Hilarowicz spent time in European biological stations – Trieste, Concarneau in Brittany, and Naples – where he conducted research on the embryology and regeneration capabilities of various invertebrates. But he also visited the Musée océanographique de Monaco in 1911 and conducted research there with the assistance of a fellow Pole, Mieczyslaw Oxner, who was on the staff of the museum. His stay in Monaco led Jules Richard, the operational manager of the Prince of Monaco's cruises and the director of the Musée océanographique, to entrust him with well-preserved fish material for anatomical studies. Nusbaum-Hilarowicz went to work right away, and a preliminary article on the light organs of the bristlemouth *Cyclothone* soon appeared (Nusbaum-Hilarowicz, 1912). In the following years Richard kept sending him more specimens of deep-sea fishes to work on. To the extent that his academic duties allowed, he pursued his investigations until the advent of the First World War.

Soon after the war broke out, Nusbaum-Hilarowicz recalled in his memoirs, the Austro-Hungarian troops on the East Front suffered setbacks that allowed Russian troops to move into Galicia and besiege the city of Lwow. Many residents fled in panic, anticipating the reprisals that followed. Nusbaum-Hilarowicz and his colleagues were harassed by Russian officers, and the campus was cut off from firewood as winter set in. The freezing indoor conditions in turn endangered the specimen collections in the Zoological Institute and also the Prince of Monaco's material on which Nusbaum-Hilarowicz was working. When he protested to the Russian governor, the latter asked: "Why the Prince entrusted you his precious material?" To which Nusbaum-Hilarowicz replied: "Apparently you have no great confidence in the Polish university and Polish scholars." The Austrians drove the Russians out of Galicia in June 1915, but the hostilities nevertheless prevented Nusbaum-Hilarowicz, who had completed his two-volume anatomical studies of deep-sea fishes, from sending the manuscripts to Monaco. He died in 1917, before the war ended, and his monographs were published posthumously in 1920 and 1923.

Nusbaum-Hilarowicz worked on a smaller set of species than Brauer, but it allowed him to do what Brauer had originally intended; that is, to expand the anatomical studies to include structures other than the light organs, such as the gonads, the digestive tract, endocrine glands, and the swim bladder, the latter an important structure for buoyancy control at depth. Between the two monographs (Nusbaum-Hilarowicz, 1920 and 1923), he treated light organs in five species: the bristlemouth *Cyclothone signata*, the hatchetfishes *Argyropelecus hemigymnus* and *Sternoptyx diaphana*, and the dragonfishes *Stomias boa* and *Photostomias guernei*. The histology was state-of-the-art and the illustrations matched in quality those of Brauer, with the added bonus of sharp colours.

Nusbaum-Hilarowicz served as a mouthpiece for many when he stated, in the introduction to his first monograph: "Of all organs of deep-sea fishes, light organs were frequently the object of refined histological investigation, because these organs always arouse the greatest biological interest" (1920). Although he acknowledged the irrefutable, that Brauer's contribution was a classic, Nusbaum-Hilarowicz argued that, given the rarity of fish material harvested from the deep sea, it was "not surprising that even in the work of such a renowned scholar there are still insufficiently resolved issues." So, the Pole felt that the main goal of his own study was to fill the gaps in Brauer's work.

Bristlemouths only possess the cup-shaped light organs of Brauer's classification. The lining of the cup is partially pigmented as usual, and the inside of the cup is filled with glandular cells organized in acini opening into a central sinus. The basal part of the cell contains the nucleus, and the large distal part abutting the canal of the acini is filled with secretory granules. Nusbaum-Hilarowicz painted a more complete picture of the reflector than Brauer, noting the complex lattice of fibres that direct the light emitted by the gland cells to the lentiform body, made of Brauer's palisade cells, through which the beam of light is broadcast to the outside. He also described the preorbital light organ, which lacks the lentiform body but, contrary to Brauer's assertion, retains a reflector. The glandular mass seems reduced to a rosette of gland cells. Nusbaum-Hilarowicz showed in a striking histological section how the aperture of the orbital organ, through which the emitted light is channelled, is oriented to shed light into the eye. In addition, the light organ and the eye are locked in a fixed position with each other by braces of stiff ligament, thus underscoring the functional importance of their relationship.

In hatchetfishes the light organs are highly developed, and Nusbaum-Hilarowicz gave them particular attention. He even provided microphotographs of the keel light organ (Brauer's *Bauschkiel*); in the early 1900s, photographing histological slides through a microscope rather than drawing was in its infancy. One can only surmise that the Prince of Monaco paid for the hefty expense of reproducing these microphotographs in the monograph. The Pole confirmed the observations of Brauer, Handrick, and Brandes and resolved divergences of opinion among them. He observed that many glandular cells contain two nuclei. Two types of gland cells are present, and he illustrated how one type is transitioning to the other. The reflector of hatchetfishes is better developed than in bristlemouths and Nusbaum-Hilarowicz was the first to identify calcified guanine in reflector cells as the source material of the reflecting mirror.

Another original and prescient contribution by Nusbaum-Hilarowicz was the observation of extracellular secretory products – from the glandular cells of the hatchetfish *Sternoptyx* – which he believed might enter into the circulation through the profusion of capillary vessels present in the light organ. Here he touched on a topic that is still unresolved today. Deep-sea fishes such as hatchetfishes contain large quantities of coelenterazine, one of the known light-emitting molecules used by many bioluminescent animals in their light organs and in other parts of their body (Mallefet and Shimomura, 1995). Although it is believed that these fishes can obtain some of their coelenterazine from their diet (luminescent prey), there is increasing evidence that they can produce their own. If the glandular cells secrete coelenterazine, then Nusbaum-Hilarowicz's observation suggests a mechanism by which coelenterazine can find its way into other parts of the body where the specific enzyme to catalyse the luminescent reaction is not present.

The barbel extending from the chin of the dragonfish *Stomias* (Nusbaum-Hilarowicz, 1923) was of particular interest to Nusbaum-Hilarowicz. By analogy with the barbels of anglerfishes, it was thought that the swelling at the tip of the barbel, from which filaments extend, was the only seat of luminous tissue in the barbel. But Brauer and Nusbaum-Hilarowitz found that glandular tissue was widespread throughout the barbel, including the peduncle, swelling, and filaments. In the swelling and elsewhere there is a simple glandular tissue composed of gland cells which, unlike those of the body organs, are narrow and very elongated. Instead of the expected sinus that glands usually possess, Brauer and Nusbaum-Hilarowitz found connective tissue enmeshed with blood vessels and nerves. Brauer thought that nerves innervated structures other than the glandular tissue, but Nusbaum-Hilarowicz observed that nerve branches invade the epithelial cells of the glandular tissue. "The presence of capillaries and nerves in the glands and the lack of gland ducts," Nusbaum-Hilarovicz wrote, "allow the conclusion that these glands function like organs of internal secretion."

Nusbaum-Hilarowitz made an astonishing discovery pertaining to these barbels: they are able to change shape in a peculiar manner. He found that they possess what he called a "pseudo-cartilaginous" skeleton, especially in the basal part of the peduncle, which ostensibly keeps the barbel upright to an extent. The barbel does not seem to have muscles, but Nusbaum-Hilarowicz's astute observations of the blood vessel walls led him to suggest that the vessels could change their volume significantly, thereby acting in conjunction with the pseudo-cartilaginous skeleton to create changes of internal hydrostatic pressure, thus allowing movements and changes of shape of the barbel. Although he was silent on the possible functional significance of his observations, here as elsewhere in his two monographs, the implication is clearly that the fish is able to control both the motility and the luminescence of its barbel through the nervous system.

If deep-sea fishes were of paramount interest to investigators of the collections from the oceanographic expeditions, the reason was simple: almost nothing was known, and curiosity-driven research moved in to fill the huge gap. But other animal phyla endowed with complex light organs were not neglected. Outstanding among these are the cephalopods and crustaceans. Cephalopod light organs were studied from specimens caught during the *Albatross* and *Valdivia* expeditions, and crustacean light organs only from specimens of the *Challenger* expedition.

William Evans Hoyle (1855–1926), the zoologist who had described the cephalopods of the *Challenger*, did not study their light organs. He was entrusted instead with the cephalopod collection of the *Albatross* expedition. Hoyle was born in Manchester, UK, and all his university education was concentrated at Oxford (Jackson, 1926). He was trained in both natural history and medicine, as were many of the investigators so far in this story. He developed an expertise in medical anatomy as well as in cephalopods, which served him well for the studies of light organs he embarked on. He was the director of the Manchester Museum at the time the *Albatross* collection fell into his lap. His description of squid light organs was inserted as an appendix to the monograph in which he provided a systematic survey of the collected cephalopods (Hoyle, 1904).

From the start Hoyle warned that the material he worked on, contrary to the *Valdivia* and *Monaco* specimens, had not been preserved with histological investigation in mind (1904), although he failed to specify what preservatives were used and what methods were applied. Despite these shortcomings, he managed to provide a vivid portrayal of the structure of the various types of light organs of two species of squid: *Pterygioteuthis giardi* and *Abraliopsis hoylei*. He had previously given a preliminary account of the light organ of another species of *Pterygioteuthis* from the Mediterranean, but it was based on a single, heavily damaged specimen (1902). Even a mere perusal of Hoyle's paper convinces the reader that squid light organs easily match those of deep-sea fishes in their sheer complexity.

Hoyle classified the light organs of *Pterygioteuthis* according to their distribution: those around the eyes (ocular), and those associated with the siphon (ventilator of the gills), the gills (branchial), and the abdomen. The spherical ocular and siphon organs are the most complex, being composed of: a pigmented capsule; a posterior cup (presumably a reflector) traversed by nerves; the inner cup, which seems to act as a light guide to direct a light beam to the outside; the central mass (presumably the light-emitting tissue) partly ensconced in a depression knob of the reflector; and the anterior cap, which may act as a lens focusing the emitted light. The gill organs are ovoid and devoid of optical accessories, and are filled with a central mass of presumed light-emitting cells (photocytes) with a granular content. Finally, the hemispherical abdominal organs are smaller versions of the ocular and siphon organs, which contain similar elements, only organized differently.

In contrast to *Pterygioteuthis*, which possesses sixteen light organs, Hoyle found that those of *Abraliopsis* are countless and are "distributed pretty freely over the ventral surface of the mantle, head, and third and fourth pair of arms" (1904). There is considerably less variety of structure among light organs than in *Pterygioteuthis*. They are all more or less spherical and consist of a lens, a pigment cup, an inner cup, a posterior hemisphere, and an internal cone. Apart from the lens, pigment cup, and internal cone – the latter reminiscent of the photocytes of *Pterygioteuthis* – even today one is at a loss to make sense of the remaining two components of the organ.

The squid material from the *Valdivia* was much more extensive and better preserved than that of the *Albatross*, and it allowed Carl Chun to gain a more rounded picture of the diversity of cephalopod light organs. He went as far as to state that there "exists a *polymorphism of the luminous organs* [his emphasis] not known to such an extent in any other organism. Unique, however, is in this respect the beautiful genus *Thaumatolampas*; I found that its 22 luminous organs are formed according to no less than 10 different structural principles" (1910). Further in the text Chun offers enchanting speculations on what perhaps lay behind that diversity:

> If we ask ourselves why there are 10 different structural types among only 22 luminous organs, we find that the only reasonable answer is that the light of the various organs differs apparently not only in intensity but also in quality ... The intensity of the light is probably proportional to the size of the luminous body and to the extent to which it is equipped with reflectors and lenses. There are a number of accessory structures, in addition to those mentioned, which suggest that the color of the light is also different. In the live animal, the middle organs of the eye emit a magnificent ultramarine-blue light, the middle of the 5 ventral organs shines sky-blue, and the two anal organs are ruby-red ... The observer who would be lucky enough to spot a living, healthy *Thaumatolampas* in all its magnificence would indeed behold a fairy-like sight.

Chun's texts on light organs, scattered throughout his 1910 monograph, celebrated this diversity but made no attempt to explain it, even in evolutionary terms. The polymorphism of the light organs of a given squid species is such that in many cases only the "luminous body" – Hoyle's internal cone containing the photocytes – remains similar in texture and cellular composition while everything else – pigment sheath, reflector, lens, fibrous layers – may vary in every respect, each according to its own "structural plan." This is particularly striking in *Thaumatolampas*, but Chun spotted similar patterns in many other squids. Even in *Pterygioteuthis giardi*, whose light organs Chun acknowledged were competently described in detail by Hoyle, he found that "the polymorphism of the luminous organs is more far-reaching than it appears from Hoyle's description." With an eye to pleasing an educated but not necessarily scientifically trained readership, Chun used a colourful analogy with gems to illustrate the diversity of the light organs of *Pterygioteuthis*:

> Specimens preserved in formol show the full sheen of the organs. One can hardly imagine a more magnificent play of colors; they shine on a dark background like pearls and precious stones, the larger organs of the eyes like blue sapphires, the anal organs like greenish or golden metallic mirrors, the organs near the gills with a flesh-pink tone, the others with a nacreous sheen. Transfer of the specimen to alcohol destroys this brilliance; the scalelike superimposed cells and concentrically stratified lamellae shrink, and the greenish metallic sheen becomes steel blue and later disappears.

Chun also improved on Hoyle's work in that he paid greater attention to the blood and nerve supplies of the light organs. Blood vessels remain ubiquitous, but Chun found innervation in almost every squid light organ examined. Although it had always been clear to Hoyle that nerves penetrate the light organs, their course inside the organ was uncertain because attempts to selectively stain nerves had failed. Chun remedied this: "After many unsuccessful attempts to stain the nerve fibrils, I finally succeeded in staining the eye organ with iron-haematoxylin after fixation with sublimate. The fibers stain deep black, in contrast to the surrounding light-colored tissue, giving a clear picture of the innervation of the luminous body." The staining technique allowed him to witness how rich the innervation was and to follow the course of nerve branches between the cells of the luminous body. Thus, he gained unequivocal evidence that the light-emitting cells themselves were the target of the innervation, and therefore are under nervous control just like the skin chromatophores responsible for the quick colour changes of shallow-water squids.

Chun added new squid species endowed with light organs. *Benthoteuthis megalops*, for example, possesses on the arms small light organs that are simple pigmented cups with no optical accessories. The path of light emission

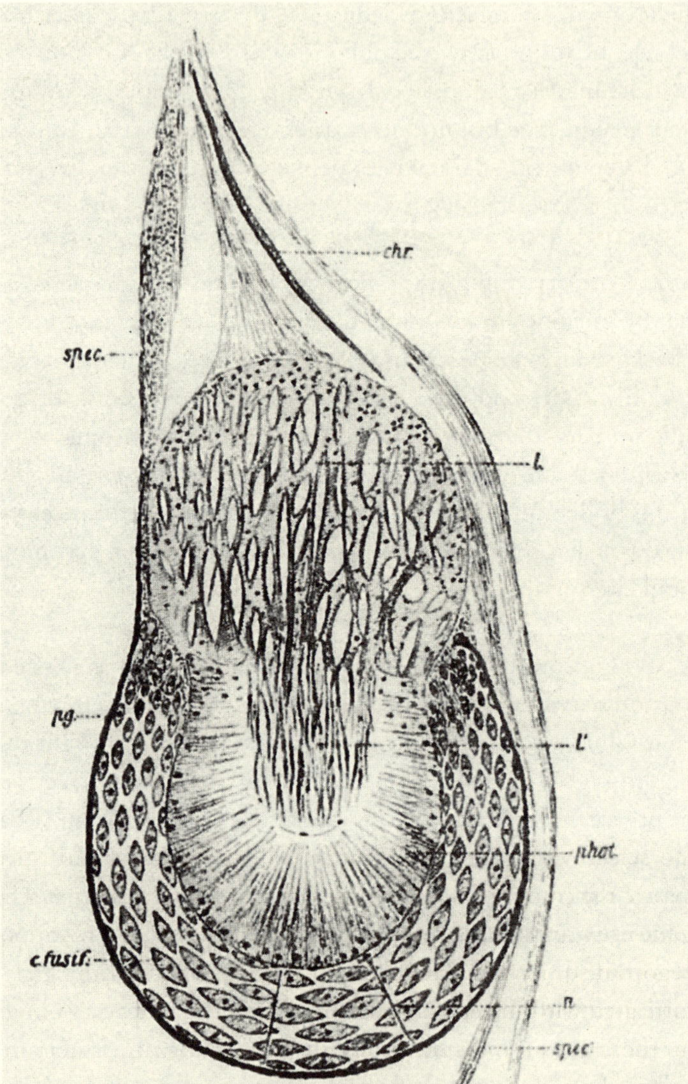

Figure 6.2 Sections through squid light organs. *Above*, body light organ of *Calliteuthis hoylei*. *Right opposite*, ocular light organ of *Abraliopsis morisii*. Note the luminous body (phot), and the various lenses (l), reflectors (spec, refl), pigment sheaths (pg), and pigment cells (chr). From Figs. 15 and 16 in Chun (1910).

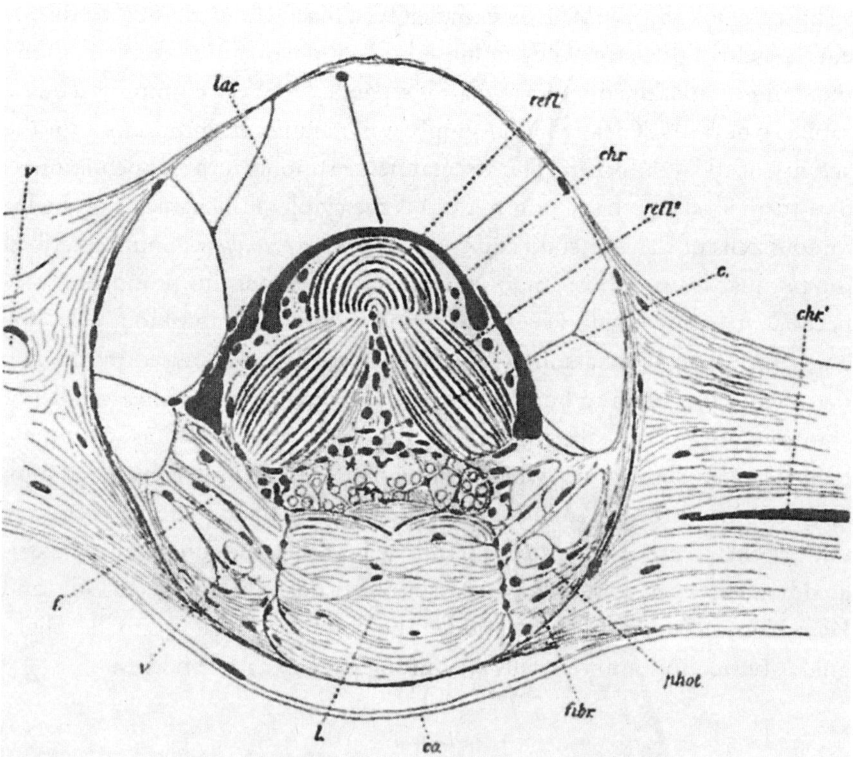

intersects with overlaying skin chromatophores which, Chun speculated, must change the colour of the outgoing light. He noted also that the "organs of *Benthoteuthis* are situated mainly on the dorsal side, while those of all other oegopsid squids are situated mainly on the ventral side. This is obviously correlated with the dorsal position of the eyes. The beam of light emitted from the organs is directed anteriorly, preventing direct illumination of the eye." In *Mastigoteuthis glaukopis* Chun saw small cutaneous light organs which Louis Joubin, the French cephalopod specialist, had previously interpreted, oddly enough, as thermal sensors (Joubin, 1893b). The chances of such organs acting as thermal sensors in a cold, unchanging environment were poor.

Another squid species with structures to which Joubin (1893c) had denied luminescent status was *Chiroteuthis imperator*. Chun drew attention especially to a pair of visceral light organs flanking the intestine and apposed against the ink sac, the structure that ejects a black cloud of "ink" to disorient a potential predator. The luminous body of the organs is peculiar among the squids in that the photocyte mass is separated into cell clumps by a dense capillary network. Clearly blood supply is of paramount importance for the activity of the light organ. The visceral nerve supplies a rich innervation.

A further oddity that Chun noticed is the cytological character of the luminous cells of the cranchiid squid *Desmoteuthis pellucida*. Chun described the peculiar feature thus: "Staining with iron hematoxylin produces a very peculiar structure which I have not found in any other luminous cells: the entire luminous body assumes a blackish shade. This is caused by the fact that each cell contains a bundle of fibers … which extend either parallel or converge toward the tapering end of the cell, like rays." Chun was totally baffled by this observation, whose significance for light emission eluded him: "I do not know the physiological function of these fibers. They are certainly not cuticular secretions of the luminous cells of the type comprising the 'striated bodies' observed in the organs of Euphausiidae [krill]." Peter Dilly and Peter Herring (1974) solved the riddle: the fibrous material is actually a light guide abutted to luminous cells containing paracrystalline bodies.

Chun's reference to euphausiids brings us to the crustaceans of the pioneering deep-sea expeditions. The first to study the light organs of euphausiids was Georg Ossian Sars (1837–1927). I mentioned Sars in chapter 4 as one of the naturalists who were instrumental in inspiring the quest for deep-sea organisms of the *Porcupine* and *Challenger*. Born in Norway to Michael Sars, a church minister and a naturalist in his own right, he studied medicine and natural history at Christiana (now Oslo) University. In 1864 Georg Sars was commissioned by the government to conduct a survey of Norwegian fisheries; in the process, he made the important discovery that cods' eggs do not settle on the coastal sea floor but float in the pelagic zone (Hestmark, 2009). He specialized in the study of crustaceans, and it was in this capacity, as he had reached the rank of professor at Christiana University, that Wyville Thomson asked him to study the euphausids collected by the *Challenger*. It was an ironic twist of fate that the man who had dredged animals at the greatest ocean depths prior to the first oceanographic expeditions should

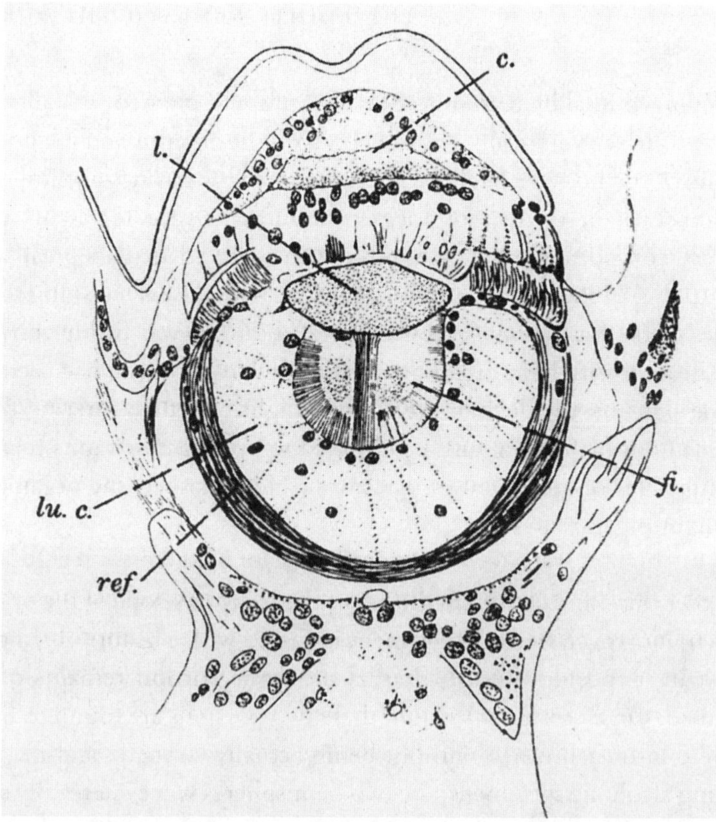

Figure 6.3 Section of a photophore of the euphausiid (krill) *Meganyctiphanes norvegica*, showing the u-shaped mass of luminous cells (lu. c) surrounding a rod mass (fi) and backed by a reflector (ref). Above are a lens (l) and a blood sinus (c). From Harvey (1952) after Valentin and Cunningham (1888).

have contributed to the reports of the *Challenger* voyage. The irony must certainly not have been lost on Wyville Thomson.

In his report on the "Schizopoda" – an obsolete designation for mysid shrimps and euphausiids –Sars devoted three pages to the light organs of the krill *Euphausia*, which he called "luminous globules" (Sars, 1875). Sars was a competent zoologist but histology eluded him. So, his approach was to dissect out light organs and observe them under the dissecting microscope, enabling him to perceive the internal structure of the organs thanks to their near transparency. Sars noted that the globules are "very conspicuous in the living animal by reason of their beautiful red pigment and glistening lustre, and are symmetrically arranged both on the anterior and the posterior divisions of the body." His illustration of the organ has no labelling so the reader must rely on the text for orientation. At first glance, the organ appears as "a very complicated structure, bearing in some particulars great resemblance to that of the eyes in vertebrates." This remark was revisiting the ambiguities of the past about "accessory eyes," but Sars was firmly convinced he was dealing with light organs; after all, Wyville Thomson had seen them emitting light on the deck of *Challenger*. Light, whether entering the eye or beamed out by light organs, is subjected to optical constraints that drive the similarities of structural organization. He described the organization of the light organ as follows:

> A rather thick and elastic cuticle forms the outer envelope of the organ, which, moreover, in fresh specimens is coated with a beautiful red pigment in its posterior half, whereas the front portion remains quite pellucid. On closer examination, these two portions are found to fit as it were into each other, without being actually connate, and on dissecting alcoholic specimens, the two hemispheres will even readily separate from each other. At the junction, a glistening ring may be seen internally, encompassing in the middle a highly refractive lenticular corpuscle. The posterior hemisphere is filled up with cellular matter, in the midst of which lies embedded a flabelliform bunch of exceedingly delicate fibres, exhibiting in fresh specimens a most beautiful iridescent lustre. To the equatorial zone of the organ, moreover, two or three thin muscles are attached, admitting, to a certain extent, of its being rolled to and fro.

Sars was struck by the analogy of movement between the shrimp's light organs and its eyes, made possible by the thin muscles. He offered no speculation as to the purpose of mobility for the light organs. However, he was acquainted with a Norwegian species of *Euphausia* that he had observed alive. He saw "that the animal is able, by varying the movements of the organs, to increase or diminish the light at will." Moreover, by crushing the organ and extricating from it the "bunch of exceedingly delicate fibres" in the middle of the organ, he could determine that this isolated fibrillary mass was the source of the emitted light.

In his monograph Sars sifted through several other species of shrimp known today to be luminescent, but either he was unaware of their luminescent capabilities or he omitted to describe their light organs. Another Scandinavian zoologist, Hans Jacob Hansen (1855–1936), partially remedied this shortcoming. Born near Odense in Denmark, Hansen earned a PhD in natural history at the University of Copenhagen in 1883 (Damkaer, 1995). From his thesis topic on insect morphology forward, Hansen remained an arthropod specialist throughout his career. In 1885 he was appointed research assistant in the entomology section of the Zoological Museum of Copenhagen and there he developed an international reputation for his work. But his outstanding scholarship failed to earn him career promotions because he fell victim to disputes between his thesis mentor and colleagues, and also because of his abrasive personality (Wolff, 1993). Angry for being passed over for a position as curator at the museum, he resigned his post in protest in 1910. Thanks to the indignant clamour of many international colleagues, he was granted a pension that allowed him to continue his research work until his death. In his own country, it didn't help that Hansen wrote polemical pamphlets leaning on extreme-right nationalism and antisemitism (Buhl, 1995).

As evidence of his standing in the scientific community, we have Hansen's contributions on crustaceans in several scientific reports of oceanographic expeditions (*Challenger*, *Albatross*, *Siboga*, and Monaco); but his only study of light organs was conducted from a new species of decapod shrimp, *Sergestes challengeri*, shrimp material of the *Challenger* (Hansen, 1903). Hansen numbered 117 light organs all over the body, nearly all of which projected their presumptive light beam downward and presented little or no variation in their internal structure. Like Sars, Hansen conducted no histological processing, relying again on teased material observed under a dissecting microscope. The organ appears complex, with two identified lenses that occupy half of the organ: an external biconvex lens and an inner, "concavo-convex" lens that "consists of two layers, is homogenous, vitreous, and slightly greyish." The two lenses reminded Hansen "of optical instruments in which the lens is composed of crown-glass and flint-glass." Behind the inner lens is a layer of "glandular" cells presumed to be the source of the emitted light. And further behind is a fibrous layer resembling a reflector.

Hansen also provided a comparative analysis of crustacean light organs (1903). The number of organs in *Sergestes*, he found, is far greater than in other crustaceans. He made the distinction between ostracods and copepods, which eject a luminous cloud, and euphausiids and sergestid shrimps, which possess what he called "real organs," by which he meant more complex organs in which the light is generated internally.

Thus ends our survey of light organs of deep-sea forms at the turn of the century. The mesmerising diversity of light organs that these researchers encountered must have raised questions as to the meaning of it all. The thought must have arisen in many minds, though unexpressed in such words, that the value of bioluminescence to the life of deep-sea animals was measured by the degree of their investment in the design and construction of their light organs.

PART THREE

OPENING UP NEW VISTAS OF RESEARCH

7

Paolo Panceri and the Italian Cohort

Having undertaken a special study of animal light, I focussed my attention on the sea dwellers with a view to search for the hitherto unknown source and conditions of this mysterious phenomenon.
–Paolo Panceri (1872)

The great oceanographic expeditions revealed to the world the existence of complex light organs in higher invertebrates and fishes, but to investigators of coastal or terrestrial fauna the luminescent apparatus of lower invertebrates was to be a far more nebulous affair. The absence of spectacular light organs made the task of these few investigators particularly difficult. But in 1870, in parallel with the groundbreaking oceanographic voyages, there began a series of investigations centred on the search for the sources of light emission in animal forms lacking conspicuous light organs, and for the physiological basis of their light emission. The instigator of this new approach was Paolo Panceri (1833–1877), an Italian naturalist whose career experienced a meteoric rise.

Panceri, born in Milan to a medical doctor, followed family tradition by going to Pavia to study medicine (Cornalia, 1877). The anatomy lessons of Professor Bartholomeo Panizza, famous among other discoveries for establishing the role of the cerebral cortex in vision, inspired Panceri to pursue comparative anatomy as his discipline of predilection. After earning his medical degree in 1856 he stayed in Pavia, where he was appointed assistant in the Department of Zoology and Mineralogy. His interest in marine animals was aroused by a collecting trip to Venice and Nice in 1858. The following year, he practised medicine briefly at the Hospital of Milan to care for the injured during Italy's second war of independence. In 1861, at the astonishing age of twenty-eight, he was courted by both Bologna and Naples to occupy their chairs of comparative anatomy. Panceri opted for Naples because he

Figure 7.1 Paolo Panceri, photographed by Francesco Gasco. Smithsonian Institution Archives, image # 85-4449, with permission.

was attracted by the rich fauna of the Gulf of Naples and thought the climate of southern Italy would be favourable to his health.

In Naples Panceri put all his energy into his teaching and research activities, and into the creation of a cabinet of comparative anatomy that would stimulate interest in the discipline from students and the educated citizens alike. Over the years he developed a good rapport with city and government officials, which stood him in good stead when Anton Dohrn (1840–1909) arrived in Naples to scout for the creation of an international zoological station. Dohrn, a German-born zoologist who espoused Darwinism under the influence of Ernst Haeckel at the University of Jena, wanted to create an independent zoological station where he could do what he liked most: help fellow scientists accomplish their research goals and foster marine biological research that would support Darwin's theory of evolution (Groeben, 1985). Dohrn arrived in Naples in 1870 and met Panceri, who helped him get the approval of the city to build his zoological station. Here is the way

Dohrn remembered their meeting in his unpublished memoirs, as quoted by Groeben (2001):

> I had the unexpected and instructive opportunity to get acquainted with a colleague for whom the significance of Darwin's theory is quite uncertain and who is a total stranger to the intellectual movement of the moment … Panceri is an autodidact and his training was based essentially on French texts and on a brief sojourn in Paris – and all the French School now demonstrates a flat refutation of Darwin and of the theory of descent … Panceri is still rooted in the tradition of Cuvier … I with my scientific theories from Jena had to be incomprehensible to him … Panceri knew neither English nor German and … all the great world of thought contained in the scientific literature of these two leading countries remained unknown to him.

Despite his diffidence with regard to Darwinism, Panceri showed an exceptionally open mind about Dohrn's project; he saw the zoological station as an opportunity to raise the scientific culture of the Neapolitans as well as provide a local centre for zoological research that worked in synergy with his own academic pursuits (Morgese and Vinci, 2010).

It was around the time of his first acquaintance with Dohrn that Panceri began his research program on luminescent animals. He seemed in part inspired by the musings of his countryman Enrico Giglioni on the phenomena of bioluminescence that he had encountered in his circumnavigation of the world aboard the three-masted schooner *Magenta* (Cornalia, 1877). Enrico Hillyer Giglioni (1845–1909) was born in London of an Italian father and an English mother. The family moved back to Italy and Giglioni's father was eventually posted at the University of Pavia, where he occupied the chair of anthropology. In 1860, when the precocious Enrico was only fifteen, several Pavia faculty members, including Panceri, helped the teenager obtain a bursary to study in London. At the School of Mines he enlisted in courses by such luminaries as the geologist Charles Lyell, the comparative anatomist Richard Owen, and the zoologist Thomas Huxley. On his return to Italy he obtained his diploma in natural history at the University of Pisa in 1864, and the following year, still only twenty, he was selected to participate in a three-year scientific mission aboard the *Magenta*.

The ship's captain was empowered by the king of Italy to negotiate commercial deals with Japan and China.

Giglioni's detailed account of the voyage was published in 1875, but just two years after his return to Italy he had assembled into one article his impressions of bioluminescent phenomena gleaned in the course of the voyage (Giglioni, 1870). The content of this paper shows a surprisingly sharp mind and scholarly ability. Giglioni identified three types of luminescent display in the oceans: (1) diffuse, homogenous, milky light; (2) scintillating, inconsistent, luminous dots; and (3) luminous discs with a generally non-scintillating, fixed light. The first type he associated with various dinoflagellates. The second he regarded as the most frequently witnessed in all kinds of seas, climates, and animal groups, among which he mentioned radiolarians, hydrozoan and scyphozoan jellyfish, comb-jellies, salps, appendicularians, pyrosomes, arrow-worms, and mysid shrimps. In the third he included scyphozoan jellyfish (*Pelagia*, *Rhizostoma*) and a shark (*Scymnus fulgens*).

Of particular interest is Giglioni's account of the luminescence of an appendicularian (probably *Oikopleura*). To this day the earliest observation of luminescence of this house-building larvacean is attributed by Galt (1978) to Lohmann (1899) not to Giglioni and the *Magenta* expedition. Galt dismissed Giglioni's account, claiming he must have just seen a show of iridescence of the tail. As the tail of appendicularians is almost synonymous with the trunk, it is likely indeed that Giglioni mistook iridescence for luminescence, as this passage suggests:

> In these small Tunicates such properties are located in the central axis of the caudal appendage, where the light appears as sharp and intense flashes, which vary in color in the same individual; to my knowledge this has not yet been recorded. I noticed for the first time in a beautiful species caught in the Southern Atlantic on 22 December 1865 … in which the trunk emitted at different intervals, a clear and vivid light, first a dark red, then blue and finally green. Many Appendicularians were encountered in the crossing from Montevideo to Batavia [Indonesia], and in almost all I confirmed this three-color phosphorescence; in a large species encountered in the Indian Ocean were the colors white, blue and green.

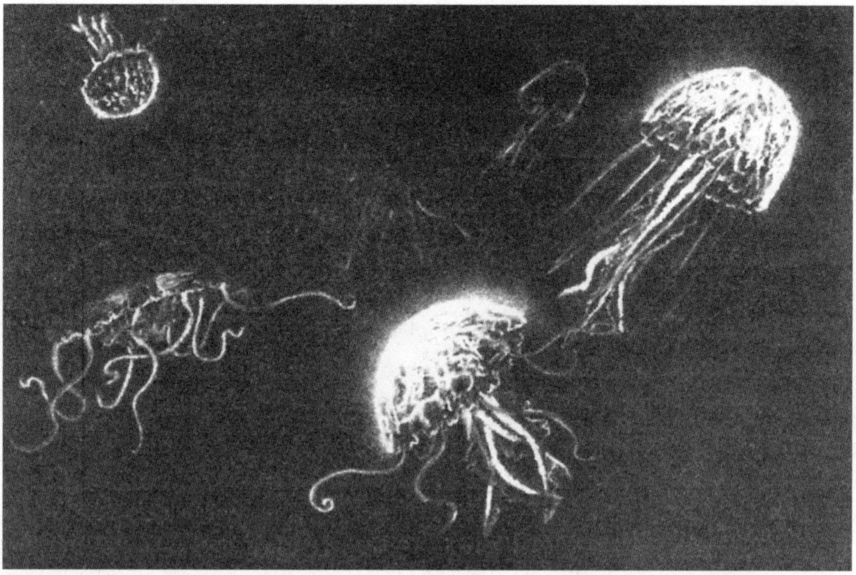

Figure 7.2 Representation of glowing jellyfish. Drawing by E. Grace White in Dahlgren (1916b).

Giglioni is unique in having observed multi-colour display in different animal groups, much of which was probably iridescence if his observation point was not initially pitch dark – a moon-lit night, for example. Green and blue figured by far as the most common (and suspect) of the colours that he observed.

Panceri's reading of these descriptions of luminescence with spectacular colour displays made such an impression on him that he embarked almost urgently on a program of bioluminescence research. To accomplish this he only had to look around in the Gulf of Naples and other nearby coasts to collect all the animals he needed: jellyfish, comb-jellies, sea pens, pyrosomes, pholads, gastropod mollusks, tube-worms, and scale-worms. Considering that Panceri had until then channelled his energy into the field of comparative anatomy, it is surprising that he so suddenly chose to venture into unknown territory – cell analysis and experimental biology – with the breezy self-confidence that is evident in his writing.

Between 1871 and 1876 Panceri devoted no fewer than sixteen papers to bioluminescent animals, not counting two early papers on the pseudoluminescence of dead fish. His work was widely and quickly disseminated;

he translated a collection of his early papers for a French journal (Panceri, 1872a), and four of his original papers were translated into English and appeared in a British journal (Panceri, 1872b,c; 1873a,b). His first contribution, on the luminescent system of jellyfish (Panceri, 1871) clearly sets out the tenets of his scientific approach. In this paper we follow the deductive process by which his experiments lead him to conclude that the luminescence of the scyphomedusan *Pelagia noctiluca* is basically a mucus secretion from specific epithelial cells, not spontaneous but elicited by touch, or by electrical or chemical stimulation. As accords with our current knowledge, he found that the light seems to turn on after mucus is ejected from the luminous cells. The cells are filled with highly refractile fine granules ranging in colour from golden to orange with red pigment mixed in. He also claimed to have seen light emanating from the "internal epithelium" (endoderm or gastroderm).

Because he was convinced, albeit without solid evidence, that the luminescence of dead and decaying fish results from the oxidation of fatty substances, Panceri carried this notion over to the luminescence of live jellyfish. "The luminous cells that are packed with the colorful granules such that the nucleus becomes buried and invisible," he remarked, "have the appearance of cells in which the content has become adipose. In fact, by their reactions and appearance, these granulations resemble fat more than anything else." In his treatment of a hydrozoan jellyfish, *Cunina moneta*, Panceri stated that the luminescent system was similar to that of the scyphozoan jellyfish *Pelagia*, including the secretory process and the fatty granules of the luminous cells. Today we know that the luminescence of hydrozoan jellyfishes is intracellular, but Panceri's interpretation that fatty substances lay at the core of the luminescent reaction pervaded many of his papers.

Next, Panceri examined the luminescence of sea pens (1873a). Sea pens (pennatulids) are anthozoans like sea anemones but, contrary to the latter, which are solitary polyps, sea pens are basically an assemblage of polyps tightly integrated together to form a colony. Apart from his rambling delusion about the fatty nature of the luminescent source, Panceri came up with valid new insights regarding the physiology of luminescent activity in sea pens. First of all, he located the luminescent sources squarely in the polyps and the zooids (modified and diminutive polyps). Second, in each polyp he identified luminous cords associated with the base of the eight tentacles distributed around the mouth, and he rightly located these eight cords in the epithelial lining aroung the pharynx; that is, in the endoderm. Third, he identified cells associated with the luminous cords; he thought the cells with "fatty content" were the light-emitting cells, but he also identified cells with elongations that are probably the cells considered since James Morin's 1974 fluorescence microscopy work to be the actual luminescent cells. And fourth, not only did he describe with great accuracy the dynamics of light emission across the colony, something Forbes only mentioned *en passant* (see chapter 3), but he designed experiments in order to understand the mechanism underlying this dynamics. The importance of his findings deserves the discussion we will now undertake.

Upon touching a polyp casually, Panceri saw a local light emission, but if he systematically touched a polyp or even an area devoid of polyps like the peduncle of the sea pen, the light emission would spread from the nearest polyp to more distal ones in succession. Electrical stimulation, he reported, brought similar results. The wave of excitation that it creates travels along the long axis of the sea pen and dies only once it reaches the other end of the animal. If the stimulus is applied to the peduncle, the wave of light travels upward to the tip of the sea pen. If the stimulus is applied to the tip, the wave travels downward toward the peduncle. So, there is no fixed polarity to the path of excitation. Even more remarkable is that if stimuli are applied simultaneously to the tip and to the opposing end on the peduncle, the two luminous waves will travel toward each other and cancel each other at their zone of meeting. This cancellation zone is explained today as the refractory zone, where excitation has occurred too recently to elicit another response so soon. Panceri also observed that these luminescent activities are highly susceptible to fatigue.

Panceri realized that the path of the luminous wave is probably the exact reflection of the underlying path of conduction of excitation. "I cannot help but call to the attention of physiologists," he wrote, "this singular property of pennatulids, *to make visible by the glow of their polyps the direction and speed of propagation of excitation, as if in these animals the internal movement of molecules which occurs during excitation put the cells of the luminous cords in contact with oxygen, a chemical action accompanied by the development of light instead of heat*" (Panceri's italics). In other words, whatever it is that conducts the excitation lights up the luminescent cells (photocytes) of the polyps along the way. "What was the nature of the conductor?" Panceri asked himself. "Could it be made of nerves?" He measured the speed of the luminous wave (5 cm/s), but found that this speed was 600 times slower than the conduction speed of frog-leg nerves. He, and Albrecht von Kölliker before him, saw "thin, pale and transparent fibres in the small partitions [septa] and muscles of polyps," but he kept his counsel until further histological investigations were conducted. If the conducting tissue turned out to be nervous, Panceri concluded, it would likely be "social," meaning that it would take a colonial nervous system, not nervous systems intrinsic to polyps, to coordinate the dynamics of sea pen luminescence.

Off the island of Capri, Panceri discovered another sea pen genus, *Cavernularia*, which was also luminous (1972a). This sea pen was caught at greater depths than the other luminous sea pens and its light emission was blue rather than greenish. Otherwise, the luminescent display presented the same characteristics that Panceri observed in pennatulids of shallower waters.

Panceri next turned his attention to another type of colonial animal, *Pyrosoma giganteum* (1873b). Pyrosomes are built around a cylindrical colonial mass upon which up to 3,000 individuals (zooids) are grafted. Luminous sparks seem to cover the entire surface of the colony, and the light output is so strong that sailors have been able to spot them from 100 meters away. Panceri was the first to accurately locate the luminescent sources for these sparks. By careful dissection and sectioning, he found a pair of cell masses on each zooid that emitted light. The cells of these structures are spherical and rather large. These suspected light organs, previously thought to be ovaries or urinary organs, are attached to the external wall but bathing in a blood sinus.

Panceri could not find any innervation to these organs, and yet when a pyrosome was stimulated by touch or other means, luminescence spread like a wave over the colony in a manner similar to sea pens, except that in pyrosomes the speed of the wave was slower. "The special studies I conducted to explain the transmission of excitation which progressively produces light in the various ascidies [zooids] of the colony," he wrote, "led me to the discovery of a peculiar social muscular system by which all the ascidies are bound to each other." So Panceri reasoned that the innervation of this colonial muscle might activate the luminous organs in coordination with the muscles. There is, however, an Italian postscript to this story. In 1911 a zoologist working at the Naples Zoological Station, Osvaldo Polimanti (1869–1947), published a paper in which he showed that the luminescence of pyrosome zooids is activated by light. He thereby discredited the theory of nervous coordination, proposing instead that once a zooid lights up following stimulation, its light activates the luminescence of neighbouring zooids, and the cascading effect results in the wave of luminescence (Mackie and Bone, 1978).

One aspect of bioluminescence research that had so far been neglected was the developmental genesis of the phenomenon. Panceri must be credited for exploring this aspect in *Pyrosoma*. He followed the formation of the embryos from the eggs (or cyathozooids) or from budding embryos forming the colony, and he found that the light organs arose in the external layer of the blastoderm. He further observed that when the embryos are hatched, the cell masses of the light organs are already functional and emit light. Thirty-five years later the Belgian zoologist Charles Julin (1857–1930), professor at the University of Liège, expanded on Panceri's contribution. Strangely, Julin found that luminescence appears twice, in different sets of cells: first, in the test (support) cells of the cyathozooid embryo, which disappear during the atrophy of the cyathozooids; and later, as Panceri had stated, in the blastoderm, where the luminous cells are gathered in the pair of what Julin calls "lateral glands of Joliet" (Julin, 1909).

The plot thickened when Julin discovered, to his surprise, that the two sets of cells, which have totally different origins, contain similar inclusions. He described the cytoplasm of these cells as being "filled by a tubing, continuous or discontinuous, displaying a very delicate reticulum of large mesh," and he identified mitochondria at the nodes of the tubing (Julin, 1909). In his long career of cytological observations, Julin had never seen anything like it. In a more detailed study, in which he added useful illustrations (1912), Julin expressed his changed belief that the "tubing" of the luminous cells included nuclein (or nuclear chromatin) rather than mitochondria. But three years later, Paul Buchner (1886–1978), a young German investigator based in Munich who was making a name for himself by proposing the existence of mycelia or bacteria inside cells – a phenomenon known as endosymbiosis – brought a new twist to the story. Buchner read Julin's 1912 paper, and its illustrations of the luminous cells immediately caught his practised eye. In a short paper (Buchner, 1915) with the provocative title "Light organs or fungal bodies," he wrote that the "tubings" inside the luminous cells of *Pyrosoma* made sense only if we looked at them as fungi. From some of Julin's descriptions he deduced that these microorganisms entered the blood sinus and from there entered the cells in contact with the sinus, and colonized the interior of the cell, from where they emitted light.

There is a fascinating symmetry in the fact that the researcher who finally solved the puzzle was an Italian who occupied the chair of comparative anatomy in Naples that Panceri had occupied over forty years earlier. Umberto Pierantoni (1876–1959) was born in Caserta, north of Naples, into a family of jurists (Salfi, 1960). After earning his doctorate in natural history in 1899 at the University of Naples, he worked in the Department of Zoology as curator of collections until 1909, when he was appointed chair of zoology at the University of Sassari on the island of Sardinia. He went on to a stint at the University of Turin before returning in Naples in 1925 to take up the chair of comparative anatomy. Around 1909 he developed an interest in symbiosis – the long-term association between organisms of different species often with mutual benefits – and a few years later turned his attention to bioluminescence.

Pierantoni's observations led him to develop the theory of "physiological and hereditary symbioses," according to which micro-organisms living in specific animal cells recognize these cells for *de novo* infection and intracellular growth, thus ensuring perpetuation of the endosymbioses from generation to generation. On reading Buchner's 1915 paper on fungal bodies, Pierantoni saw in the luminous cells of pyrosomes tangible evidence of the validity of Buchner's theory. So he conducted studies to confirm Buchner's hunch. In the paper reporting his findings (Pierantoni, 1921), he reconstructed the entire cycle of spore release from luminous cells and *de novo* infection of the next generation of test cells, and confirmed that the micro-organisms are luminous bacteria. Pierantoni (1918) also discovered the symbiotic light organs of shallow-water squids such as *Sepiola* and *Rondeletia*, in which luminous bacteria are lodged in special organs, but not intracellularly. In later developments, Giuseppe Zirpolo, who worked in the Bacteriological Laboratory of the Military Hospital of Caserta, cultured and characterized the luminous bacteria of the squids

Sepia (Zirpolo, 1918) and *Rondeletia* (Zirpolo, 1919) obtained at the Naples Zoological Station (Zirpolo, 1918).

~~~~

Returning to the 1870s, we find Panceri pressing on with his bioluminescence research program, this time tackling the pholads. As mentioned in chapter 1, this unusual mollusk, which bores holes in clay or soft rock of shorelines, had been on the radar of observers ever since Antiquity. The clam *Pholas dactylus* may have been the most observed luminous animal over the longest span of centuries, but, although many believed the whole epithelial surface of the clam secreted a luminous "liquid," by Panceri's time no one had yet ascertained the source of its luminous secretion. Panceri was the first to address the problem with the necessary scientific rigour (1873b). "To solve this problem," he wrote, "I only had to use a small trickle of water which, falling in the dark on the animal, whose mantle and anterior siphon had been exposed, swept away the largest part of the luminous mucus, and allowed me to see the *luminous organs* of these mollusks" (Panceri's italics). In the illustration that Panceri provided, the distribution of the glandular tissues is clearly marked: in an arc running from the superior rim of the mantle halfway down the valves of the shells, in two triangular patches near the entrance to the anterior siphon, and in two long, sinuous cords running parallel to each other along the anterior siphon. Surgical removal of these organs eliminated light emission entirely.

Following a suspicion that the organs are controlled by the nervous system, Panceri traced the innervation of the triangular patches and cords. The branchial ganglion sends out two nerve trunks that form two small ganglia from which nerves ramify and send tiny nerve branches into the light organs. Turning to the description of the organs themselves, which lie on the mantle, he found a simple structure in which the secretory epidermis rests over a crest of dermis. The secretory cells are ciliated – they possess hair-like projections from the surface – and are filled with granules that discharge to the outside at the slightest touch. Panceri further observed that "the luminous content of the cells of the luminous epithelium dissolves in ether and alcohol," thus seemingly reinforcing his notion that the luminous substrate is a fatty substance. Air and oxygen sustained the luminescence of the luminous
~~~~

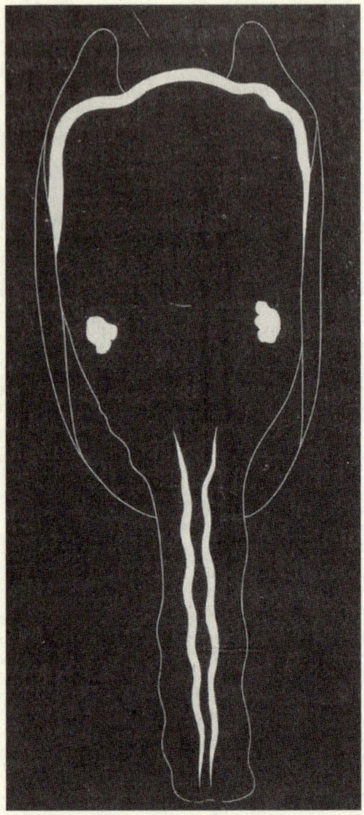

Figure 7.3 *Right* Representation of the clam *Pholas dactylus*, highlighting the luminescent zones. From Panceri (1872c).

Figure 7.4 *Opposite* Representation of the transparent body of the gastropod mollusk *Phyllorhoe* (left) and the distribution of its light sources (right). From Panceri (1873c).

material well into the animal's putrefaction, a clear indication that an oxidation process is involved. Finally, with the assistance of British zoologist E. Ray Lankester, he was able to determine that the emitted light is monochromatic in the blue range.

Another mollusk, the pelagic sea slug, interested Panceri as well. *Phyllirhoë bucephala* pushes transparency to the point that all its internal organs are visible to the onlooker. Given such a glassy surface, it is no wonder that the many naturalists who had happened on this gastropod did not in the least suspect the presence of light organs or its luminescent capability. Panceri, ever the perceptive observer and by now attuned to looking for signatures of luminescence, witnessed bright scintillating light coming from thousands of dots all over the body upon chemical stimulation (1873c). These dots in fact correspond to small clusters of large cells sunk under the epidermis. Heinrich Müller (1820–1864), a German anatomist affiliated with the Uni-

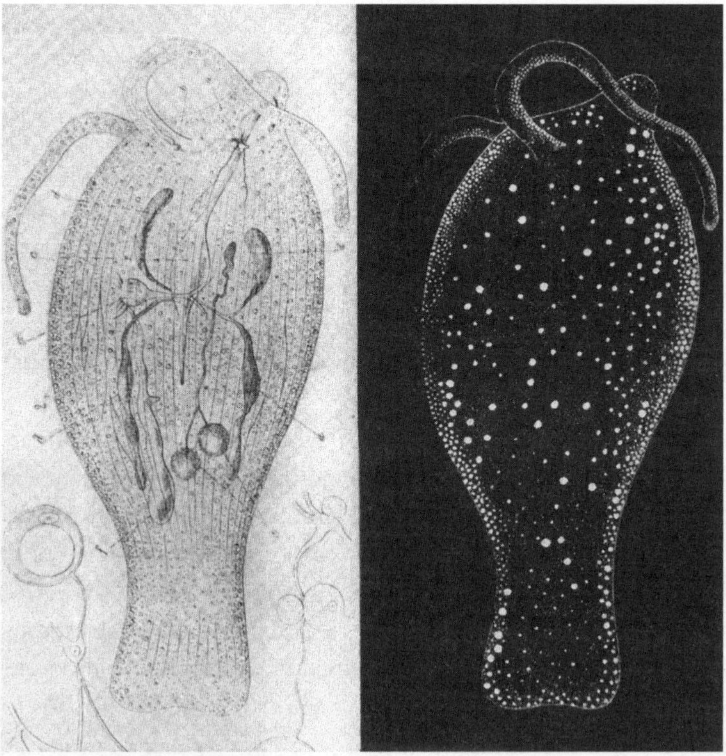

versity of Würzburg, described "spherical cells with sharp dark outlines, lying over fine nerve branches, which contain not only a nucleus but also a spherical body of varying size, yellow and refringent" (Gegenbaur, Kölliker, and Müller, 1853). Müller had no way of guessing that these cells could emit light, but Panceri, in honour of the German who first described them, named them Müller's cells.

Panceri thought that nerve cells in the tentacles of the sea slug were also a source of light emission, but time has proven this assertion wrong. His own experiments with tactile stimulation, intended to prove that nerve cells emit light, showed that the emission of light was too slow to start, develop, and extinguish to be assigned to nerve cells. But that did not deter Panceri from insisting that the tentacle nerve cells are luminescent. Even to this day, the identification of Müller cells as the source of luminescence is not entirely settled (Herring, 1978).

The next project that engaged Panceri was an investigation into assorted species of comb-jellies (1873d). Prodding one of the eight "ribs" (ciliated comb rows) sends a wave of light travelling along the rib. If the stimulus is applied at the anal (or aboral) pole of the gooseberry-shaped body, the light wave shoots along the rib to the oral pole; and the reverse process also applies. So, as in sea pens, the direction of travel is unpolarized. With tactile contact over a larger surface of the gelatinous body, Panceri managed to light up all eight ribs, thereby suggesting that graded sensory stimulation can recruit luminescent tissue from more ribs. The excitation is susceptible to fatigue after a short period of stimulation, but it recovers after fifteen minutes at rest.

Panceri's quest for the source of the luminescence of comb-jellies met with less success. He correctly determined that the light originates in "particular matter" surrounding the gastrovascular canal of the ribs. Where he erred was in thinking that acellular elements (vesicles) were involved because he could not associate nuclei with these structures. From Allman's contribution (see chapter 3) it was known that solar light inhibits the luminescence of comb-jellies. Panceri further determined that solar or artificial light completely eliminated luminescent capability and that luminescence was restored fifteen to thirty minutes after returning the comb-jellies to total darkness. Even moonlight had a partial inhibitory effect on luminescence.

It is not clear what caused Panceri's last papers to appear posthumously in 1878 (1878a,b). He had completed an important paper on annelid worms and other "worm-like" organisms of the Gulf of Naples in 1875, but its publication was postponed until 1878 (1878a). One reason might be that he was caught up in a swirl of career activities: scientific expeditions, the official opening of the Naples Zoological Station in which he had invested so much, the revamping of the zoological curriculum on campus, public lectures, and other extracurricular events (Cornalia, 1877). Panceri was an exceptionally energetic personality, but at the same time his health was fragile and had been the reason for his settling in Naples instead of Northern Italy to pursue his career. On 29 February 1877 he gave a public lecture on the luminescence of animals which had cost him days of feverish preparation. On 4 March he fell ill, suffered respiratory distress as pneumonia took over, and died on 12 March 1877, only forty-three years old.

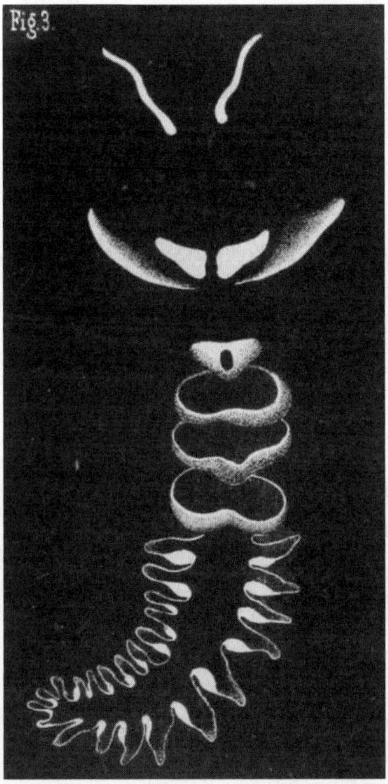

Figure 7.5 Drawing highlighting the luminescent zones along the specialized body segments and tentacles of the tube-worm *Chaetopterus*. From Panceri (1872a).

A medley of luminous organisms was covered in Panceri's 1878a paper: the tube-worm *Chaetopterus*, the terebellid worm *Polycirrus*, the fire worm *Odontosyllis*, several scale-worms (polynoids), the acorn worm *Balanoglossus*, and finally the brittle star *Amphiura* (now *Amphipholis*). Among his contributions here was his careful analysis of the luminescence of the tube-worm, which superseded Johann Friedrich Will's superficial study (mentioned in chapter 3). Panceri appended an illustration that provides a vivid and accurate portrayal of the distribution of the light-producing glands all along the body: antennae, at the base of the wing-like notopods, on the suctorial disc, at the edge of mid-body flaps, and on the serial notopods of the posterior region. Only in contact with freshwater do all the light-producing surfaces light up almost simultaneously and reveal the exudation of bright mucus that ensues. Tactile or electrical stimulation only produced local luminescent responses. Panceri investigated in greater detail the wing-like notopods, which produce the brighest luminescence and the most copious mucus. Under the microscope he identified as the light sources in this gland multiple layers of large mucous cells packed with granules. Strangely, he did not address the paradox of a luminescent animal living permanently inside an opaque tube.

The luminescence of the fan worm *Polycirrus*, Panceri noted, is of a beautiful violet shade, and it comes out in flashes by touching the light-producing parts of the body, which include most of the body in the two species examined, and a spectacular display in the shock of fine hair-like tentacles in one of the species. Panceri traced the luminescence to spherical mucous cells in contact with the exterior via a duct piercing through the external cuticle, thus strongly suggesting an external secretion of luminous slime.

The fire worm *Odontosyllis* stages one of the most spectacular displays of luminescence on the sea surface. Before Panceri, the famous Swiss zoologist René-Edouard Clarapède (1832–1871) who, like his Italian counterpart, was professor of comparative anatomy (in Geneva) and died prematurely, had earlier described the luminescence of this worm.

> When I placed the animal on a glass strip, around noon, by a splendid summer day, I was struck by the sparkling, emerald green dots lining the sides of the animal. These dots shone a few moments and gradually disappeared. Realizing that it was a phenomenon of phosphorescence, I placed the animal on a dark surface for improved observation. Every acute irritation with a needle caused the appearance of two green shining bands on the sides of the animal. After a few moments, these bands were reduced to two series of bright dots which seemed to correspond to the base of the feet. More than once phosphorescent Annelids, marine or terrestrial, were recorded, but I believe never before has been reported a light bright enough to explode like this in the middle of day, under the noon sky. (Clarapède, 1864)

Panceri confirmed Clarapède's observation, but he added microscopic observations of the light sources, which turn out to be epidermal mucous cells strikingly similar to those of *Polycirrus*. Panceri then moved on to scaleworms, which also exhibit a green light emission. He determined that the only source of light was the elytra, which cover the dorsal surface of the worms like scales on the sides of fishes – hence the designation "scale-worms." These elytra, he claimed, are made of a cuticle, a hypodermis, and a profusion of nerves that have "special endings." If the head is stimulated, luminescence spreads from elytrum to elytrum toward the tail, and vice-versa if the tail is stimulated. If one elytrum is removed, the spread of luminescence along the body ceases. Panceri argued that the unusual crowding of nerves in the elytra, issuing from an elytral ganglion, points to their nodule-like endings as the probable source of light emission. In this, as in the case of the sea slug *Phyllirhoë*, Panceri was misled to assign the light source to nerve cells in the same way as Quatrefages (as we saw in chapter 3), erred in seeing the source of scale-worm luminescence in muscle cells.

The worms discussed so far are annelids. The acorn worm has nothing to do with classical worms, but belongs rather to the hemichordates, which are more akin to echinoderms. Panceri was the first to record its luminescence. If he agitated the water in which *Balanoglossus* bathes, the whole surface of the animal glowed with a pale blue light. Under the microscope he assigned the light source to pear-shaped mucous cells extending through a narrow neck to the surface as in *Polycirrus* and *Odontosyllis*. From these cells "gushes a yellow material that also has the appearance of fat droplets, which are moving in the water without dissolving and disperse as if they were sparks."

Panceri also studied a brittle star as Quatrefages had done a generation earlier (see chapter 3). In *Amphiura squamata* Panceri observed that all five arms of the echinoderm are luminous, and on closer examination he saw the source of the greenish light in a pair of organs in each consecutive segment of the arms. When these are touched, repetitive flashes travel along the arms, followed by an escape response. This motion of course suggests that the luminescence serves a defensive purpose. Panceri did not share Quatrefages's belief that the light comes from muscle cells, but he did not suggest alternatives. Considering that Panceri was so keen to associate luminescence in the sea slug with nerve cells, it is ironic that the light-emitting cells of *A. squamata* turned out much later to be a form of nerve cells of the radial nerve running in each arm (Brehm and Morin, 1977).

The swan song for Panceri's opus of works was a paper dedicated to the luminescence of the hydroid colony *Campanularia flexuosa*, which he found in abundance along the Amalfi coast (Panceri, 1878b). Edward

Forbes had earlier touched on the subject of hydroid luminescence (see chapter 3), but Panceri deserves the distinction of providing a more detailed description and, as with other forms he investigated, of attempting to unveil the cellular source of the light. Hydroids include *Hydra*, a solitary polyp, but the majority are colonial, with the polyps – gastrozooids (or feeding polyps) and gonozooids (or reproductive polyps) – budding off a colonial stem tissue called the stolon. The stolon branches give the colony a tree-like appearance.

Panceri had made it his mission to trace the sources of light at the cellular level in animals in which these cells are not housed in organs, in contrast to fishes, cephalopods, and crustaceans. He was not always on the mark, but his success rate was reasonably good. He credited his success to a simple technique – exposure to fresh water – which stood him in good stead. But the technique could also be used to screen for luminous animals in the field. He explained that finding hydroids such as *Campanularia* settled on rock algae "will not be so easy to succeed, because these are often small and their light intermittent and fleeting. But if you remember the power fresh water has to fix the light of marine animals for some time … immersing the clumps of algae in it will make it easier to collect the small organisms that are the cause of the observed scintillation" (Panceri, 1878b). In the laboratory, again using fresh water and placing the hydroid under the microscope, he saw that the light originated in ectodermal cell clumps along the stems and stolons, and in the pedicels supporting the polyps. Perhaps Panceri put too much confidence in his technique, for James Morin and George Reynolds (1974), using fluorescence microscopy, showed that luminescence is endodermal and less widespread than in Panceri's account. The nearly spherical cells have few distinguishing features and their content is nondescript. Disturbances triggered multiple flashes of light throughout the colony. As a legacy to his physiological investigations, Panceri proposed the universal principle that all luminous marine forms, from dinoglagellates to radiolarians, from coelenterates to fish, only luminesce from provocation by stimuli of any sort.

While Italian investigations were centred around Naples and were dominated by Panceri's drive, northern Italy also contributed a distinguished researcher a few years after Panceri's death. Carlo Emery (1848–1925) followed a path similar to that of Panceri, but in reverse. Born in Naples to Swiss parents, he completed medical studies but soon turned to the natural sciences, the field in which he earned his doctorate (Wheeler, 1925). Having started his career in Sardinia as professor of zoology at the University of Cagliari, in 1881 he was called to fill the chair of zoology at the prestigious University of Bologna. Although he was considered one of the most distinguished entomologists of Italy, he also published on other zoological subjects, notably the light organs of the lanternfish (Emery, 1884a). (This paper predates by a few years the series of monographs on the light organs of deep-sea fishes discussed in the previous chapter.) Emery investigated both the small pearly organs on the lower (ventral) sides of the body, and the caudal-dorsal "glands," which were somewhat neglected or misconstrued by the oceanographers. He found them a technical challenge because of the brittle texture of the associated scales and reflectors located next to soft tissue; sectioning the organs for histological observation brought the risk of tearing off soft tissue in which the luminous cells are likely to be housed. He used paraffin and celloidin as embedding material to minimize tissue stress during the slicing.

Emery's specimens of the lanternfish *Scopelus* (= *Hygophum*) *benoiti* and *H. elongatus*, preserved in alcohol, were shipped to him from the Naples Zoological Station. His representation of the histological slides, stained with a carmine-based solution, gives an accurate picture of the structural organization of these light organs. The elongated organs possess a relatively large lens abutting against a transparent opening of a covering scale. A deep scale lies in front of the reflector. The back of the organ is lined by a pigment layer and a reflector (tapetum), and between this screening in the back and the superficial lens lies a suspended tissue mass made of connective tissue in which is encased "a specific mass or rather stack of flattened cells," which are the suspected luminous cells. A secondary arrangement of pigment and reflector layer adjoining the dorsal half of the lens ensures that any light emanating from the luminous cells is not directed sideways to the fish's flank, but at a downward angle. The dorso-caudal glands (or caudal luminous organs) differ from the pearl organs by virtue of their

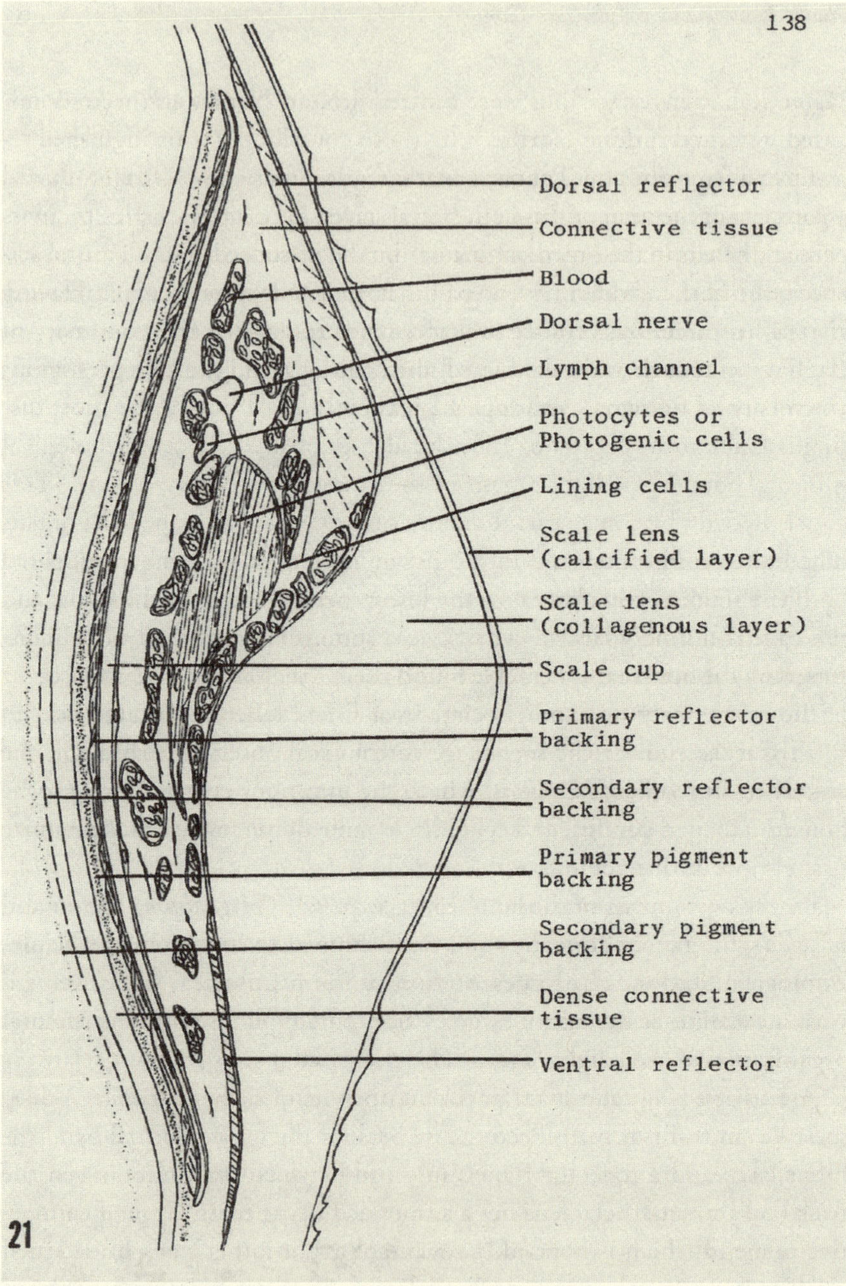

Figure 7.6 Schematic drawing of a section through a photophore of the lanternfish *Stenobrachius leucopsarus*. Note the blood and nerve supplies, and the complex accessory optics (multiple reflectors, scale lens), which ensure a specific orientation for the emitted beam of light. From Figure 21 in O'Day (1972).

larger size, the absence of a lens, and a much larger mass of the flattened, heavily stacked luminous cells.

Emery's contribution is of particular interest for his effort to follow the development of the pearl organs in the larval stages. He concluded that luminous cells and others in the organ originate from the mesoderm germ-layer, and he specified that the ectoderm has no part in it. Emery depicted a cupula in which the nascent luminous cells are oval and at a later stage elongate and become flat. Elements such as the associated scale, the lens, and the dermal elements (pigment and reflector layer) appear later.

~

The Italian connection to early studies of lanternfish light organs does not end here. Fifteen years later a young researcher, Michele Gatti (1880-1904), expanded on Emery's work. Gatti is a mysterious figure about whose life no biographical information can be found except that in 1899 he published a preliminary paper on the light organs of fishes (Gatti, 1899). A note appended at the end of this paper suggests that Gatti was at the time a doctoral student at the Sapienza University of Rome under Giovanni Battista Grassi, an entomologist famous for having been denied – unfairly, it is claimed – a Nobel Prize in Medicine or Physiology for his work on the life cycle of the malaria-transmitting mosquito (Capanna, 2008). The note also announced a forthcoming monograph by Gatti on his comparative analysis of fish light organs. This monograph appeared posthumously in 1904; what caused his untimely death at the age of twenty-four (*Revista bibliografica italiana* 10, 1905, p. 282) is not known.

Already in his preliminary paper Gatti developed the theory that there are two general types of light organs in fishes: the glandular type and the electric type. The glandular type is an organ in which the luminous cells are gland cells amenable to secreting a luminous substance. In keeping with the anatomists of the oceanographic voyages, he found this type of cell in the stomiid fishes, but he noticed them also in the luminescent coastal fishes. The electric type is found exclusively in lanternfishes, where "light production is undoubtedly due to a special substance which is consumed inside the cell's granules." He saw Emery's stacks of flattened cells in his own lanternfish material, and their resemblance to such stacks in the electric organs of fishes – the electrocytes or electroplaques – struck him particularly. The fact that both lanternfish light organs and electric organs also have a rich innervation and numerous blood vessels made the analogy even more compelling to Gatti, who must have been aware of the work of his fellow countryman Carlo Matteucci on the subject of the electric organ of the torpedo fish (Matteucci, 1844). (This is the same Matteucci discussed in chapter 3 in relation to his observations on the firefly.) Gatti concluded that histologically the specific body of lamellae (flattened cells) resembles an electric organ, but he stopped short of stating that it functions like an electric organ.

In his definitive and large-scale study (1904), Gatti described in detail the light organs of glandular type in *Maurolicus*, *Gonostoma*, *Stomias*, *Bathophilus*, and *Porichthys*, and those of electric type in the lanternfishes *Diaphus*, *Lobianchia*, *Electrona*, *Ceratoscopelus*, *Lampanyctus*, *Hygophum*, and *Myctophum*. And, like Emery, he examined closely the development of the light organs of lanternfishes. While agreeing with the light organ organization depicted by Emery, Gatti disagreed with his compatriot on some points, mainly the development of the organs. Gatti believed the organs have a mixed origin: mesodermal and ectodermal. The luminous cells of the "specific body," he wrote, arise from an infolding of the ectoderm. The covering scale of the organ is the first dioptric element to appear, and in fact these scales are the first to appear on the body of lanternfishes. The lens and the deep scale derive secondarily from the covering scale. The pigment layer and reflector are mesodermal structures and appear last.

Gatti mustered all the available power of the microscopes of his time to provide a surprising amount of detail on these strange lamellar cells, which he and Emery identified as the luminous cells. For a start, how do these cells flatten? The central cells of the forming specific body flatten first; once they have flattened into thin lamellae the nucleus is pushed at the edge and is connected only by a narrow neck to the rest of the cell. Gatti believed the lamellae are multi-nucleated (that each cell possesses more than one nucleus). The ventral parts of the stacked lamellae are more granular than the dorsal parts and extend into many interdigitations. In some lanternfish species, these extensions cause anastomoses between lamellae, which may have led to their multinucleation. Why such a complex cellular organization and what does it contribute to the process of light production? Either Gatti was at a loss to answer or he did not bother to speculate on the functional significance of what he observed.

After his lanternfish work, the entomologist in Emery attracted him to the light organs of fireflies, in particular the local species *Luciola italica*. Emery was an accomplished polyglot (Wheeler, 1925), so it is not surprising that he published his first firefly paper in German (Emery, 1884b) and his second in French (1886). Like all his contemporaries who examined firefly light organs, Emery was unaware that Peters had seen the rosette organization of the lantern as early as 1841 (see chapter 3), even though he cited Peters's paper! Nevertheless, Emery's contribution is critically important. He did not content himself with describing the anatomy of the light organs, but sought to do so while the constituent cells were emitting light, so as to trace the source of the light. The motivation for Emery's research was a paper by Heinrich Ritter von Wielowiejski on lampyrid fireflies (Wielowiejski, 1882), which was based on the latter's doctoral dissertation at the University of Leipzig. Wielowiejski investigated the complex tracheole (respiratory) system of firefly light organs. He concluded that the tracheal end cells – located at the beginning of the tracheoles – were not the luminous cells as was then believed, but were involved instead in controlling the access of oxygen to the parenchymous cells around the tracheoles, which he thought, without experimental evidence, were the true luminous cells.

In his 1884 paper Emery noted the rosette pattern of the light organ, in which the circular centre represents the tracheole viewed in cross-section, the parenchymous cells around and between tracheoles tending to form a concentric spread of "petals." By gluing the elytra of the insect to a substrate, pressing a glass cover on the abdomen and placing the preparation under a low-power microscope, he was able to see a ring of light emission around the tracheoles of the rosettes. As the luminous ring seemed to him at the approximate location of the tracheal end-cells, Emery concluded that Wielowiejski must have been mistaken and that the tracheal end-cells were indeed the luminous cells. Emery hypothesized that the luminous substance is formed in the parenchymous cells but is transferred into the end cells, where it gives rise to the light emission.

Emery revisited the problem two years later (1886). Conscious that his experimental setup was far from ideal, what with the spastic movements of the insect which caused the rosettes to come in and out of the microscopic field of view, he decided to refine the conditions of observation. He removed the abdomen, treated it with osmic acid and secured it on a slide for microscopic observation. He found that osmic acid only dimmed the light emission but didn't extinguish it. Now, aided by a stable field of view, he was able to see that the light emission was not restricted to a narrow ring around the tracheole cylinder, but spread farther, to where the parenchymous cells are. So, Emery had to give up his former theory and rally to Wielowiejski's. Over eighty years later new imaging techniques for low-light microscopy revealed the microsources of luminescence in the rosette, showing that the parenchymatous cells are indeed the photocytes (Hanson et al., 1969) and that Wielowiejski had been right after all.

Emery never returned to fireflies as an object of study. He enjoyed many years of research on his favourite insects, ants, until 1906, when he suffered a stroke that left him permanently paralysed on the right side (Wheeler, 1925). He learned to use his left arm to continue experimenting on ants, draw, and write papers. He died in 1925. In the years of Emery's firefly studies (1884–86), a new investigator of bioluminescence emerged who was to revolutionize the field and put the French on the map.

8

Raphaël Dubois and the Chemistry of Living Light

> Thenceforth it is evident that the luminous phenomenon is the result of a reaction of a chemical nature. The notions obtained from the previous experiments allowed us to extract from the luminous parts of *Pholas dactylus* two substances which, when mixed in the presence of water, determine the appearance of light.
> –Raphaël Dubois (1888)

If one regards Paolo Panceri as the scientist who opened up the experimental approach to the identification and characterization of luminous cells, then Raphaël Dubois must be considered responsible for unlocking the chemical Pandora's box of the luminescent material inside these cells. In so doing, Dubois dispelled two notions that had kept scientists shackled for the greater part of the nineteenth century. One was the idea – untested – that light production results from the oxidation of fat inside the granules of luminous cells; we have seen that Panceri himself, despite his often-clever experimental observations, was so entranced with this notion that it became an obsession. The other was the view, prevalent in all other physiological fields as well, that processes such as biophotogenesis, as bioluminescence was often called, are intrinsic to living tissues and cells, and cannot be reproduced outside the living environment of the organism. Observations of light production ceasing after the death of the luminous organism, or explaining away the enduring luminescence of dead bodies or secretions as examples of cells and tissues surviving the organism for a while longer, had helped that notion gain currency. The course of action whereby Dubois refuted these notions constitutes nothing less than a game-changing scientific exploration, the introduction of a new paradigm that has determined the path of scientific inquiry in the field to this day.

For a scientist of Dubois's stature it is surprising that no formal biography, even in French, has ever been published. The following depiction of his life, career, and personality is a composite picture based mostly on bits of unpublished accounts and on archival material. Christian Bange, a retired historian of science at the Université de Lyon, and his wife, Renée, have pored over the scientific archives of Raphaël Dubois, which are preserved in various locations. In addition to readily accessible published papers, they have produced two invaluable documents: a painstakingly detailed inventory of the contents of the collections held at the Institut Michel-Pacha in Tamaris-sur-mer (hereafter named "Bange-Tamaris") and the text of an essay on the scientific archives of Raphaël Dubois communicated at the Annual Conference of the Société d'histoire et d'épistomologie des sciences de la vie held in Milan in 2005 (hereafter referred to as "Bange-Conference"). I am deeply indebted to them for allowing me access to these precious resources.

Raphaël Horace Dubois was born in Le Mans on 20 June 1849 into the family of a pharmacist. He attended the Lycée du Mans, enrolled in the Medical School of Tours and, after an interruption necessitated by the Franco-Prussian war of 1870, resumed his studies in Paris (Philippe Jaussaud, *Raphaël Dubois et la bioluminescence*, posted on the internet). He became a fully licensed pharmacist in 1875 and a medical doctor the following year. He practised medicine for a few years, during which he actively militated against alcoholism and "social scourges" (Bange-Conference). Did he perhaps feel that his idealistic vision of social medicine was getting him nowhere? What we know is that he abandoned his practice in 1882 to become a research assistant to Paul Bert at the Sorbonne. Bert, a student of the famous physiologist Claude Bernard, was spurred into entering politics by the rise of the Third Republic after the Franco-Prussian war and in 1881–82, just prior to the arrival of Dubois in his laboratory, was appointed minister of education. Bert brought important reforms to the French educational system, making it universally accessible and reducing the influence of the clergy in schools. Back on the bench, Bert had Dubois assist him in implementing a research program on the effects of anaesthesia.

At the time, Bert and a zoologist colleague had organized a marine laboratory at Le Havre (Normandy), to which Dubois soon migrated to run experiments on the effects of anaesthesia and "poisons" on mollusks. In a serendipitous turn of events, a child hanging around the port brought him an exotic insect that turned out to be a luminous click beetle (*Pyrophorus*) (Jaussaud); it had travelled to France from the West Indies in the wood cargo

Figure 8.1
Raphaël Dubois.
Courtesy of Musée d'histoire de la médecine et de la pharmacie, Université de Lyon 1.

of a ship (Bange and Bange, 1994). Dubois conducted a thorough dissection of the single specimen and was so intrigued to learn more about these insects and their luminescence that he managed, thanks to a singleness of purpose that was to be a trademark of his career, to enlist the manager of the Le Havre laboratory and a horticulturist in Guadeloupe to import hundreds of these beetles for a comprehensive study.

This work determined the course of Dubois's career; he placed "biophotogenesis" at the centre of his research and never looked back. The experimental model he had so fortuitously encountered was the click beetle *Pyrophorus noctilucus*, an elaterid beetle distantly related to the lampyrid beetles, which include the fireflies most people identify with. As a model, this species proved ideal: the light organs emit a beautiful and very bright green light and their "great strength, an exquisite sensitivity and a sufficiently high grade of organization do make it possible to analyse successfully all the expressions of their organism susceptible to exert influence on the luminous

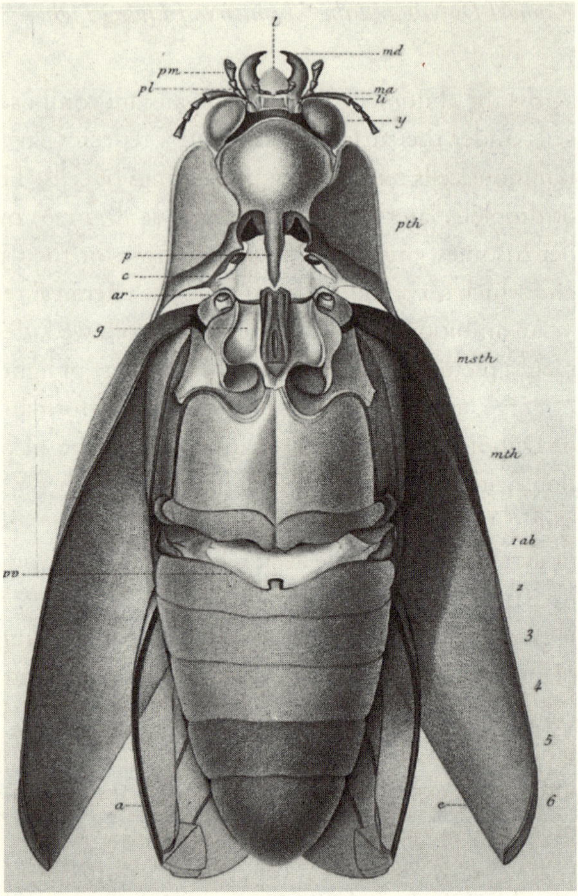

Figure 8.2 Dubois's click beetle, viewed from its ventral surface to show the two bands of light organ in the thorax and abdomen. From Dubois (1886a).

function" (Dubois, 1886a). In this research project, which was of an unprecedented scale for a work dedicated to a single species, Dubois showed himself to be a great experimentalist. He published two short preliminary articles on the work in progress (1884, 1885); and the monograph of the completed work (1886a), which was in fact his doctoral dissertation at the Sorbonne, contained enough material for two theses.

Dubois first described the embryology and anatomy of the beetle and its light organs. From him we learn that the unfertilized egg is already luminous, as are the embryo in the egg and the hatched larva. This larva has light sources on the prothoracic segment, and the second moult larva shows in addition strings of minute light sources on the abdomen. The adult has a pair of light organs placed wide apart on the prothorax and a "ventral plate" across the abdomen. The light organs are composed of a mass of luminous cells under the surface cuticle and a reflector layer underneath. The large luminous cells tend to pile up as strings of cells. They contain oily granules or droplets, later identified (Hanna et al., 1976) by electron microscopy as peroxisomes, organelles that are known for their role in catalase reactions and which serve in fireflies to store the chemical reactants of luminescence.

An ambitious series of experiments followed, designed to obtain as complete a picture as possible of the physiology of light production and the role of nerve, air, and blood supplies in the performance of light organs. Some of Dubois's experiments sought confirmation of previous firefly investigations with regard to the role of oxygen and other gases, and of various chemicals. One interesting experiment revisited Jean-François Macaire's finding of 1821 that nitric oxide excites luminescence. Dubois found it counterintuitive that a gas acting as an anaesthetic should excite rather than extinguish light emission. But he confirmed Macaire's finding and was at a loss to explain it. (The role of nitric oxide will be discussed in chapter 17.) Dubois's experiments suggested that oxygen was carried to the light organs by the blood sinuses, not by the respiratory tracheole system. Dubois incorrectly interpreted experimental results to suggest that light emission was dependent on muscle activity; what was really happening was that nerve activity excited muscles as well as the light organs. His experiments involving ablation and destruction of nerve centres showed that the brain was necessary for light production; however, he did not extend these experiments to the peripheral ganglia, from which the lantern nerves proceed.

Careful mapping of the zones lit by the light organs in the insect's surroundings, Dubois argued, would help assess the role of the luminescence. He concluded that the light was used for visual guidance when walking (prothoracic light organs) or flying (abdominal organ) after dusk. He also showed by spectroscopy that the dominant colour of the prothoracic light was the hue of green matching the canopy where the insect lives.

Dubois felt strongly about refuting the theory that combustion was a mechanism of light production. He made an analogy with a blacksmith fanning a simmering hot iron, requiring that the luminous cells be kept in a subdued glowing state but with no visible light, which would be kindled by oxygen bursts. Classical physiology, he insisted, was inadequate to settle the issue; to dismiss the theory, one had to reduce the light-producing function to a chemical phenomenon. This was a bold step never really considered before. Here Dubois was heeding the reductionist precept of Bernard, according to which "the work of the physiologist ends when a biological phenomenon is reduced to the level of a physico-chemical phenomenon." One of his preliminary papers (Dubois, 1885) touched on these experiments, but the latter are described fully in his 1886 monograph.

First, Dubois ground dried light organs in such a way that all cells had emptied out their contents. Adding water to the resulting extract restored light production. If he immersed the insect in boiling water and repeated the extraction, no light appeared by adding water. Now he ground the tissue and let the extract in contact with water luminesce until no more light could be elicited. When he put this extract in the presence of another, which had been boiled previously and was not giving any light, light production was restored. "There is a striking analogy," Dubois wrote, "between the physico-chemical phenomenon which provokes the appearance of light in the luminous cell, and what happens within the hepatic organ for the glycogenic function." In both cases, Dubois suggested, there is a catalytic action of a heat-sensitive "diastase" (enzyme) on a heat-resistant substance, resulting in one case in sugar regulation and in the other case in the release of energy in the form of light. These simple, astute experiments were to have an enduring impact, leading to future biochemical and molecular studies of bioluminescence.

Taking a break after his thesis work to learn new ideas in the laboratory of the celebrated physiologist at the University of Heidelberg who first introduced the word "enzyme," Wilhelm Kühne (1837–1900), Dubois reported on the luminescence of two arthropods captured in the vicinity of Heidelberg. In the first paper (Dubois, 1886b), he revisited the myriapod, whose luminescence Jules Richard had described the year before (see chapter 4). The luminescence of *Scolioplanes crassipes* was thought to radiate from all over the body, but Dubois found to his surprise that it originated in specialized cells in the wall of the gut and that the secretory product of these cells was flushed out through the anus. The little we know today about myriapod bioluminescence was reviewed by Jörg Rosenberg and Victor Meyer-Rochow (2009), who concluded that myriapods use their luminescent secretions for defence. After this discovery Dubois (1886c) turned to the springtail *Anorophorus*, whose luminescence was first observed by Allman

(1851, see chapter 4). Another surprise awaited Dubois: the luminescence of this insect had its source in nodules near the urinary tract (tubules of Malpighi), but the luminescence was internal; no secretion was observed.

Prominent French academics, including his mentor Paul Bert, were so impressed by Dubois's scientific accomplishment that they championed his application for the vacant position of chair of General and Comparative Physiology at the Université de Lyon. When the chair became his, he was thirty-seven and had vaulted to the highest pinnacle of academia without climbing the ladders of the professorial hierarchy. And he accomplished this in the third largest city of France, after Paris and Marseilles. No wonder he aroused the envy and bitterness of colleagues. His personality added to the sulfurous mix. A biographical note published after his death mentions that "his imposing head, framed by a magnificent mane of hair, gave an impression of energy, intelligence and self-will" (Pourcel, 1929). However, his brazenness, in conjunction with his tremendous industry and singleness of purpose, predictably tended to sour relationships with many of his peers.

From the moment of his appointment in Lyon, Dubois wanted to conduct research on marine bioluminescence and he committed all his fabulous energy and drive to organizing a marine station that would facilitate his ambition. In spite of his personal connections, however, he failed to obtain the necessary support (Bange and Bange, 1994). So, he had to settle for conducting his studies at the biological station of Roscoff on the coast of Brittany. There he had access to the common piddock *Pholas dactylus*, the mollusk whose luminescence had already been observed by the Ancients and was of late studied by Panceri. Dubois's motivation was simple: he wanted to generalize to all luminous marine animals the results of the *in vitro* experiments he had begun with the terrestrial click beetle. Apparently for the sake of posterity he made sure to enlist the director of the Laboratory of Experimental Zoology at Roscoff, Yves Delage, and others as witnesses of his experiments (Dubois, 1887a).

Dubois basically replicated his previous experiments in *Pholas*, producing the two extracts from the luminous glands and pouring them each into their assigned test tube after filtering. Then in the dark he mixed the content of one tube into the other and watched the mixture turn brightly luminescent. The historical importance of this finding is such that what Dubois wrote next deserves to be quoted in full:

> One of these substances was obtained in a crystalline state: it exhibits very unique optical properties which confer to the photogenic tissues we examined a peculiar opalescent shimmer previously described in *Pyrophorus* and other luminous animals. It is soluble in water, poorly soluble in alcohol, soluble in petroleum spirit, benzine and ether. We propose to designate it with the name *lucifèrine*, pending an analysis to determine its elementary composition and chemical function.
>
> The second substance is an active albumin-like compound, like those known as *soluble ferments* (diastases, zymases, etc.), with which it shares all the general characters. We will call it *Lucifèrase*.
>
> These two substances are necessary and sufficient to produce *in vitro* the phenomenon of animal luminescence, improperly called *phosphorescence* and the mechanism of which has been explained until now by more or less likely hypotheses, based on insufficiently broad experimental studies. (Dubois, 1887a)

Thus were coined the generic names of the two classical reactants of the bioluminescent chemical reaction, and these names have stuck to this day. Dubois never explained how he decided on these names. The derivation from the Vulgate Latin word *lucifer*, meaning "bringer of light," seemed at once appropriate and self-explanatory to Dubois, but he never disclosed the mental process that led to their choice. Beyond these semiological considerations, it is important to stress that Dubois was at the avant-garde of the test-tube biochemistry that developed in the years to come, and that his pioneering contribution in this regard also had an impact beyond the field of bioluminescence (Bange and Bange, 2010).

While Dubois was pursuing his research on pholads, which culminated in another sizable monograph (Dubois, 1892), he tried again to arouse interest in a marine station with a physiological vocation. This time he rallied to his cause a marine officer, Commander Marius Michel (1819–1907), who had been commissioned to build a sea port in Constantinople and a network of lighthouses along the coasts of the Ottoman Empire (Bange and Bange, 1994). The businesses stemming from these ventures had made Michel very wealthy, and he was rewarded by the sultan Abdulhamid II with the honorary title of Pacha of the Ottoman Empire. Michel Pacha, as he became known, was working on a seaside resort project in Tamaris-sur-mer near

Toulon when Dubois met him. And so it transpired that Michel Pacha provided all the funds for a marine biology station built in the Moorish style, which was completed in Tamaris in 1890 and bore the name of Dubois's benefactor: L'Institut de biologie marine Michel-Pacha.

Despite this good fortune, as Bange and Bange (1994) explained, personality difficulties persisted: "This much originality and cheeky luck in a man reaching his goal so rapidly invited lasting enmities against Raphaël Dubois and his station." One contentious point, of course, was that the zoologists who had founded the pre-existing French biological stations had had to peddle their projects for many years and raise money in trickles to bring their brainchildren to fruition. But the bitterest pill to swallow was that French zoologists, who had a low regard for marine physiology as a field of investigation (Fontaine, 1980), had fought hard to keep physiology at bay in their stations. Then, along came brash Dubois founding a station dedicated to the study of the physiology of marine animals! It rankled even more that the Tamaris station was such a success that Prince Albert I of Monaco (see chapter 4) was in 1910 inspired by Dubois's example to set up his Institut océanographique in Paris, where marine physiology figured prominently (Fontaine, 1980).

Dubois concentrated his teaching in Lyon and his research in Tamaris. His broad research interests included anaesthesia and the biological underpinnings of commercial products such as pearls and silk – Lyon being the world capital of the silk industry at the time. But bioluminescence remained his major field of research. He followed up his *Pholas* luciferin-luciferase paper with another (1889), in which he did experiments purporting to demonstrate that luminescence in *Pholas* was part of a reflex circuit. Beaming a light on *Pholas* caused a series of siphon muscles to contract, thereby spewing a luminous cloud in the surrounding water. Since *Pholas* lacks eyes, Dubois assumed that epidermal light-sensitive cells connected to nerve cells were part of the reflex circuit. In his monograph on *Pholas* (1892), in which all aspects of the anatomy and physiology of the mollusk are covered in addition to luminescence, Dubois drew a misguided analogy between the *vision dermatoptique* of the skin and the retina of the eye; he failed to draw a critical distinction between simple light receptive cells and an image-forming organ.

In his monograph Dubois criticized Panceri for having missed the glandular epithelium of the inner surface of the siphon. This tissue, as Dubois showed by various experiments, is the biologically relevant light organ. He established the role of the visceral ganglion in controlling the luminescence of the siphon: direct stimulation of the ganglion induces luminescence, and severing the pallial nerves issuing from the ganglion abolishes the light emission. At the level of the gland of *Pholas,* Dubois described a subcutaneous organization in which nerve endings enter into contact with muscle fibres criss-crossing the spaces between and under the mucus cells of the luminous gland. From this arrangement, he deduced that neural control is achieved by inducing the local muscle cells to contract and thereby press on the mucous cells to expel their content. Then, if appropriately stimulated, the siphon itself contracts to fan the already secreted mucus out into the surrounding water, where it will form a luminous cloud concealing the mollusk from a potential predator.

Curiously, however, in this 1892 monograph Dubois had little to add to his paper (1887a) on the chemistry of the *Pholas* luminescent system in which he had identified the luciferin and luciferase extracts. In fact, he was totally silent on the subject. But he did mention that chemicals causing the coagulation of albumin-like substances (today's proteins) destroy luminescence – by which he meant that it cannot be restored. It would seem that Dubois's notion of the luciferin-luciferase duality was evolving in his mind, but that he was uncertain about the direction his thoughts were taking. In a paper on a luminous centipede the following year (Dubois, 1893), we are given a glimpse of this evolution. Almost *en passant* he mentioned that, contrary to his previous idea of two distinct substances acting on each other, "in reality there are only two successive states of the same substance modified by oxygen and water, and for which I will keep the name *luciferin* until its atomic structure is determined." In retrospect, it is easy to see that Dubois was caught off guard when his idea of "one chemical model fits all" for luminescence garnered no clear-cut experimental support.

In hindsight, it is easy to understand Dubois's confusion. His method led him to think that luciferase was the protein in the luciferin-luciferase model (the luciferins discovered later in the latter half of the twentieth century are non-proteinic, small molecules). But in fact, as the team of Anthony Campbell at the University of Wales found out, *Pholas* has a photoprotein – pholasin – to which a luciferin is bound, and pholasin reacts with *Pholas* luciferase and oxygen to generate light (Dunstan et al., 2000). It is an unusual case, where the luciferin-luciferase reaction involves two proteins.

It wasn't until five years later that Dubois finally explained himself openly, using a textbook of general and comparative physiology to clear the air (1898). He was so enamoured with the field of bioluminescence that it took up 40 percent of the book, a clear case of scholarly imbalance. Dubois reviewed his own research thus far, cursorily discussed all luminous organisms known at the time, and, only toward the end, in the twenty-third lesson of the textbook titled *Conclusions générales relatives à la production des radiations lumineuses et chimiques par les êtres vivants*, did he arrive at the heart of the matter. He confessed that the importance of oxygen in the luminescent chemical reaction had eluded him, but that he had too few click beetles left to address his omission with extracts. He thought that the chemical work with the readily available *Pholas* would correct that, but instead it led him astray into thinking that "the light emission was the result of the transformation of a colloidal, albumin-like substance into a crystalline compound in the presence of air and water and in conditions where life can be exercised." The fact that he had abandoned the idea of two substances interacting with each other demonstrates that Dubois failed to recognize the heat-sensitive component not only as an enzyme, but as one that catalyses an oxidation. He also confided that many of his peers had reproached him sharply for his about-face, and he defended himself by pleading guilty only to having underestimated the technical challenges of the chemical research.

Now that he had recognized his error, he conducted a series of experiments designed to carefully monitor the conditions under which his extracts were obtained (Dubois, 1898). More rigorous chemical work allowed him to take into account the importance of the alcohol extracts and the critical acid-base environment of the extracts. He then obtained consistent luciferin-luciferase reactions, and his belief that the original assessment of the chemical mechanism in 1885 and 1887a was the correct one was restored. He revisited the problem in a later paper (1911) in which he now considered the luciferin of *Pholas*, contrary to that of *Pyrophorus*, to be an enzyme (zymase) oxidized by a peroxidase to produce the light emission. Dubois's script here came close to anticipating the concept of photoproteins such as pholasin, which reconciles the confusing results that confronted him.

Dubois also made another important observation in this 1911 paper. He tried in vain to detect luciferins in non-luminous tissues of various mollusks and crustaceans, but he found luciferase activity in many non-luminous tissues and in the blood of these organisms. It is not clear if he was detecting luciferases specifically or oxidases in general. The question of the distribution of luciferins and luciferases in organisms was to occupy biochemists late into the twentieth century.

Once he had made a name for himself and his career was far advanced, Dubois seems to have given thought to marrying. He had a relationship with a young woman – Lucile Julienne Pignet, born in 1873 – with whom he had a son, Jean, born in 1898, and a daughter, Rose, born in 1899. The boy was born out of wedlock, as Dubois and Lucile married on 30 September 1899, and the girl may have been born shortly afterward. I am grateful to Jean-Claude Autran, retired researcher of the Institut national de la recherche agronomique, for sharing these details about Dubois's private life that he painstakingly retrieved from official documents (census, birth certificates, Act of Marriage). Dubois was fifty when he married; thus far his scientific, academic, and administrative activities had left him no time to envisage founding a family. Since no surviving biographical document has mentioned his family life, it is a matter of conjecture whether he was truly in love with Lucile or married out of a sense of responsibility for having fathered her children.

In the years between his first luciferin-luciferase paper in 1887 and the early twentieth century, Dubois had associated the luciferase fraction with cell organelles he called "vacuolides." While sampling the luminous sea water of Menton on the Mediterranean coast near Italy, he had seen these organelles whenever the sea water was luminous, and the luminescence could not be attributed to dinoflagellates such as *Noctiluca* (1887b). He discovered that he could reproduce the harvesting of these vacuolides from different luminous organisms by cell lysis of their luminous organs. He described the organelles as roundish or ovoid, membrane-bound, and containing refringent crystalline structures and a small vacuole. He later concluded that these vacuolides are universally present in all cells (1919a).

Dubois's discovery of vacuolides, first observed in light organs, is important for the history of biology because it involves a priority dispute over the discovery of mitochondria. Dubois, a proud and ambitious man who saw himself as a gentleman and expected his peers to behave reciprocally, was very sensitive to colleagues seemingly encroaching on his claims to priority of discovery. The archives of Raphaël Dubois provide ample evidence of this side of his nature (Bange-Tamaris), and the vacuolide controversy stands out in them (Bange-Conference). In a book on symbiosis, Jan Sapp (1994) touched on this controversy:

> In this regard, the Lyonnaise physiologist Raphaël Dubois entered the debates to claim priority for the discovery of mitochondria and their physiological role in the cell. In *Les Symbiotes*, Portier [1918] had mentioned Dubois after Altmann [1890], stating that in 1896 Dubois offered more evidence for the existence of these organelles, which he called "vacuolides." Dubois [1919b] pointed out that in fact it was in 1887 that he first proposed this term to designate the small bodies which captured his attention in his studies of luminous insects, and which he found later in all cells. He believed that they were identical in structure and function to many other elementary granules to which one had given many names around the same time – plastidules, bioblasts, leucoplasts, and so on. As he saw it, that his vacuolides later became known under the neologism mitochondria, named by German histologists, was simply due to xenophilia (love of strangers) … He asserted, in no uncertain terms, that all the work conducted since the time in which he wrote simply amounted to confirmations of his own research and what he had been teaching for more than twenty-five years.

Dubois bitterly resented being overlooked by his own countrymen in favour of foreigners (Germans). Was he reaping the bitter fruits of the enmities he had garnered this far in his career? Did the fact that Germany's standing as the scientific champion of Europe, and the position of the German language as the language of scientific communication with English, allow the Teutons to muscle in on the French? This debate cannot be settled here. But, if one examines the evidence, carefully reviewing his original papers, it does seem that Dubois had a case and was unfairly treated in the adjudication of the original discovery of mitochondria. As recently as thirty-five years ago a historical analysis of discoveries related to mitochondria (Ernster and Schatz, 1981) failed to mention Dubois at all. But Dubois's case is not entirely airtight. His vacuolides seem also to include cell organelles other than mitochondria. He claimed to associate luminous dots with vacuolides, but we know today that mitochondria themselves, although they may be metabolically involved in luminescence, are not sources of light. As cell lysis does not discriminate which organelle type is released, Dubois's vacuolide populations were probably contaminated with other cell organelles.

La vie et la lumière, Dubois's last book-length statement on bioluminescence came on the eve of the First World War (Dubois, 1914). It recapitulated the survey of luminous organisms and touched again on the physiological importance he attached to vacuolides. Dubois also discussed at length his work on cultures of luminous bacteria. He undertook rigorous experiments to assess the optimal culture conditions for growing bacteria with strong luminescence. He made the interesting observation that making the culture broth slightly alkaline restores luminescence in cultures where luminescence has faded. The most original chapter in his book dealt with potentially practical applications of bioluminescence.

Dubois had some talent for entrepreneurship, as was demonstrated by his stewardship of the Biological Station of Tamaris-sur-mer, and he also had a flair for showmanship. He put this talent to good use on the occasion of the Exposition universelle of Paris in 1900. In the basement of the Palais de l'Optique he created a dazzling light show for the public by uniformly coating the interior surface of 25-litre glass jars with cultures of luminescent bacteria. Not only was this the first time in history that bioluminescence had been offered to a general audience as entertainment but it was its first appearance as a promotional tool for a potential new technology for lighting up homes. In the wake of this public success, Dubois created smaller "living lamps" along similar lines. Naïvely, as posterity showed, in extolling the potential of his gadgets, he got carried away by his eloquence:

> The fact is, besides the advantages physiological light offers for the eye and for the preservation of the lit objects, it has the enormous advantage of being cold, which is welcome in hot countries, where the incandescent lamp is already burdensome. For the same reason, it cannot set things on fire; it would already be useful in places where the dangers

> of explosion are real, such as mines, powder magazines, etc.; it shines day and night and for this reason ignition is avoided; and last, there are no wires, pipes, machines or repairs: the costs are insignificant, maintenance nonexistent. One can carry my [living] lamp everywhere, for it is very portable and resistant to rain or wind. It is truly the ideal lighting in all respects and it is certainly cold light which owns the future. If the practical goal is not reached by myself, it will be by others; to me perhaps will remain the merit of having showed the way to produce this great economic revolution.

Of course, his dream was never realized, and the conventional lighting industry, if it ever truly feared the competition, has no cause for worry today. Physiological light has never been a viable business proposition, and the usefulness of bioluminescence as a technology took a different direction. But it took six decades for tangible applications to appear.

As his archives clearly show, Dubois was a convinced pacifist. The draft of a book intended to provide a biological explanation for war was found among his papers (Bange-Conference). Dubois actually used this material for a published lecture in which he declared that biological principles and cosmic (physical) forces were what had determined the course of human invasions that resulted in bloody wars throughout history (Dubois, 1918). He pleaded for the supremacy of the rationality that science provides in trying to prevent future world conflicts. He had a low opinion of diplomats, whose secrecy and machinations he blamed for the outbreak of wars; diplomats, he wrote, were like alchemists prior to the advent of rational scientists. The text is replete with half-baked notions that have no currency today. His suggestion for preventing another war was to occupy the east bank of the Rhine; we know what happened when this idea was implemented a few months after his lecture.

Always on the watch for grandiose projects to conceive and propose, Dubois was a mover. Tamaris-sur-mer was never enough. He devoted his considerable energy to promoting a school for applied marine fisheries (École pratique des pêches maritimes) and an oceanographic institute. At every turn he was frustrated by the lack of resources and support. We saw that the Oceanographic Institute became the brainchild of the Prince of Monaco. Pourcel (1929) deplored the fact that Dubois's requests for the necessary funds to the relevant government ministries were always denied: "The worries of material life and the lack of encouragement certainly affected his personality and contributed to a premature discouragement." Fallout from the later years of his research career, such as the lack of recognition by his peers and challenges to the originality of his discoveries, fanned the flames of his bitterness after the war.

Dubois kept a wary eye on the burgeoning career of American biologist E. Newton Harvey, whose studies of bioluminescence had started around the time those of the Frenchman were entering a decline. This is not the place to discuss Harvey's life and contribution to the field, to which chapter 10 is devoted. Suffice it to say that Dubois was not so much jealous of Harvey's incipient aspiration to succeed him as the great man of the bioluminescence field, as he was irritated by the way Harvey went about stepping on his toes in regard to acknowledging his life's work.

In a letter to Charles Richet, editor of the *Journal de Physiologie,* dated 19 January 1918, Dubois complained that his work was not discussed in the journal's pages by his French peers (Bange-Conference). "No analysis of my works has appeared in the latest issue [of *Journal de Physiologie*]," Dubois moaned. "In contrast, I find there four analyses … of publications by the American Newton Harvey, a genuine piece of scientific gibberish. The author is fishing in murky waters and has come up with an ingenious trick to downplay my discovery by changing the French names of 'Luciférine' and 'Luciférase' to 'Photogénine' and 'Photophéline.' I have unmasked the stratagem of this foreigner and shown the nature of his errors, deliberate or not, but obvious errors."

The paper in which Dubois purported to "unmask" these errors (Dubois, 1917) is actually a hodgepodge of responses to several investigators who had slighted him by ignoring or misrepresenting the content of some of his papers and books on bioluminescence. The controversy embroiling Dubois and Harvey stemmed from the poor state of knowledge on the biochemistry of bioluminescence at the time. Dubois had concluded that the luciferin of *Pholas* was basically a coenzyme interacting with luciferase – in fact it is a photoprotein – whereas Harvey, working on the crustacean *Cypridina*, had found, correctly, that luciferin was a small, thermostable substance. That was the American's justification for suggesting the two new names for the interacting molecules. Both antagonists believed that there was only one chemical type of bioluminescence, which was applicable to all luminescent organisms, so they were bound for a collision course. But beyond the science, Dubois, if not in so many words, was accusing Harvey of bad faith in crediting other researchers (Mölish and McDermott, of whom more later) ahead of Dubois for the discovery of chemical substances involved in bioluminescence. It could simply have been a case of youthful neglect of the French scientific literature on Harvey's part.

Early in the twentieth-century world of science, priority of discovery tended in Europe to be a matter of gentlemanly agreement, but in the United States it was more a predatory tool of the publish-or-perish pressure cooker. The lack of a common language for publications gave American scientists, very few of whom were polyglots, excuses for not citing papers in languages other than English or German (the two leading languages in science at the time). Carlo Emery, for example, could write in German as well as in French and Italian, but the fact that Dubois could write only in French (although he could read English and German) denied him considerable critical exposure. Whether the non-polyglot Europeans could benefit from exposure to Anglo-Saxon scientists depended on the good will of British periodicals such as *Annals and Magazine of Natural History* or the *Quarterly Journal of Microscopical Science* to translate and re-broadcast their results. If one adds to this the fact that American biologists of the era had a better command of German than of French scientific literature, as a result of completing their training in German universities, then the neglect of French papers is compounded.

Even by French standards, Dubois's attitude to Harvey's faux-pas was inordinately bilious. Witness his off-the-record remarks and marginalia. Among the documents unearthed in Dubois's archives (Bange-Tamaris) is a paper by Harvey in the margins of which Dubois had penned cursory comments: "has not mentioned my shipment of Pholades"; "does not mention my spectrophotometric measurements … about the energy (which is all in light energy, he mentions Langley and Fery but not me"; "bastard! writes many pages to prove what is already established by my thesis." His tone of indignation did nothing to de-escalate the feud. Perhaps Dubois should have heeded the quip of the Nobel Prize–winning physicist Richard Feynman on the topic of priority: "Always give the bastards more credit than they deserve" (Dyson, 2011).

That said, it would be difficult not to agree that Dubois was unfairly treated by some of his peers. He was ambitious, energetic, and highly productive, and he broke new ground with his chemical theory of living lights. He deserved better treatment and recognition from the scientific community. As Christian and Renée Bange rightly put it, Dubois possessed "a fierce attachment to a sentiment of scientific honour which reminds us that science, wishing to be truthful, started in the beginning of modern times as an activity of gentlemen. All his life Dubois conformed himself to this selfless ideal: in 1927 he was still pursuing his work on biophotogenesis and kept the hope of refining a lamp based on this process. He was to die a little more than a year later, on 22 January 1929, as he was entering his eighty-first year, three days after having sent a last communication on the reflexes of the praying mantis to the Société de biologie" (Bange-Conference).

9

Bioluminescence Spreads Further Afield

> At first, the observations were superficial, but later the organization of living beings producing light, their photogenic function, the cause of their luminosity, etc., were studied more and more in depth.
> –Gadeau de Kerville (1890)

While the career of Raphaël Dubois was unfolding, a large cast of investigators from around the world were making contributions to the field of bioluminescence. Adding to the many participants from the usual nations such as Great Britain, Germany, and the United States, a new player in the field now signed in from the East. Imperial Japan made its entry into bioluminescence research at the same time as it embraced academic scientific research at large. In 1890 the Japanese minister of education expressed his imperial government's new awareness when he declared that the "flourishing or decline of a country has much to do with the flourishing or decline of its science" (Bartholomew, 1978). By 1894 university chairs modelled after the German academic system had been created in Kyoto, Tohoku, Kyushu, and Hokkaido, in addition to Tokyo. Applied science was the favoured child of this policy, which sought to spur industrial growth and a technocracy for imperialistic purposes (Mizuno, 2009), but basic research benefited indirectly by receiving some of the funds showered on university campuses. The founding of the Misaki Marine Biological Station in 1886, followed by others in the early 1900s (Inaba, 2015), allowed Japanese researchers easier access to marine luminous organisms.

Later in this chapter, we summarize original turn-of-the-century contributions to the study of bioluminescence in different animal groups. It seems timely here to take stock of the essays and reviews by some of Dubois's contemporaries that were then current in the field, including those of the new contributors from the East.

The first such contribution was by a compatriot of Dubois. Henri Gadeau de Kerville (1858–1940) was the son of an industrialist and member of an old French aristocratic family. At twenty he decided not to involve himself in the family business but rather to devote his life to natural history, and his parents helped him in such a way that he was financially self-sustaining throughout his life. He sought no official university diploma, trained himself as an autodidact, and did his research and writing from his home in Rouen, Normandy (Clément-Guyader, 1992). An amateur scientist, his bioluminescence-related work included a compilation of studies on luminous insects spiced with a few of his own observations (Gadeau de Kerville, 1881), and a general survey of bioluminescent organisms (Gadeau de Kerville, 1890). This survey alone deserves treatment here.

Although Gadeau de Kerville's 1890 survey was the work of a popularizer of science, the penultimate section of his book was of special interest to both specialists and the educated public. In what he called the "natural philosophy of luminosity," Gadeau de Kerville addressed questions that had begun to percolate among the initiated and would be raised repeatedly in the century to come: (1) What is the origin of the luminescent capability of the currently known luminous organisms? (2) Why is it that the number of non-luminous organisms far exceeds the number of luminous organisms and that marine luminous species far outnumber those of terrestrial ones? (3) Why is it that both luminous and non-luminous species belong to the same genus? and (4) Why do light organs of various complexity coexist within the same phylogenetic group?

To the first question the Frenchman, a convinced Darwinian, gave a correct evolutionary answer: luminescence capability must have originated in the ancestral, primordial organisms from which the current organisms derived by heredity. Although he did not specifically mention Ernst Haeckel, Gadeau de Kerville obviously borrowed from the illustrious biologist's recapitulation theory for the justification of his answer. He noted that the eggs and embryos of many of the luminous organisms are also luminescent; therefore, if the embryological development of an organism recapitulates the evolutionary development of the group to which it belongs, then the primordial organism must also have been luminous.

His answer to the second question was a clever take on Darwin's concept of the struggle for existence. As the luminous ancestors evolved, new species were faced with new environmental conditions in which the advantages of luminescence (visual assistance, bait to attract preys, sexual attraction) were offset by disadvantages, such as becoming visible to predators. In such cases luminescence in these groups disappeared gradually. A case in point is the evolution toward active life during daylight (diurnal lifestyle). This would help explain why bioluminescence is so predominantly associated with marine life. Terrestrial habitats, Gadeau de Kerville argued, tend to have a larger ratio of diurnal versus nocturnal species than in marine habitats, so the tendency to nocturnality in marine life has favoured a greater retention of bioluminescence there.

To the third question, as to why one finds luminous and non-luminous species in the same genus, Gadeau de Kerville suggested two possible scenarios: that natural selection either works gradually – from generation to generation – to remove luminescence capability in species where its advantages have vanished in an evolving environment; or it works as a result of sudden "accidental variations," which would create a non-luminescent species in a fast-track evolution. It is interesting that Gadeau de Kerville here anticipated the theory of mutations of the Dutch botanist Hugo de Vries by over a decade.

And finally, the gradation of complexity in the light organs within a phyletic group, Gadeau de Kerville explained, is caused by the highly modulated trade-off between the usefulness and harmfulness of improving the organization of these organs in the species involved.

This was the first time in the bioluminescence field that such questions had been addressed so directly and comprehensively. Unfortunately, Gadeau de Kerville's seminal essay on evolutionary mechanisms of bioluminescence has been overlooked to this day by successive generations of bioluminescence experts. Even Harvey, in his *History of Luminescence*, had failed to read, or see the importance, of this segment of the Frenchman's 1890 book in his cursory mention of the book's existence. Gadeau de Kerville was addressing a biological issue quite distinct from that of his compatriot Dubois, who was investigating the proximate causes of bioluminescence by applying functional analyses. The zoologist and historian of biology Ernst Mayr has remarked that "changes of concepts have a much stronger effect on the development of biological sciences than the discovery of new facts" (Haffer, 2007). This was certainly the case with Dubois, who revolutionized his field whereas many of his contemporaries were just content to add new observations. But Mayr also drew a distinction between immediate and ultimate causes in biology: "Functional biology deals with direct, proximate causes, which concern the phenotype and poses 'How?' questions, whereas evolutionary biology investigates the history of the genotype of organisms and poses 'Why?' questions" (Haffer, 2007). Gadeau de Kerville was a pioneer of the latter approach, whereas Dubois championed the "How?" questions.

Gadeau de Kerville also touched on a subject rarely discussed in his or subsequent times: the use of luminous organisms by non-luminescent animals. This was not a matter of recruiting symbiotic bacteria to substitute for one's own bioluminescence, as in squids and fishes. To illustrate his point, he uses a story told by natives from Southeast Asia according to which the male and the female of the baya weaver bird mix fireflies into the clay of their nest to light it up. Gadeau de Kerville adds that snakes and rats are their greatest threat to the nests and that the diffuse glow emanating from the nest keeps these nocturnal, photophobic predators away. However, he based his story on a myth originating from India, not on a scientific report (Davis, 1973); the bird was credited with too much cleverness.

~

In Dubois's promotion of the use of bioluminescence for human purposes, he had a predecessor who deserves mention. As early as 1890, ten years before Dubois's demonstration of "living lamps" at the Paris Exposition universelle, Samuel Pierpoint Langley (1834–1906), an American, had wanted to test whether animal luminescence could become the cheapest and most energy-saving form of lighting. Langley was many things: astronomer, physicist, inventor, and aviation pioneer alongside the Wright Brothers (McCullough, 2015), but a biologist he was not. Not that it mattered, because all Langley wanted to do was make physical measurements of the light emission of a bright insect – Dubois's click beetle (*Pyrophorus*) – to demonstrate what was already suspected; that is, that bioluminescence yields insignificant heat radiation (Langley and Very, 1890). Langley did not trust Dubois's own measurements because Dubois's apparatus was not in his opinion sensitive enough. For his purpose Langley used an instrument of his own invention, the bolometer, which was ultrasensitive in detecting infra-red radiation. He and his assistant, F.W. Very, confirmed that the light emission is indeed accompanied by infinitesimal heat radiation, and concluded:

> Resuming then what we have said, we repeat that nature produces this cheapest light at about one four-hundredth part of the cost of the energy which is expended in the candle flame, and at but an insignificant fraction of the cost of the electric light or the most economic light which has yet been devised; and that finally there seems to be no reason why we are forbidden to hope that we may yet discover a method (since such a one certainly exists and is in use on the small scale) of obtaining an enormously greater result than we now do from our present ordinary means for producing light.

Langley's paper caught the interest of the Japanese zoologist Shozaburo Watasé (1864–1929). A native of Tokyo, Watasé obtained his bachelor of science at the College of Sapporo in 1884 (Watasé, 1890). After studying zoology at the Imperial University of Tokyo, he decided in 1886 to travel to the United States and enter a PhD program at Johns Hopkins University in Baltimore, under the supervision of William Keith Brooks (1848–1908), a prominent zoologist who trained many future luminaries of American biology. The topic of Watasé's thesis was the morphology of the compound eyes of crustaceans (1890); there is no evidence that he ever conducted bioluminescence research, but we know he was keenly interested in the phenomenon. He spent time at the Marine Biological Laboratory (MBL) in Woods Hole, Massachusetts, during his PhD program, and it is assumed that he somehow became affiliated with the MBL in the following decade. His musings on animal luminescence were delivered in the form of two lectures at the MBL, which the Japanese zoologist intended, but failed, to publish with other material in a monograph. Watasé eventually returned to Japan to take up a professorship at the Imperial University of Tokyo (Li, Su, and Wu 2010).

Watasé's first lecture on the physical basis of animal luminescence (Watasé, 1895) reflects a reductionism as stringent as Dubois's. He points out that "it is easily conceivable that the animal that produces heat, as all animals can, may just as well produce light under certain circumstances, for both are but the manifestations of the same energy, and can be produced by essentially the same physico-chemical antecedents." After a discussion of Langley and Very's paper, Watasé turned suddenly to the various colours manifested by animal light emissions. Previous investigators had been at a loss to explain how these colours can be produced. Watasé's originality was to propose a simple, and very plausible, physico-chemical mechanism by which colour differentiation could be accomplished:

> The difference in color, when exhibited by different organisms, is probably due to some slight chemical differences in the light-giving substance. It may be supposed that under the influence of oxygen, the molecules of the given photogenic substance are set in vibration, the rate of vibration depending on and being characteristic of the particular species. And in those cases where a series of colors are displayed in succession by the same and one organisms, it may be supposed to be due to either of two causes: (1) the same photogenic substance is agitated with different degrees of frequencies at different periods in the life of the organism, or (2) a series of photogenic substances are produced, each one of the series representing a stage in the chemical metamorphosis of the substance.

While the above paper may have shown prescient insight, Watasé's second lecture (1898) was a throwback to the days of Quatrefages and the latter's obsession with establishing a link between muscle contraction and light emission (see chapter 3). Watasé recognized three types of animal luminescence: luminescent glands discharging their content externally; light organs in which the secretory products remained internal; and animals "in which the definite photogenic organ as such does not exist, but the light-giving material formed by the secretory process of the protoplasm accumulates along the course of muscle-fibres or other contractile protoplasmic material, the light manifesting itself in sparks or scintillations along the course of the fibres at the time of their contraction." Of the latter type he gave as examples the eggs of comb-jellies and the dinoflagellate *Noctiluca*. His explanation of the underlying mechanism of this third type was a bit strained. "The heat-producing particles and the light-producing particles," he wrote, "objectively considered, may not be very different from each other. *They may be variations of similar chemical substances, as the resulting energies, the products of their oxidation, are the variations of the same radiant energy* (his italics)." Thus, stimuli "which induce combustion of the thermogenic molecules, may also be presumed to incite combustion of the photogenic molecules as well, which exist side by side."

An essay by the American Charles Cleveland Nutting (1858–1927), another contemporary of Dubois's, returned a year later to the much-debated subject of the uses deep-sea animals make of their bioluminescence. Nutting made his career at the State University of Iowa, where he was curator of its Museum of Natural History (Calder, 2004). Having participated in one of Alexander Agassiz's oceanographic expeditions, he had become acquainted with many luminous organisms. This experience no doubt prompted him to reflect on the ecological function of deep-sea luminescence (Nutting, 1899). Nutting believed that there is enough ambient light in the deep-sea, some of which is provided by luminescent animals, for the colours of deep-sea animals to serve as advertisements for intraspecific recognition or sexual attraction. Could deep-sea luminescence serve similar purposes?

Being a Darwinian fanatic, Nutting had "scant sympathy with those naturalists who delight in demonstrating, to themselves at least, the falsity of the good old Darwinian dictum that every character possessed by an animal is of use to the species, or was of use to its ancestors." Nutting clearly signalled where he sided. But in what ways is luminescence useful to the bearer? In two ways, he suggested, using the coloration analogy: "one to the effect that the light attracts the mate and thus serves the purpose of *attractive coloration*, the other that it attracts the prey and serves the purpose of *alluring coloration*." Owing to his belief that comb-jellies possess "eye-spots," further into his paper he added a third role: their luminescence "may serve to keep them together, and thus effect the same end as 'directive coloration' among vertebrates and insects." All the examples he drew from in allocating these roles are free-swimming animals, so what happens with "fixed" or sessile animals like the sea pens and soft corals? The prevailing view that luminescence serves a protective role in these bottom-dwellers rested on the assumption that luminescence is a warning signal to the predator that the potential prey is unpalatable owing to its stinging nematocysts. But Nutting dismissed this opinion on the ground that to him the stinging cells of sea pens and gorgonians are relatively harmless compared, say, to jellyfish. Instead he thought the light attracts preys in a way "analogous, perhaps, to what is known as 'alluring coloration,' and its function in decoying prey." Future investigations, as we shall see later, proved him wrong in favour of the protective role. The impact of his paper, however, was aided by a positive review in the journal *Science* (C.L. Franklin, *Science* 11: 954, 1900), which ensured a broad dissemination of his thoughts.

~

As the field moved into the twentieth century, two reviews of bioluminescence were published in Germany. In the first, August Pütter (1879–1929), then an assistant in the Laboratory of Physiology at the University of Göttingen who had just completed his Habilitation thesis, produced a synthesis of the major investigative themes to date (Pütter, 1905). These included: the distribution of luminous organisms; the location of the sources of light emission; the physico-chemical conditions for obtaining luminescence; the physiological mechanisms controlling light emission; and the ecological significance of bioluminescence. Pütter was the first to use the concept of "ecology" in relation to bioluminescence:

> In one respect, the wide distribution of luminescence in the organic kingdom appears such as to suspect that it was used for specific ecological purposes. But this assumption from the distribution patterns is not necessarily correct, because luminescence could indeed be a random property of certain nutrient cycles, in which case the emergence of luminescence ability would be indifferent from an ecological point of view. Since, as we have already emphasized in the introduction, each chemical reaction could be connected to the production of light rays, luminescent activities are not as isolated as they may seem at first glance.

In Pütter's view the "diffuse" luminescence of bacteria and dinoflagellates would be ecologically neutral, whereas animals possessing luminous glands or optically complex light organs (photophores) would have a greater impact on their environment. Pütter himself never conducted research on bioluminescence, and his paper has the feel of a series of lectures or essays. His command of the literature was superficial; for example, he only cited the 1892 *Pholas* monograph of Dubois, who had by 1905 published a great deal more.

The review by Ernst Mangold (1879–1961), in contrast, was systematic, and supported by numerous illustrations and a full, up-to-date bibliography. Mangold, the son of a schoolteacher, was born in Berlin. He studied medicine and science in Giessen, Leipzig, and Jena, and at the latter university he received his medical licence and doctorate in 1903 (Wormer, 1990). He was habilitated in 1906, at which time, under the guidance of the great Ernst Haeckel in Jena, he pursued work on invertebrates at the Naples Zoological Station. From his stay in Naples he produced two papers on bioluminescence. In the first (Mangold, 1907), he studied the luminescence and light organs of the pearlside fish *Maurolicus pennantii* (Sternoptychidae). Although he did not observe spontaneous light emissions, he quickly elicited luminescence from light organs proximal to the site of mechanical or electrical stimulation. Immersing the fish in freshwater caused a long-lasting glow from all light organs, whereas irritant chemicals such as dilute sulfuric acid had no effect. In Mangold's second paper (1908), he examined the luminescence of an echinoderm, a brittle star. Spark-like luminescence occurred in response to vigorous mechanical stimulation. In Harvey's (1952) rendering of Mangold's observations, stimulation of an arm "sets off an impulse which moves along the arm in either direction as a light wave and will pass by way of the nerve ring in the disc to other arms." Mangold found that cocaine, muscarine, and pilocarpine – the latter two mimics of the neurotransmitter acetylcholine – were strong stimulants of brittle-star luminescence.

When Mangold published his exhaustive review of bioluminescence (Mangold, 1910), he was thirty years old and a lecturer at the University of Greifswald (Wormer, 1990). His original publications on bioluminescence may have spurred him to the daunting task of the review, but a more compelling motivation was likely to promote his career. Being called to write a chapter on bioluminescence in a pioneering and prestigious monograph series, *The Handbook of Comparative Physiology*, amounted to being recognized by his peers as an authority on the subject and gave Mangold the visibility he needed to rise on the academic ladder; he moved to a fast-track position at the University of Freiburg the following year. The First World War momentarily halted his ambitions while he served as a physician in military hospitals. Whatever his motives, Mangold's contribution to the handbook did not disappoint. He systematically covered all the groups of luminous organisms, from protists (bacteria and dinoflagellates) to fishes, and followed with general sections on various aspects of the field, in which he covered the characteristics of light emission, the factors affecting bioluminescence, and the theories of light production.

To my knowledge, Mangold (1910: 323) was the first to use the word "bioluminescence" (*Bioluminescenz* in German) for the phenomenon of luminescence in living organisms. The word "luminescence" itself was first coined by the German physicist Eilhard Wiedemann (1852–1928) in an attempt to sort out the terminology for the different forms of light emission in nature (Wiedemann, 1888). Under the umbrella of "luminescence" Wiedemann included all forms of light not resulting from incandescence; in other words, cold light. *Photoluminescence* is produced when a physical body is excited by lights of narrow- or broadband wavelengths, as in fluorescence and phosphorescence. *Thermoluminescence* results from a body excited by a rise in temperature; *electroluminescence* occurs in gases subjected to an electrical field, *crystalloluminescence* takes place when a solution turns into crystals and *triboluminescence* when crystals are fractured or crushed. Wiedemann's sixth and last type – *chemiluminescence* – is of special interest to this book; this form appears in the course of some chemical reactions. Dubois's luciferin-luciferase reaction qualifies as an example of chemiluminescence; it therefore follows that bioluminescence is a special case of chemiluminescence.

After Wiedemann's publication, the word "luminescence" was slowly and gradually substituted for phosphorescence or luminosity in the scientific literature on living lights. The trend climaxed with the introduction of the word "bioluminescence" in 1910. However, the word had no foothold in the literature until it resurfaced in 1916 in E. Newton Harvey's scientific articles and in 1920 in his book *The Nature of Animal Light*, but with no acknowledgment of Mangold's priority in coining it. Harvey's 1952 classic book *Bioluminescence* finally consolidated the word's widespread usage to this day.

In his review Mangold largely revisited the themes raised by Pütter in addition to others. In his ecological perspective, he embroidered on the reflections of August Brauer and Carl Chun already discussed in chapter 6. Mangold emphasized that non-luminous fish species overwhelmingly outnumber luminous fish species in the oceans, but ignored the fact that many luminous species are represented by so large a number of individuals worldwide that the luminescent biomass belies the apparent marginality of luminous species. He also pointed out the importance in the marine ecosystem of animals using their luminescence at night or in the twilight zone, but underestimated the role of luminescence in depths of permanent or near darkness. Mangold's thoughts on mechanical and electrical stimulation of luminescence in relation to control are among the most original that he offered. The fact that these stimuli can elicit not only local luminescence but also luminescence far from the point of stimulation suggested to him that nervous control is at play. In lower invertebrates Mangold favoured the intervention of cutaneous nerves, but acknowledged that in fishes and insects the central nervous system may be involved. He urged future experimentalists to investigate the role of the spinal cord in the coordination of light organ activity in fishes. He also noted that strong stimuli may excite light organs or luminous cells directly, whereas luminous responses to milder stimuli are likely to reflect the mediation of the nervous system. These thoughts turned out to be prescient.

In which directions did the understanding of bioluminescence progress around the core years from 1890 to 1920? Mainly in the accumulation of new information on the phenomenon in already-identified luminous species, but also in records of newly discovered luminescent species. More detailed analyses of the histology of light organs also appeared, as well as a greater recognition of the role of nervous control. And new investigators entered the field, although for most of them their dedication to bioluminescence research was but a passing phase in their career. Before proceeding to Part Four, on the American ascendency in the field of bioluminescence in the twentieth century, let us take a moment to survey this progress from the turn of the century, as Mangold had done, in group-by-group phylogenetic order.

Ctenophores and Cnidarians

Some comb-jellies, many of which look like transparent gooseberries, emit dynamic flashes that run along their comb-rows, as first demonstrated by George Allman (chapter 4) and Paolo Panceri (chapter 7) on *Beroe*. Now, a young American investigated a newly discovered luminous species of comb-jelly, *Mnemiopsis leydyi*. Amos W. Peters was a doctoral student in zoology

Figure 9.2 Representation of comb-jellies, highlighting the luminescent comb-plates. Drawing by E. Grace White in Dahlgren (1916b).

at Harvard College at the time of his research. Peters (1905) confirmed Allman's observations that luminescence is present in the early developmental stages as well as in adults and that luminescence is inhibited by light. His major new findings were that early embryos produce flashes, but not the one-cell egg, and that flash activity follows a daily rhythm of low flash activity in daytime and peak activity at night. Because the rhythm persists in animals constantly maintained in the dark, it would be considered an intrinsic or circadian rhythm by today's criteria. Peters also established that bioluminescence in *Mnemiopsis* is entirely independent from the ciliary locomotor activity of the comb-plates.

The other investigation of significance in this group was conducted by Peters's mentor, George Parker, on sea pansies. A full discussion of the biography and scientific career of Parker can be found in *Dawn of the Neuron*, a book on the pioneers of research on early evolved nervous systems (Anctil,

2015) and will not be repeated here. Suffice it to say that Parker, by then a full professor at Harvard, took advantage of a stay in 1919 at the Scripps Laboratory in La Jolla, Southern California, to investigate the sea pansy *Renilla amethystina* (now *Renilla köllikeri*). The sea pansy is a member of the sea pen family, whose bioluminescence was studied by Paolo Panceri (see chapter 7). But it is the wayward cousin of the family, resembling a kidney or the pansy flower rather than an old-fashioned pen. Because of its shape Parker viewed the sea pansy as more easily amenable to experimentation and to the study of colonial organization than the sea pen. The polyps (autozooids and siphonozooids) resemble a forest with a flat canopy cover, the latter forming the rachis or colonial tissue mass.

"The phosphorescence of Renilla," Parker (1919) observed, "is limited to the superior surface of the rachis, and when this surface is scrutinized closely under a hand lens, it is found that the phosphorescence is not a property of the whole surface, but appears only in certain almost microscopic white granulations. These occur around the openings in the common flesh through which the autozooids emerge and particularly on the siphonozooids." Gently poking these granular locations produces a soft greenish glow. But if a stronger mechanical or electrical stimulation is applied at any point on the colony, "waves of phosphorescence sweep from this area as a center over the whole of the superior face of the rachis. These waves succeed one another at such a rapid rate that the whole superior surface seems to be covered with a rippling glow emanating from the region of stimulation. After the application of the stimulus the luminous response quickly subsides."

Parker's description of his experimental results recalls Panceri's on sea pens, and Parker came to a similar conclusion: "It is difficult to understand how these successive activities are induced unless it is assumed that the luminous points are all controlled by a nerve-net whose form of transmission is reflected in the outward moving circles of light." Parker was not aware of Panceri's paper at the time, however; but as Panceri's article had appeared forty-seven years earlier, Parker had no easy excuse for overlooking it. Parker noticed that the luminescence of the sea pansy, like that of comb-jellies, is inhibited by light. In a later, more detailed paper, Parker (1920) measured the speed of the luminescent waves over the rachis and arrived at the same value as Panceri before him – around 5 cm/second.

Figure 9.2 Representation of scale-worms shedding their luminescent scales when attacked by a crab. Drawing by Bruce Horsfall in Dahlgren (1916e).

Annelids

Most luminescent annelids are polychaete worms, which are exclusively marine forms, but some terrestrial annelids (earthworms) are also luminous. The polychate species of interest here are scale-worms, the Bermuda fire worm, bristle worms, and a parchment worm, all of which have been introduced to varying extents in previous chapters. Scale-worms were of particular interest because of their abundance along European coasts and the bright flashes emitted by their "scales" (elytra). A student of these scales, William Aitcheson Haswell (1854–1925), was working at the Australian Museum in Sydney when he published a paper in which he presented the histological organization of the elytra (Haswell, 1882). Haswell noted the upper and lower cell layers under the cuticle of the scales, and the nerve entering the scales through the scale tubercle, which attaches the scales to the dorsal side of the body. His discerning description of the scale-worm's activities related to luminescence led him to propose a role for these organs:

> When certain species of *Polynoë* are irritated in the dark a flash of phosphorescent light runs along the scales, each being illuminated with a vividness which makes it shine out like a shield of light, a dark spot near the centre representing the surface of attachment where the light-producing tissue would appear to be absent. The irritation communicates itself from segment to segment; and if the stimulus be sufficiently powerful, flashes of phosphorescence may run along the whole series of the elytra, one or more of which then become detached, the animal meanwhile moving away rapidly and leaving behind it the scale or scales still glowing with phosphorescent light. The species in which the phenomenon of phosphorescence occurs are species characterized also by the rapidity of their movements, and also by the readiness with which the scales are parted with; and it seems not at all unlikely that the phosphorescence may have a protective action, the illuminated scales which are thrown off distracting the attention of an assailant in the dark recesses which the Polynoidae usually frequent.

A more precise picture of the organization of the elytra was provided a few years later by Étienne Jourdan (1854–1930), a lecturer and later professor at the Faculté des Sciences de Marseille (Jourdan, 1885). He more accurately called the cell layers "epithelia." He noted that the epithelial cells projected "fibrils" across the thickness of the scales, from the upper to the lower epithelia and vice-versa. He confirmed Panceri's description of the rich innervation inside the scales (chapter 7), but refuted the Italian's claim that the source of luminescence resided in this innervation. He saw the source of the light in the cells of the lower epithelium, which are more granular than those of the upper epithelium, but incorrectly assumed they were gland cells. Jourdan was the first to notice the presence of a small ganglion inside the scales.

Although scale-worms had until then been the focus of interest in luminescent marine worms, other polychaetes had attracted the attention of zoologists. The spectacular display of the Bermuda fire worm (*Odontosyllis enopla*) was investigated fully for the first time by Thomas Walton Galloway (1866–1929), at the time professor of zoology at the James Milliken University in Decatur, Illinois, and his student Paul Smith Welch (1882–1959), who would later distinguish himself as the author of the first textbook of limnology in the United States. They conducted their field research at the

Bermuda Biological Station, whose director, Edward L. Mark, had been Galloway's (and George Parker's) thesis supervisor at Harvard. From the content of the paper one surmises that Galloway actually did the field work in Bermuda in 1904, and conducted the morphological/histological work years later on campus with Welch's assistance.

Galloway and Welch (1911) found that fire worm luminescence occurs at the sea surface and that it is associated with reproduction, which takes place shortly after the full moon over several months (lunar periodicity). The precision goes so far as to enable prediction of the onset of spawning around an hour after sunset for each full moon. Females are twice the length of males and attract the males with bright continuous glows, whereas the males emit only brief flashes. The timing is important so that the males can fertilize the eggs shortly after their release; if the males fail to show up, Galloway and Welsh observed, the females stop glowing after ten to twenty seconds. Here is the authors' description of their mating choreography:

> In mating, the females, which are clearly swimming at the surface of the water before they begin to be phosphorescent, show first as a dim glow. Quite suddenly she becomes acutely phosphorescent, particularly in the posterior three-fourths of the body, although all the segments seem to be luminous in some degree. At this phase she swims rapidly through the water in small, luminous circles two or more inches in diameter. Around this smaller vivid circle is a halo of phosphorescence growing dimmer peripherally. This halo of phosphorescence is possibly caused by the escaping eggs, together with whatever body fluids accompany them. At any rate the phosphorescent effect closely accompanies ovulation, and the eggs continue mildly phosphorescent for a while. The fact that the luminosity is known at no other time is further suggestive that it is produced by the material which escapes from the body cavity. If the phosphorescent glands are external, as the histology of the epidermis at least suggests, the discharge of the glands is closely correlated with ovulation.

The male approaches the female and joins the balletic *pas de deux* of the intended synchrony:

> The male appears first as a delicate glint of light, possibly as much as 10 or 15 feet from the luminous female. They do not swim at the surface, as do the females, but come obliquely up from the deeper water. They dart directly for the center of the luminous circle and they locate the female with remarkable precision, when she is in the acute stage of phosphorescence. If, however, she ceases to be actively phosphorescent before he covers the distance, he is uncertain and apparently ceases swimming, as he certainly ceases being luminous, until she becomes phosphorescent again. When her position becomes defined he quickly approaches her, and they rotate together in somewhat wider circles, scattering eggs and sperm in the water. The period is somewhat longer on the average than when the female is rotating alone; but it, too, is of short duration.

The eyes of the males are larger than those of females. Galloway and Welch did not dwell on this sex difference, but it is presently known that the large eyes of some deep-sea fishes tend to be more light-sensitive (collect more light) than small eyes. The onus of luminescence detection rests on males, who have to spot ovulating luminescent females from a distance, so sharp visual detection is important to males. Interestingly, many firefly species also display a sex difference in eye size in favour of males, and in this case it bestows an advantage in visual acuity to males (Case, 1984). Galloway and Welch tentatively identified as the source of the luminescent secretion a type of gland cells inside which the secretory material is "twisted," but they offered no supporting evidence.

One might quip that there is a fascinating historical twist to these worms as well. In 1935 a naturalist in charge of sponge fisheries investigations in the British West Indies, Lionel Ruttledge Crawshay (1868–1943), published a paper in which he proposed that the mysterious lights that Christopher Columbus and members of his crew saw prior to their first landing in the New World in 1492 was a bioluminescent display of spawning fire worms (Crawshay, 1935). He based his theory on Galloway and Welch's descriptions of the bioluminescent events in synchrony with the full moon, his own observations of another Caribbean species of *Odontosyllis*, and the date of the sighting by Columbus – 11 October – which would have been a likely moment for spawning. Unfortunately, this is one hypothesis that cannot be experimentally tested, but it makes for interesting speculation.

Another family of polychaete worms was added to the list of luminescent annelids in the 1880s. The zoologist who introduced them, Richard Greeff (1829–1892), was already a seasoned professor of zoology and comparative anatomy at the University of Marburg when his two papers dealing with the luminescence of bristleworms (*Tomopteris*) appeared (Greeff, 1882, 1885). Greeff travelled extensively to gather the material for his researches (Hess, 1904). It was while screening the pelagic fauna off the coast of the African colony of Guinea in 1880 that he came across tomopterid worms. He noticed "rosette-like" organs on the appendages (parapods) of these worms, which he took at first for glands but which, upon examination of living material, he concluded were light organs (Greeff, 1882). He observed a bright yellowish luminescence emanating from these dot-like spots at the base of the parapods. The rosettes are innervated, so Greeff inferred that they are under nervous control. In a subsequent paper, Greeff (1885) zoomed in on these organs, which appeared to be associated with the genital organs of these worms.

The parchment worm *Chaetopterus variopedatus* turned out to be a tricky puzzle to solve. It was clear from Panceri's work that the source of the light emission was a luminous secretion (see chapter 7), but the epidermis of the whole worm is full of mucous cells of various types, especially those secreting the mucus that participates in building the parchment tube in which it lives. In short, the worm is a mucus machine. How, then, can one distinguish the luminous gland cells from the other types of gland cells? The first to address this conundrum was Emanuel Trojan, then assistant professor at the German University of Prague, using material collected, again, at the Zoological Station of Trieste (Trojan, 1913). It stood to reason that the matter would be resolved by a process of elimination, on the presumption that the luminous gland cells should overwhelmingly outnumber all other kinds of cells in the patches of luminous epithelia distributed in specialized appendages along the body. But Trojan's *a priori* proved to be mistaken. So in the end, feeling compelled to choose mucous cells present in slightly larger numbers than the other types, he picked cylindrical or pear-shaped cells spanning the entire epidermal layer, which were filled with large granules and opened to the exterior through a fine pore. For lack of sound scientific ground to stand on, Trojan's choice had all the trappings of guesswork.

In one of those unfortunate twists of fate surprisingly common in the history of biology, another investigator of the parchment worm showed up, but at the wrong geopolitical time and place. Anna Krekel spent all her university years in Heidelberg, starting in 1912 according to enrollment archives. She began her doctoral studies around 1914–15 under the supervision of the then-aging Otto Bütschli (1848–1920), the charismatic and world-famous protozoologist whose liberalism toward women in science and readiness to accept female students in his laboratory were recently documented (Anctil, 2015). The First World War influenced Krekel's fate in three ways. First, as men were called to the front, women stood better chances of being accepted as graduate students. But then, as lines of communication became disrupted, the war made it difficult for scholars to keep pace with the scientific literature, and Krekel became acquainted with Trojan's work long after she had conducted her histological research in 1915 and 1916. Had she known of his paper when she began her doctoral project, she likely would have chosen a different topic. And finally, war and the turmoil of its aftermath in Germany apparently postponed the publication of her thesis research, which appeared only in the year of Bütschli's death (Krekel, 1920).

Nevertheless, Krekel not only gave a more precise description than Trojan of what she considered to be the two contenders for the status of luminous gland cells but she also came up with a sounder basis for their identification. She also provided colour illustrations of histological slides stained with eosin and haematoxylin. She found that both gland cell types are interspersed between ciliated sensory cells.

> Two types of glandular cells can be distinguished clearly. The first is characterized by its strong staining with eosin. These unicellular eosinophilic glandular cells are roughly flask-shaped. Basally they extend a fine projection. Their nuclei are flattened and are usually at the lower end of the large mucous mass. Its content consists of either more or less numerous small grains, or they have merged into a single homogenous secretory ball. These cells do not possess a particularly differentiated secretory pore … [The second] are slimmer and fusiform and extend almost through the entire thickness of the epidermis. Basally they also end with a fine projection. They empty their secretion in the same manner as the eosinophilic cells, but through a slimmer

> excretory pore. The blue-stained mucous cores hardly stand out from their surroundings, are also parietal, and lie in the deepest part of the basal slightly enlarged cells.

So the red eosinophilic cells contrasted sharply with the blue basophilic cells stained by haematoxylin. She noticed that in certain cases there was an abrupt boundary line between luminous and non-luminous patches whereby the basophilic cells gave way to eosinophilic cells, respectively. This she considered a critical observation for assigning a luminescent role to the basophilic cells, thus supporting Trojan's original assignment. Future investigations proved both Trojan and Krekel wrong: Nicol (1952a) tentatively identified the eosinophilic cells as the luminescent gland cells, and Anctil (1979) confirmed it by fluorescence and electron microscopy.

Over the years, luminescent earthworms have attracted less interest than polychaete worms, and yet they are similar to parchment worms in that they also burrow in the substrate (the soil) and eject a luminous secretion. Emilia Rota (2003, 2009) made an exhaustive review of the literature, but only a few salient contributions are highlighted here. The first is an observation by the British entomologist Allen Harker (1847–1894) of numerous small luminous earthworms in a peaty soil, which he tentatively identified as belonging to the genus *Enchytraeus* (Harker, 1887). "In a darkened room," Harker wrote, "a single worm on being gently rubbed glowed like a fine streak of phosphorus."

Another Englishman, William Blaxland Benham (1860–1950), who spent most of his career as an earthworm specialist in New Zealand (Miller, 1952), wrote about a local species, *Octochoetus multiporus*: "[It] has a milk-coloured coelomic fluid of very great tenacity; it can be drawn out into strands, and soon hardens on exposure to air. In the dark, when the worm is handled, this fluid is discharged abundantly from the dorsal pores and from the mouth, which it reaches through the 'peptonephridia' [kidneys] opening into the buccal cavity" (Benham, 1899).

John D.F. Gilchrist (1866–1926), a Scottish expatriate to South Africa, reported at length on a genus (*Chilota*), which expels a luminous, viscous fluid from the mouth and anus (Gilchrist 1919). The light emanates from the coelomic fluid, and Gilchrist identified small granules in chloragen cells and/or amoebocytes as the source. He intuitively assigned a defensive role for the luminescence.

Stanislaw Skowron (1900–1976) was a young Polish biologist benefitting from a travel grant of the Rockefeller Foundation (Jordan 1987) when he studied the earthworm *Microscolex* in Naples (Skowron, 1926). From crude experiments, he came to conclusions similar to Gilchrist's, except he found that the luminous slime of *Microscolex* emanates mainly from the anus. Two years later, at the time of his Habilitation, which opened to him an assistant professorship at the Jagiellonian University of Cracow, he produced another paper, in which he obtained chemical extracts from the luminescent material of *Microscolex*. From these extracts Skowron surmised that, to produce the light, an oxidation involving two substances was involved. A professor of zoology at Charles University in Prague, Julius Komárek (1892–1955), took Skowron's approach in his study of the earthworm *Eisenia submontana* (Komárek, 1934).

Mollusks

In the slime category, the piddock *Pholas dactylus* is a worthy competitor to the parchment worm. The German investigators who dealt with it, Bernhard Rawitz (1857–1932) and Johannes Förster (1888–?), had to face the same murky problem when it came to identifying the luminous gland cells. Rawitz was an assistant professor (*Privatdozent*) at the University of Berlin when he produced his study of clams, which included the histology of the luminous glands (Rawitz, 1892). Rawitz's descriptions are confusing, however. Förster was critical of Rawitz's work, and the great invertebrate neuroanatomist and Arctic explorer Fridtjof Nansen (1861–1930), commenting on an earlier paper by Rawitz, complained of his unreliable methodology and shabby scholarship (Nansen, 1887). Förster's work, in contrast, stands up to scrutiny.

Ernst Johannes Rudolf Förster studied zoology at the University of Leipzig, and his doctoral thesis, under the guidance of Carl Chun (of *Valdivia* fame), described his work on the luminescent glands and the nervous system of *Pholas*. The luminescent epithelia, Förster remarked (1914), contain long ciliated cells between which the glandular cells are inserted. Their ciliae generate a current that moves the luminous mucus away, once it is exuded. The glandular cells are organized in two layers, a superficial one made of mucous (slime) cells, which stain lightly (eosinophilic), and a deeper layer made of the luminous gland cells, which stain darkly (basophilic). The luminous gland cells are filled with highly refringent granules; because they lie deeper in the epithelium, these cells possess a long neck to bring the mucus to the surface. The notion that luminous gland cells are basophilic, as seemed to be the case in annelids, was thus generalized. Förster was the first to observe the maturation of the secretory process over time; however, he failed to provide convincing supporting evidence for his identification of the luminous gland cells. He considered *Pholas dactylus* to be the only member of the genus to emit light, but a study by Yô Kaname Okada (1891–1973), a Japanese zoologist who was spending six years as visiting scholar in Europe (Okada, 1994), showed that two other species (*P. candida* and *crispata*) are also luminous (Okada, 1927b).

In thinking of luminescent mollusks, cephalopods come foremost to mind for the sheer number of their luminescent species. In an excellent review of luminous cephalopods, Samuel Stillman Berry (1887–1984) claimed that "it can be truly said that no other class of animals can compare with the cephalopods in the complexity, diversity, beauty, brilliancy – in brief, the high specialization of organs devoted to the production and utilization of that form of energy which to our human faculties finds expression as light" (Berry, 1920a). Born in Maine and educated at Harvard and Stanford, where he earned his PhD in 1913, Berry worked at Scripps Institution for Biological Research in La Jolla, California, until 1916, when he abandoned academia to pursue a business career in Redlands, a town near San Bernardino, Southern California (Roper, 1984). Meanwhile, he had become an authority on cephalopods and he remained so, making them a lifelong hobby. His interest in bioluminescence was a collateral effect of his investigations.

Berry rightly noted that, despite the large number of cephalopod species that possessed light organs, the actual light emission of very few squids had ever been observed. The main obstacle to observing them was the inacccessibility of their mostly deep-sea abodes. But by the dawn of the twentieth century, reports of several sightings of their luminescence were published. In Japan, Shozaburo Watasé, discussed earlier in this chapter, discovered a new species of squid, the *hotaru-ika* – the firefly squid – whose luminescence he quaintly described:

> When seen in daylight they [light organs] appear to be small black spots, but in the night all these spots shine with a brilliant light like that of the stars in heaven … When these spots (while the *hotaru-ika* is alive) are viewed under the microscope, they are very interesting. When the animal is about to produce the light, the membranes [chromatophores] covering the spots will concentrate and remove themselves, thus opening a way for the light. The light is so brilliant that it seems like a sunbeam shot through a tiny hole in a window curtain. Again, when the *hotaru-ika* wishes to shut off the light, the membranes will expand and cover the spots. (Watasé, 1905, translated from Japanese in Berry, 1920a)

Clearly the pigment cells (chromatophores) serve as a kind of shutter, an eyelid of sorts. Berry himself gave the firefly squid the scientific name *Watasenia scintillans*, in honour of its Japanese discoverer. At Watasé's urging, a staff member of the Chiba Medical School on the Tokyo Bay, Sotoo Hayashi, conducted a histological study of the firefly squid's light organs (Hayashi, 1927). The light organs have a well-developed reflector and are remarkable for their rich blood supply. The numerous rod-like organelles in the luminous cells, though suggestive of bacteria, were not regarded as such by Hayashi. All attempts to culture bacteria from these organs failed, and it took an electron microscopic study decades later, by the same Okada who had worked on *Pholas* in 1927, to discover that the rod-like structures are paracrystalline organelles (Okada, 1966), the likes of which, as it turned out, are found in the luminous cells (photocytes) of light organs from different mollusks and scale-worms (Bassot, 1966).

Another witness of the wonders of squid bioluminescence was Werner Theodor Meyer (1882–?), a student of Carl Chun, a squid specialist in addition to his other attributes. He poked at *Heteroteuthis dispar* at the Naples Zoological Station and "it shot rapidly through the water, and spurted through its funnel a luminous secretion which floated in the water as separate globules, these being drawn out by the currents into shining threads, a pyrotechnic display (*Feuerwerk*) which it was able to repeat many times"

Figure 9.3 Representation of a school of luminescent squids moving in deep pelagic waters. Drawing by E. Grace White in Dahlgren (1916d).

(Meyer, 1906). Despite their splendid fireworks, however, the structure of the light organs appears unremarkable (Meyer, 1908).

Berry (1920b) was struck by the diversity of structure of light organs in cephalopods – what he called the "polymorphic nature of photogenic organs." This diversity was previously remarked upon by Chun and Hoyle (see chapter 6), but Berry was the first to realize its significance. His analysis suggested that, "whereas about a third of the genera cited each possess photophores belonging to a single general type, nearly as many have strongly dimorphic photophores, and an even greater number have trimorphic or polymorphic organs." These statistics led him to "a general statement of fact that those species having a relatively abundant development of integumentary photophores distributed over the body generally fail to evolve a great variety of other types." To the questions: "Are our present [cephalopod] species descended from an ancestral photogenic stem, some branches of which have now yielded up the function? Or has photogenesis arisen several times in this class of animals?" Berry found that the burden of evidence favours the polyphyletic origin of light organs.

The luminescent system of the nudibranch *Phyllirhoë* came to the attention of Emanuel Trojan, who refuted Panceri's notion (see chapter 7) of luminous cells as neuronal derivatives (Trojan, 1910). He distinguished two modes of light emission: a low-level, diffuse light spread over the entire body, which he associated with individual mucous cells; and little spots of light of more limited distribution, which he associated with small luminous glands. The latter looked like bundles of mucous cells clumped together like bulbs of garlic. Trojan also followed individual nerve fibres that end on some of these mucous cells.

I end this section with a short note reporting the discovery of a rare freshwater bioluminescent snail in 1890; until then only marine and terrestrial luminescent forms had been known. The discoverer was Hans Heinrich (Henry) Suter (1841–1918), a Swiss-born malacologist (mollusk specialist) who had moved to New Zealand with his family in 1886 (Hedley 1919). Suter eventually joined the staff of the Canterbury Museum in Christchurch, but he was living in Wellington when he happened on the limpet *Latia lateralis* (= *neritoides*) in a creek (Suter, 1890). "I was greatly astonished," he recalled, "when at night-time I found all the animals highly phosphorescent. The margin of the mantle showed a violet light, and this was intensified by a touch with a needle. The secreted mucus was also phosphorescent for some time. I do not know of any other fresh-water shell showing this phenomenon, though it is well known in many marine shells, especially in the Pholadidae." Strangely, Sutter seems to have gotten the colour of the light wrong, as it is known to shine green (Meyer-Rochow and Moore, 1988, 2009). This error again illustrates how unreliable the human dark-adapted eye can be in discriminating colours.

Crustaceans

The juvenile stages of the euphausiid *Meganyctiphanes norvegica* live near the ocean surface, whereas full adults inhabit deep waters. Previous investigators of its light organs had not taken this into account (see chapter 6), but Rupert Vallentin (1858–1934) and Joseph T. Cunningham (1859–1935) at Oxford University noted their location in their description of the "photospheria," as they call the light organs of the adults (Vallentin and Cunningham, 1888). These light organs, present near the eyes, and on the thorax and

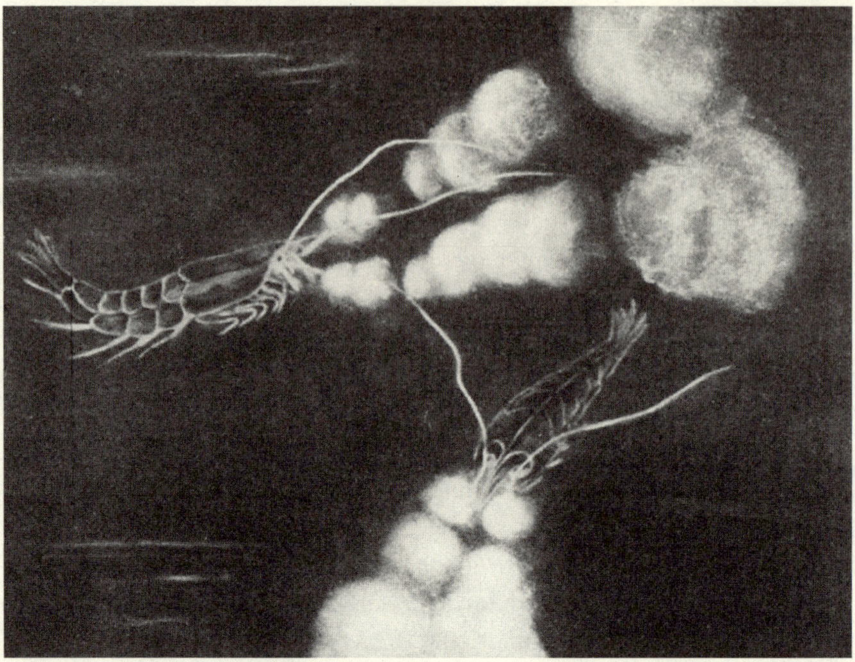

Figure 9.4 Representation of shrimps spewing out large luminescent clouds. Drawing by E. Grace White in Dahlgren (1916f).

abdomen, are remarkable for their optical accessories: a thick reflector behind the large luminescent cells, the array of rigid fibres, which Bassot (1966) called "rodlets" and may serve as optical guides to collect the light from the photocytes and direct it to a lens constricted by concentric lamellae. Using young specimens in which the reflector was not yet developed, Wilhelm Giesbrecht (1854–1913), a Prussian zoologist and crustacean specialist working at the Naples Zoological Station, sought to determine the cellular source of light emission in the photospheria. Unfortunately, he missed his target in stating that the rigid fibres (rodlets) are the source of the light (Giesbrecht, 1896). Trojan (1907) disagreed with Giesbrecht and confirmed the large cells on the inner surface of the reflector as the source of light.

Rupert Vallentin and John Cunningham, trying to understand the role luminescence played in animals, made observations on the luminescent behaviour of krill. As far as undisturbed behaviour is concerned, they found that in the dark, "the animals swimming about in a glass jar of sea-water gave out short flashes of light from time to time. Each flash was of short duration, but sometimes lasted longer than at others; when several animals gave out light simultaneously or in rapid succession, the effect was very brilliant and beautiful, but nothing like continuous luminosity was ever observed." When prodded, the krill emitted a series of flashes as they flapped their tails in an escape reflex. Vallentin and Cunningham concluded that the organs must be under nervous control, but a new study a century later (Herring and Locket, 1978) suggested that nervous control is effected indirectly, by controlling blood flow to the photocytes.

Decapod shrimps, in their turn, were investigated by Stanley Kemp (1882–1945), a London-born marine biologist who worked in Ireland and had just been appointed to the Indian Museum (Calman, 1947) when he published his paper on *Sergestes challengeri*, *Acanthephyra* (*Systellaspis*) *debilis*, and *Hoplophorus* species (Kemp, 1910). As an illustration of diversity, Kemp observed that within decapods the photophores of *Sergestes* differed as much from those of *Acanthephyra* and *Hoplophorus* as decapod organs differed from those of euphausiids. The large luminous cells with a granular content are immersed in a blood sinus and are sandwiched between a reflector layer and a complex lens. Arata Terao, of whom nothing is known except that he was affiliated with the Zoological Institute of the Tokyo Imperial University, confirmed Kemp's findings in another species of *Sergestes* in which he counted no fewer than 150 photophores (Terao, 1917).

Terao also made interesting observations on decapod light emission. He noticed that "the photophores emitted dim greenish-yellow light in an intermittent way, each time starting suddenly and vanishing with as much promptness after a longer or shorter period of illumination. Frequently, after dark intervals of varying length, the lighting up of different photophores in the same body occurred one after another in serial succession, beginning with those at the head end and thence progressing posteriorly, to finish up at the tail end." Having had the chance to observe both euphausiids and decapod shrimps in the same catch, Terao was adamant that the light emission of the decapods was much weaker than that of euphausiids.

Up until this time, it was thought that in shrimps, with the exception of *Hoplophorus*, light is emitted directly from inside the organs, but in all the luminescent copepods and ostracods (except halocyprids) an expelled secretion was involved. Giesbrecht's work on copepod luminescence is admired to this day; as Herring (1988) remarked, "All subsequent investigations have drawn heavily on this classic account, though few have matched its detail and accuracy." Giesbrecht (1895) reported that the luminescence emitted by marine copepods is produced by small cutaneous glands on the legs and elsewhere that exhibit a pore through which secretion is exuded. Females, he specified, possessed more of these glands than males. He found in *Pleuromamma abdominalis* sub-cuticular cells to which he attributed the production and secretion of luminous material; these cells contained a greenish-yellow secretion. He also found that their luminescence capability varied according to season.

The last crustaceans featured here are of utmost importance for the history of bioluminescence, as we will have occasion to discover in the next chapter. These are the ostracods. Their story begins with a German naturalist, Franz Martin Hilgendorf (1839–1904), whose doctoral thesis work at the University of Tübingen on fossil freshwater snails caught the attention of Charles Darwin in the sixth edition of *On the Origin of Species* for its evolutionary implications (Yajima, 2007). As part of exchange agreements between Germany and the Meiji Era Imperial Japanese government, Hilgendorf came to Japan in 1873 as a "Westerner employed in the modernization of Japan" (Yajima, 2007). He was credited with introducing the teaching of Darwinian evolution in Japanese universities. When he returned to Germany in 1876, Hilgendorf brought with him a collection of specimens, which were stored in the Zoological Museum of Berlin. Among them were ostracods. Years later Gustav Wilhelm Müller (1857–1940), an ostracod specialist based at the Zoological Museum of Greifswald, found Hilgendorf's ostracod specimens and assigned them the name *Cypridina hilgendorfi* in his honour.

Gustav Müller used the last pages of his large paper on cypridinids (Müller, 1891), to discuss the luminescence of ostracods. He compared the ejection of a luminous cloud by these ostracods to that of the ink by the cuttlefish or, more strikingly, to the luminous tail produced by a comet. The amount of ejected luminous fluid seemed disproportionate in relation to the tiny size of the animal. Because ostracods swim as they eject the fluid, the streaming luminous cloud appears to issue from the tail region. But Müller soon realized that the source of the cloud is the "upper lip," which has an opening to the outside independent of the mouth. It is in fact the opening of the maxillary gland. In this gland he noticed three distinct groupings of gland cells, each displaying a different morphology in their granular content. From his detailed observations Müller suggested that two secretory substances are channelled by the squeezing action of local muscles in separate ducts opening out through distinct pores. The two secretory products thus only mix once they have been ejected into the surrounding water, and this mixing is necessary for the production of the light.

Other investigators soon followed up on Müller's seminal paper. Hiroshi Watanabe (1897) noted that *Cypridina hilgendorfi* was known locally as the marine firefly (*umi-hotaru*) and he confirmed the presence of two secretory products. He also assumed, although without providing evidence, that the luminescence serves to frighten potential predators. Naohide Yatsu (1917) and Yô K. Okada (1927a) added details of organization of the maxillary gland and corrected a few errors of observation or interpretation on Müller's part. Müller's idea of the coming together of secretory products in a chemical mixing was soon to flourish in the first half of the twentieth century.

Insects

Research on insect bioluminescence intensified, and at the turn of the century progress in understanding how it functions was palpable. The era saw an upsurge of contributions by American investigators, a trend that continued well into the twentieth century. The bulk of the research was concerned with the structure and development of the light organs, but other aspects such as physiological control and luminescent behaviour were also examined.

Max Verworn (1863–1921) made a single contribution to the field of bioluminescence, but it stands out as an off-the-beaten-path foray into realms untrodden by his peers. Born in Berlin and educated at the University of Jena where he earned his medical degree, Verworn rose to a distinguished career as the foremost cell physiologist in Germany, but also as a controversial figure for his unorthodox philosophical ideas (Fröhlich, 1923). At the time of his paper on the firefly *Luciola italica* (Verworn, 1892), he was assistant professor in Jena, but his originality and experimental acumen were evident. Where others had only seen series of flashes emitted by fireflies, Verworn discerned a pattern of rhythmicity that set him thinking along the lines of a control by some pacemaker brain centre. To test this idea, he embarked on a series of experiments (decapitations and chloroform anaesthesia, for instance), which led him to locate an "automatic centre" (pacemaker) in

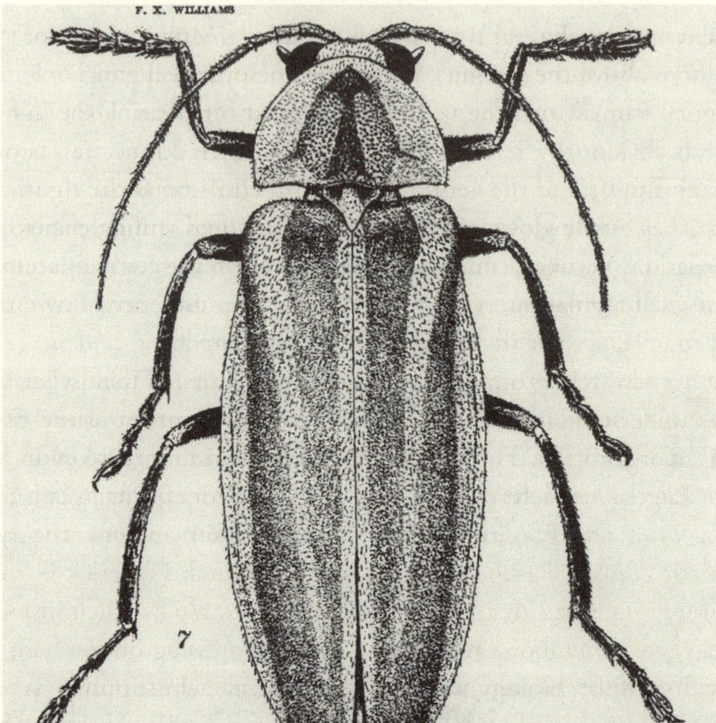

Figure 9.5 A female firefly (*Photuris pennsylvanica*, top), and the arrangement of the light organ (shaded area) on the abdomen of a male (bottom left) and a female (bottom right). Plate 2 in Williams (1916). Courtesy of John Wiley & Sons.

the circumoesophageal ring, which includes the brain (supraoesophageal ganglion, above the foregut), and the suboesophageal ganglion (under the foregut), from which the ventral nerve cord innervating the light organs extends posteriorly. Severing or anaesthetizing the connection between the central ganglia and the ventral nerve cord eliminated the rhythmicity of the flashes, while glows of varying duration could still be elicited by local stimulation. Verworn concluded that neurons in the central pacemaker fire impulses at regular intervals, and that these impulses travel down the nerve cord to impose their rhythmicity on the light organs.

Other researchers of a physiological bent contented themselves with trying to understand luminescence control downstream, at the level of the firefly light organ itself. The German Johannes Bongardt (1856–?) in 1903, the Swiss Eugen Steinach (1861–1944) in 1908, the Americans Joseph H. Kastle (1864–1916) and F. Alex McDermott (1885–1966) in 1910, the American Elmer J. Lund (1884–1969) in 1911, and the Dutch Frederick C. Gerretsen (1889–1966) in 1922 are the most outstanding examples. Their investigations did away with Dubois's notion that muscles pressing on the lymph of the light organs, by analogy with our diaphragm, help supply oxygen to the photocytes. They tended instead to emphasize the role of the tracheoles in allowing rapid access of oxygen to the luminous tissue. Although the view prevailed that the fine tracheoles pass between the photocytes, Lund, availing himself of the best microscopic resolution available, found that tracheole branches interdigitate inside the photocytes, thus creating more intimate areas of gas exchange.

To the theory that muscle-powered air pressure is the ultimate control mechanism of light emission in fireflies, also promoted by the German Carl Heinemann in 1886 and by Watasé in 1895, was opposed the neural control hypothesis. Using high-power microscopy, Lund (1911) had pinpointed the photocytes with their granules as the light source, but he erroneously regarded the "crystalline" granules as by-products of the oxidation reaction producing light. His microscopic observations of live luminescence showed, however, that light was emitted in the photocytes, but initially in the vicinity of the tracheal end-cell. From this zone, luminescence spreads further afield upon repeated stimulation.

Similar but rather crude experiments made earlier by the American entomologist Alpheus S. Packard (1839–1905) pointed in the same direction

(Packard, 1896). And Bongardt (1903) had observed nerve endings on these tracheal end-cells, thus suggesting that innervation excites the tracheal end-cell, which in turn excites the photocyte. To add support to the neural control hypothesis, Steinach had found that a summation of sub-threshold stimulatory impulses was necessary to induce luminescence, an observation compatible with what is known of spatio-temporal summation of synaptic inputs in neurobiology. Further support by Kastle and McDermott was the excitatory effect of the neuro-humoral agent adrenaline on firefly luminescence.

Histological studies brought little new to what was already known of the organization of firefly light organs. Anne B. Townsend (1904) and Alex McDermott and Charles Crane (1911) gave great emphasis to the tracheal system in their descriptions, which presented only small variations in organization between species. The most original contribution here is that of Francis X. Williams (1882–1967), a student at Harvard of the great American entomologist William Morton Wheeler (1865–1937), who revealed to the world the social behaviour of ants (Parker, 1938). Williams produced a doctoral thesis in which he followed the embryological development of the firefly light organ (Williams, 1916). "There are, in general, two views as to the origin of the photogenic organs in the Lampyridae;" Williams explained: "according to one they are developed from the ectoderm; according to the other they are related to the fat-body and are therefore mesodermal." The first view was championed by Dubois and a few others, the second by Bongardt and the majority of firefly students. Williams's observations put him decidedly in the camp of the majority. He showed in larvae how the light organs are "derived directly from the fat-body by the expulsion of its cells, which migrate to the ventral hypodermis [epidermis under the cuticle]." These cells proliferate and form a mass, which becomes invaded by the tracheoles and nerves. Walter N. Hess (1890–1974), an entomologist at Cornell University, corrected and improved Williams's developmental observations with more detail (Hess, 1922).

Tropical fireflies, and especially elaterid beetles, were examined by Carl Heinemann, a German medical doctor living in Veracruz, Mexico (1872, 1886), and by Erich Geipel (1915), a student of Carl Chun in Leipzig. Unfortunately, their contributions were overshadowed by the large-scale work of Dubois on elaterids.

During the period covered here a new luminous insect species was introduced to the world that eventually became a major tourist attraction in New Zealand. The entomologist who performed the introduction was George Vernon Hudson (1867–1946). British-born, Hudson moved to New Zealand with his father when he was fourteen (Salmon, 1947). Already an amateur entomologist, the teenager turned semi-professional the moment he set foot "down under." But to make a living he eventually found employment at the post office in Wellington, where he rose to chief clerk. Hudson, also an amateur astronomer, is believed to have been the first to submit a proposal for daylight saving time (Hudson, 1898). In 1886, aged only nineteen, he published a paper in which he described the cave-dwelling glowworm *Arachnocampa* and corrected misinformation on the insect by a British entomologist (Hudson, 1886). Here is how he described the glowworm's display:

> When carefully examined with a bull's-eye lantern and pocket lens, this light is found to proceed from a large glutinous knob, situated at the *posterior* extremity of the larva, a fact I have verified by repeated investigations: but the insect's curious habit of occasionally travelling backwards has doubtless led to this mistake. It inhabits irregular cavities in the bank, where it hangs suspended in a glutinous web, which also appears to envelop the body, large quantities of sticky mucus being periodically shot out of the mouth of the larva … I think [the light] may often assist the larvae in escaping from enemies, as when disturbed they nearly always gleam very brilliantly for a few seconds, suddenly shutting off the light and retreating into the earth.

Hudson asserted that the larva belongs to a gnat of the mycetophilid family. In 1926 he added more information, pointing out that, "When it is in a small cave, the light also reflects on the pendants of [glutinous] beads, thus lighting up the whole of the cave." This is the spectacle that would later be exploited in the Waitomo Cave by the New Zealand tourist industry. In the intervening years Hudson changed his mind about the role of the larva's luminescence, stating, "It is my belief that the web is formed to entangle insects, which are attracted by the light" (Hudson, 1926). Wheeler and his student Williams, mentioned earlier in this chapter, studied the light organ of the glow-worm on the occasion of a visit to New Zealand (Wheeler and Williams, 1915). Close examination "showed the luminous organ very distinctly in the dilated ovoidal terminal segment of the body." A bluish-green light came off the dorsal surface of the segment. A surprising discovery was that the light organs are in fact part of the Malpighian tubules, the kidney structures of insects. As these structures are ectodermal in origin, so must be the light organs, which at the time distinguished them from the mesodermal origin attributed to firefly light organs.

Other mycetophilids also produce light. Hans-Jürgen Stammer (1899–1968), a student of the discoverer of endosymbiosis Paul Buchner (see chapter 7), produced a landmark paper on the fungus-eating gnat *Keroplatus testaceus* one year after his Habilitation at the University of Breslau (Stammer, 1932). He observed the weak luminescence of the whole body of larvae as well as the pupa, whose light shines through the cocoon. Stammer attributed the light source to the insect's fat body, with some of the fat cells let loose in the hemolymph. He was perplexed that the light source of *Keroplatus* differed so strikingly from that of its New Zealand cousin, *Arachnocampa*. Forty-five years later Jean-Marie Bassot (of whom more in chapter 14) identified giant secretory cells in the hemolymph of the Appalachian mycetophilid *Platyura fultoni*, whose exuded secretion causes the long-lasting, blue luminescence of this species (Bassot, 1978). Kotaro Osawa and collaborators (2014) confirmed in the Japanese species *Keroplatus nipponicus* that the larval body fluid is luminescent, as expected from exuded secretions.

Echinoderms and Tunicates

The brittle stars *Ophiopsila* and *Amphiura* were investigated by Ernst Mangold (1908) and Emanuel Trojan (1909). Both agreed that their light emission originates in gland cells deep inside the arms, but also that these cells send a thin tubular extension to the epithelial surface. They believed the luminescence to be intracellular and under nervous control. Upon stimulation, flashes like lightning travelled along the arms. The suspected luminous gland cells of the deep-sea star *Brisinga*, a form then believed to be intermediate between starfish and brittle star, are similar in appearance to those of *Ophiopsila* and *Amphiura* (Thust, 1916).

Among the tunicates, the appendicularian *Oikopleura* was studied in detail by Hans Lohmann (1863–1934). He obtained his specimens in the harbour of Kiel in Northern Germany. These peculiar animals live inside a mucous "house" (tunic) which they build themselves. Hans Lohmann (1899) followed three developmental stages of *Oikopleura*, each represented by a different type of luminescence. As Harvey (1952) well summarized Lohmann's findings: "In the free-swimming tunic-less form, Lohmann observed a bright greenish flashing on stimulation, of short duration, that came only from the trunk, not from the tail … Second, at the time the tunic unfolded, the light of the whole trunk with the exception of the gonadal cavity became brighter and white and was continuous. The luminescence came from the oikoplast epithelium that secretes the tunic. Third, after the tunic had unfolded and the animal had rested, the continuous light disappeared and flashes of light again occurred on stimulation."

Fishes

The great hold on the public imagination that deep-sea fishes conjured up in the wake of the oceanographic expeditions pushed luminescent shallow-water fishes into the shadows. But progress was being made all along, if only because these fishes were more easily captured alive than deep-sea fishes and also easier to experiment with in seaside laboratories. Of particular interest to us here are sharks, flashlight fishes (Anomalopidae), the knight fish (Monocentridae), and the midshipman fish (Batrachoididae).

Leopold Johann, whose biography is unknown except for the fact that he was pursuing doctoral studies at the University of Rostock, Germany, at the time, fired the first salvo with his histological investigation of the skin light organs of the shark *Spinax niger* (Johann, 1899). He described small organs recessed in the skin, largely surrounded by dark pigment, with photocytes and secretory masses inside but no obvious innervation. The next year Rudolf Burckhardt (1866–1908), a professor of paleontology and comparative anatomy at the University of Basel, Switzerland (*The Geological Magazine*, vol. 5, 1908, p. 144), who was enjoying a sabbatical leave at the British Museum of Natural History, greatly expanded the number of known shark species that possess light organs while collecting in the Bay

Figure 9.6 Representation of the shark *Spinax niger*, highlighting the light-emitting ventral patches. Drawing by Bruce Horsfall in Dahlgren (1917e).

of Naples (Burckhardt, 1900). Ten more species belonging to different genera were added to the list. As for structure, Burckhardt had nothing to add to Johann's descriptions.

A detailed study of the Japanese sharks *Etmopterus lucifer* and *E. frontimaculatus* was performed in 1911 by Hiroshi Ohshima (1885–1971), then a student of Watasé at the Imperial University of Tokyo. Ohshima, based at Kyushu University in Fukuda, would later rise to prominence and train many of the best Japanese zoologists of the twentieth century (Baba, 1974). Ohshima witnessed "whitish" light emanating from the ventral side of these sharks, which appeared only upon stimulation. When rubbing the shark, "the luminescence was not immediately called forth but became apparent after some minutes, the light gradually appearing or vanishing or attaining maximum intensity here and there at different places." Such observations tended to support Johann's assertion that the light organs are not under nervous control. Ohshima proposed a control mechanism based on the heavy pigmentation of the light organs:

> The facts that no sudden change of luminosity takes place and that there is such a local difference in the intensity of the emitted light may, in some cases, be due to the action of the pigment cells which form what I have called the "iris." When contracted, they allow the exit of the light produced in the photogenic body lying underneath, while their expansion makes the iris act as a screen that shut[s] in the light.

Charles Frederick Hickling (1902–1977), a young investigator in the Department of Oceanography at the University of Liverpool, made a clever observation about the light organs of *Spinax niger* which induced him to offer a – maybe too clever? – hypothesis on the biological role of shark luminescence (Hickling, 1928). He noted that: "The complex lantern-like structure of each individual organ seems designed to throw out a parallel beam of light, and to prevent scattering of the rays; the arrangement of the axes of all the organs parallel to the median vertical axis of the fish, seems to aim at precisely the effect described above, namely, that the luminescence will only shine upon objects immediately beneath the ventral surface." Hickling assumed that a flash of light would allow at once seeing a prey just beneath the shark and for the shark to snatch the prey. But descriptions by others have stressed the long duration of the light emissions, and Hickling, like others before him, had missed the opportunity to invoke counterillumination – the camouflage of the ventral silhouette – as the ultimate function of the luminescence.

The flashlight fishes, originally described by Weber (see chapter 4), came under the close scrutiny of Otto Hermann Steche (1879–1945). The son of a wealthy and cultured family – his grandmother was a classical singer and a friend of Franz Liszt – Steche was born in Leipzig and graduated in medicine at the University of Freiburg, where he also obtained a doctorate in zoology (Dudek, 2013). After garnering a second doctorate in Leipzig in 1906, Steche embarked on a trip around the world that brought him around the Malay Archipelago and into contact with the flashlight fishes *Anomalops* and *Photoblepharon*. Upon his return to Leipzig he published an extensive paper on the histology and physiology of their subocular light organs (Steche, 1909). As explained in chapter 4, the luminescence of these organs is continuous, but the light blinks because in *Anomalops* the dark pigment over the organ contracts or expands, by analogy with the shark *Spinax* just

Figure 9.7 The flashlight fish *Photoblepharon palpebratus* with its conspicuous subocular light organ (top), and a section through the light organ (bottom). Plate 19 in Steche (1909).

mentioned, or an eyelid-like structure slides over the organ, as in *Photoblepharon*. In *Anomalops* Steche clocked ten seconds of light alternating with five seconds of dark and he considered that the luminescence served as a searchlight to attract prey, but also that the blinking light served to confuse prey.

Steche's histology revealed an organ packed with slim and parallel tubes spanning the depth of the organ and with blood capillaries running between them. A reflector layer covers the back of the organ. Near the surface of the organ tubes fuse with each other to create a reservoir zone, and a pore leads to the outside. By analogy with deep-sea fish light organs in which acinar tubes were assigned a glandular function, Steche saw the tubes of flashlight fishes as a highly organized glandular tissue. The thought that the luminescence might be of bacterial origin never seemed to have occurred to him. The organ is innervated, but Steche could not resolve the final destination of the nerve fibres; he assumed they reach the tubes, but strangely he failed to grasp that the pigment cells or the eyelid-like muscular structure must have nerve endings in order to effect their blinking action.

Flashlight fish material was not the only thing Steche brought back from the Malay Archipelago. His anthropological fixation on the Javanese races and half-casts planted the seed for an ingrained racism which, fuelled by personal setbacks – the loss of his first wife to Spanish influenza in 1918 and dissatisfaction with his academic career – led to his full embrace of Hitler's national-socialism (Gissing, 2003; Dudek, 2013). The Nazi party put him in charge of an educational program to promote "racial hygiene" in schools. Taken prisoner at the end of the war, he died of an infection in a military hospital.

To the Australian ichthyologist David George Stead (1877–1957) we owe the first mention of the luminescence of the knight fish *Monocentris gloriamaris* (Stead, 1906). Stead could not help remarking, almost anachronistically, on the "quaint-looking creature … clad in a strong coat of mail, formed by the large bony scales." His description of the light organ could hardly be shorter: "On each side of the head near the mouth are peculiar luminous discs, which are probably of service to the fish in assisting it to obtain its food." The Japanese species (*M. japonicus*) was examined by S. Yoshizawa (1916, cited by Okada, 1926). Both Yoshizawa and Okada noticed short acinar tubes in the light organs, and Okada assumed, as Steche before him, that they constituted a glandular tissue. It took a bacteriologist, Yoshiwo Yasaki, to demonstrate unequivocally that the luminescence of the knight fish originates in symbiotic luminous bacteria (Yasaki, 1928).

Another addition to the list of luminescent fishes was the midshipman fish, *Porichthys notatus*, a member of the toadfish family found in coastal waters of the western coast of North America. The man who brought this fish to the attention of his peers was Charles Wilson Greene (1866–1947). Born in Indiana, Greene belonged to the first cohort to graduate from the newly founded Stanford University in 1892 (*The Physiologist* 14: 1–2, 1971). After completing a master's degree at his alma mater, he obtained his PhD in physiology at Johns Hopkins University in 1898. When he published the results of his investigation (Greene, 1899), he had recently been appointed

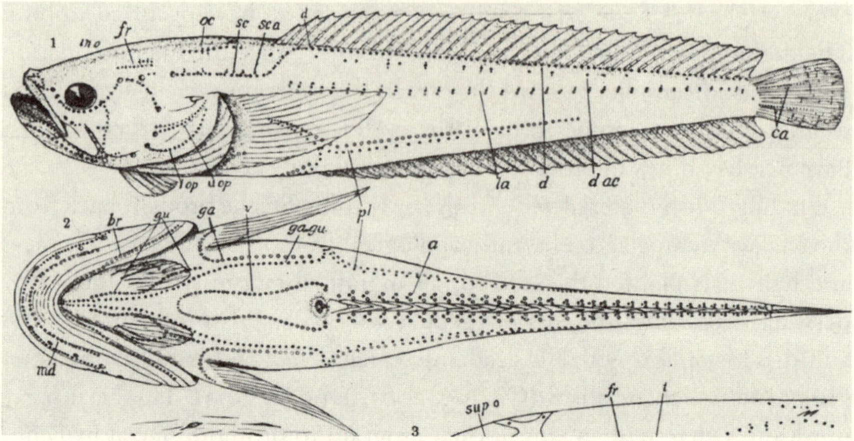

Figure 9.8 Schematic diagram of the midshipman fish *Porichthys notatus*, showing the distribution of the minute photophores from side and ventral views. Plate 38 in Greene (1899). Courtesy of John Wiley & Sons.

assistant professor at Stanford. On the skin of the midshipman fish, Greene noted, "the phosphorescent organs appear as bright silvery spots distributed in lines or rows over the surface of the body of the fish." He counted a whopping seven hundred such spots over the body.

The organs lie deep in the skin. They contain a large, pear-shaped cellular lens, and the photocytes are clumped at the distal face of the lens; behind the photocyte mass – which Greene called the gland – lies the pigment-coated reflector through which nerves and blood vessels penetrate. He also observed that the organs start to appear late in the embryonic stages of the fish. If he rubbed the belly of an already luminescing fish, the light emission became brighter. Applying electric current likely to excite the light organs directly or the local cutaneous nerves elicited a dramatic response: it "called forth a brilliant glow of light from apparently every well-developed organ in the body. All the lines on the ventral and lateral surfaces of the body glowed with a beautiful light, and continued to do so while the stimulation lasted. The single well-developed organ just back of and below the eye was especially prominent. No luminosity was observed in the region of the dorsal organs previously described as rudimentary in structure."

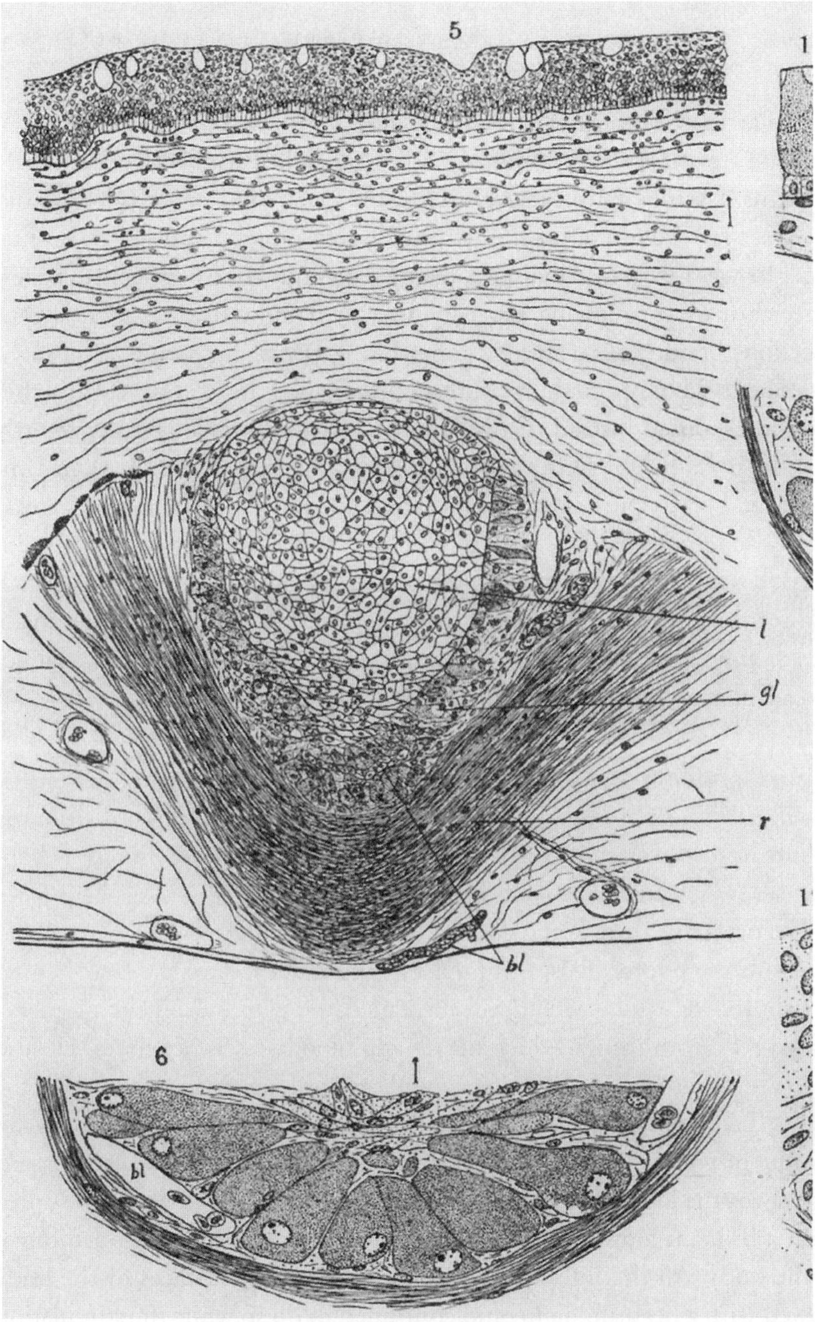

Figure 9.9 Section through a ventral photophore of the midshipman fish (top), and high magnification of the mass of luminescent gland cells (bottom). (l), lens; (gl), luminescent gland; (r), reflector; (bl), blood vessels. Plate 39 in Greene (1899). Courtesy of John Wiley & Sons.

A quarter of a century later, Greene, now a professor at the University of Missouri, revisited the midshipman fish while doing field work at Stanford's Hopkins Marine Station in Monterey. In this second paper (Greene and Greene, 1924) he re-examined the issue of physiological control. In his first paper he was lukewarm about endorsing a nervous control for the luminescence. But now, following experiments which showed that subcutaneous injections of adrenaline slowly induced long-lasting light emission from a widening field of photophores, Greene abandoned the nervous control theory and espoused instead the involvement of a hormonal control. Shortly after the publication of Greene's first paper, chemists at Johns Hopkins University had partly purified adrenaline from adrenal glands, and the pharmaceutical company Parke-Davis put it on the market in 1903 (Valenstein, 2005). The only role entertained for adrenaline at the time was hormonal, so Greene's understanding of adrenaline's effect on the photophores made sense. Only in the 1930s was it suspected that adrenaline (or noradrenaline) also acted as a neurotransmitter.

PART FOUR

THE AMERICAN ASCENDANCY

10

E. Newton Harvey and the Princeton Laboratory

> Physiology was a direct approach to the workings of the cell itself, unit of all living things; an attempt to explain functional activity of organisms in terms of chemistry and physics, what seemed to me a rational approach to every problem of biology.
> –E. Newton Harvey (quoted in Johnson, 1967)

The First World War was a watershed for American science. During the second half of the nineteenth century, American scholars had few if any domestic models to go on by in the way of promoting institutional scientific research of a quality on a par with the great European centres of learning. Their perusal of the scientific literature from across the pond made it all too clear how far behind their disciplines stood in the United States, especially the biological disciplines. The response of young investigators, either urged on by their local mentors or on their own volition, was to travel to Europe to learn how science was done by the luminaries of the great European universities and to get up-to-date on the latest scientific methods, recording instruments, and results. Closing the knowledge gap seemed the best way to cure a lingering inferiority complex. As Robert J. Frank, Jr, recorded in his historical analysis of the American cohort of visiting physiologists between 1865 and 1914 (Frank Jr, 1987), the main take-home lesson to learn was which model for running a successful and competitive research lab should best fit the American way. Budding physiologists looked at British, French, and German institutions and gradually came out in favour of the German model, on the basis of heavily subsidized universities, recruitment by merit, and the dismantling of hierarchical barriers in favour of individual ambitions and top-quality student training.

The Americans had already started to narrow the gap with Europe in the decade before the war. But the war was a catalyst, in that the drain on human and material resources during the war and the social chaos in Germany after the war crippled that country's scientific enterprise. Meanwhile America prospered and American research institutions increased in number and quality as they adopted a German-inspired yet home-grown management style. One difference with the European models was the American predilection for privately funded universities as a way of promoting the best quality and secure a competitive edge, prime examples being the motivations leading to the creation of Johns Hopkins University and the Rockefeller Institute of Medical Research (later Rockefeller University). Thus the scales were tipped in favour of the United States, and American science never looked back.

One of the beneficiaries of this emerging American hegemony was a central and pre-eminent figure in the future development of bioluminescence research: Edmund Newton Harvey (1887–1959). The two major sources on Harvey's life and career are a biographical memoir written by his student and collaborator Frank H. Johnson (Johnson, 1967), in which excerpts from unpublished autobiographical notes by Harvey are cited, and Harvey's voluminous correspondence, preserved at the American Philosophical Society in Philadelphia. The following account relies heavily on these sources.

Johnson (1967) aptly remarked that Harvey "was the acknowledged Dean of Bioluminescence; no one ever has matched, and perhaps no one ever will, his widely encompassing, scholarly contributions to our knowledge of the mystifying natural phenomenon of the emission of visible light by living organisms, which stimulated his greatest interest and research activity, his boundless energy, and his contagious enthusiasm." The man behind this aura was born in suburban Philadelphia on 25 November 1887, the son of a Presbyterian minister who died when Harvey was only six. He was raised with his three sisters by "his devoted, tolerant, and understanding mother" (Johnson, 1967). Tolerant she was of a boy whose curiosity about local living things and fixation on ingenious mechanical contrivances created havoc at home but also anticipated a passion for biology and a gift for designing instruments of experimentation. As he was growing up, his skepticism displaced the religious background of his family. On the missionary zeal of Christianity, he had this to say: "I could not see how conversion to Christianity would help the Chinese or Japanese, who already had a highly developed system of ethics, which suited very nicely the elaborate culture and dense population of these nations." His agnosticism and aversion to organized religion remained with him for the rest of his life.

Figure 10.1 E. Newton Harvey in 1928. Courtesy of Marine Biological Laboratory Archives.

In 1905 Harvey enrolled at the University of Pennsylvania, "where he distinguished himself for his versatile activities in biology and other sciences and where in his junior year his life's interest in General Physiology was kindled in a course on this subject given by Professor Ralph Lillie" (Johnson, 1967). General physiology was a young discipline founded by Jacques Loeb (1859–1924), a German-born and trained biologist who emigrated to the United States and taught at the universities of Chicago and Berkeley before founding a department of experimental biology at the recently created Rockefeller Institute of Medical Research in New York (Pauly, 1987). For Loeb "experimental biology" and "general physiology" were interchangeable terms to define the new discipline which he intended as a sharp break from medical physiology. The mission of general physiology, Loeb urged, should be to study "the phenomena of life which are common to all animals and plants," including "the energetics of life phenomena" and "the constitution of living matter" (quoted in Pauly, 1987). Ralph Lillie and Harvey's favourite teacher at the University of Pennsylvania, Edwin G. Conklin (1853–1952), were strong proponents of general physiology and they won over their pupil in this regard. Loeb created an outlet for the promotion of his pet discipline, the *Journal of General Physiology*, in which Harvey was later to publish results from some of his work. As Harvey himself admitted, "Loeb's definition of living organisms as 'chemical machines, automatically capable of maintaining and reproducing themselves' appealed greatly to me, and is still as good a definition as can be found."

After obtaining his first degree in 1909, Harvey moved to Columbia University to pursue postgraduate work in the laboratory of Thomas Hunt Morgan (1866–1945), who was in the early stages of the fruit fly genetics work that earned him the Nobel Prize. But Harvey's PhD subject was outside the realm of Morgan's research interests: the permeability of cells. Left pretty much to his own devices, Harvey successfully completed his PhD program in just two years. His old teacher Conklin had left Pennsylvania in 1908 to head the Department of Biology at Princeton University. So, shortly before being granted his PhD Harvey was invited to deliver a talk on his doctoral research at Princeton, where his reputation had preceded him and where he made a good impression. He was appointed Instructor of Biology there at the age of twenty-three and remained affiliated with Princeton from 1911 until his death in 1959.

The impression Harvey made on his peers was of "a tall, thin but athletic, quiet, dignified, slow-speaking but fast-thinking, well-poised, polished-mannered, handsome young man, young enough in fact (as he once recalled to the writer [Frank Johnson]) occasionally to be mistaken for a freshman on the campus." The quaint, well-groomed Princeton campus in the early 1910s projected an image in sharp contrast to the larger research-oriented American universities such as Harvard, Johns Hopkins, and Chicago, to name but three. Steeped in the humanities, Princeton attracted dapper young men like F. Scott Fitzgerald, who enrolled there in 1913, men who exuded style and took in its relaxed atmosphere. Harvey voiced his disappointment "that scientific studies were on a small scale in comparison to his accustomed environment." Rather than try to emulate the scientific programs of other elite universities, Woodrow Wilson, who had taken the reins of Princeton in 1902, went in the opposite direction, trying to implement a preceptorial teaching system predicated on the British "Oxbridge" model, which suited the humanities. Fortunately for Harvey, Wilson resigned the year before his own appointment. The new leadership steered a new course that pleased Harvey, who remarked in his autobiographical notes that "Princeton University gradually developed its scientific departments and now ranks with the foremost in science studies. After a few years I became so loyal a Princetonian that no offer could have enticed me away, and I came to regard the town as the ideal place to live."

We saw in chapter 8 that Raphaël Dubois got hooked on bioluminescence when he watched a stowaway insect on a French dock which turned out to be luminous. What was Harvey's trigger for devoting the larger part of his career to bioluminescence research? If Dubois's insect came to him, Harvey had to travel far to witness the bioluminescent phenomena that determined the future course of his life. Johnson (1967) writes of "the rugged life [Harvey] had spent on expeditions, his love of travel and the great outdoors." Among his early travels was an excursion in 1909 to the Tortugas in the Florida Keys to assist Alfred G. Mayer, director of the Department of Marine Biology at the Carnegie Institute of Washington, in his work on jellyfish. Harvey's first scientific publication resulted from this adventure (Harvey, 1909). Harvey repeated the experience in 1910 and 1911. Mayer was so impressed with his charge that he invited Harvey on yet another expedition in 1913, when Harvey was already well settled in Princeton. This time he travelled all the way to the Great Barrier Reef, and it was in its vicinity that something activated in him at the sight of a luminescent display. While "he does not mention any particular incident or type of organism that sparked such consuming dedication to this field; perhaps it was a spectacular 'phosphorescence' of the sea, of a kind he had observed earlier, or perhaps it was various luminescent representatives of the new marine life he encountered, and undoubtedly collected, there" (Johnson, 1967).

Prior to his first effort on bioluminescence, Harvey had published mainly on the physical-chemistry of cell membrane permeability, on artificial parthenogenesis – a favourite subject of Jacques Loeb – and on the speed of transmission of nerve impulses in a jellyfish. Now in 1913, a short note appeared on the temperature limits of bacterial bioluminescence. But a more significant first contribution came the next year on the subject of fireflies collected around Princeton (Harvey, 1914). In this pivotal article Harvey positioned himself as chiefly interested in understanding the chemical mechanism of light production. He wanted to test the oft-repeated claim that the luminous material is a fatty or fat-like substance such as albumin or lecithin. He already knew that an oxidase is involved which, as an enzyme, is destroyed by heat, but what of the other component, which is heat-resistant? He wrote:

> I can state definitely that the "luciferin" of the common fire-fly is not a true fat or any fat-like body such as lecithin. The dried material may be extracted with anhydrous ether and the ether extract evaporated to dryness. On adding water or a watery extract of luminous organ (to add an oxidizing enzyme) or potato juice (to add an oxidase) to the residue no phosphorescence took place; on adding water to the original ether extracted material brilliant phosphorescence occurred. The same results were obtained with anhydrous chloroform, ethyl alcohol, acetone and carbon tetrachloride. The material is therefore insoluble in fat solvents.

An eye-catcher in the above passage is the quoted word "luciferin." Where did Harvey pick up the word? His article cited no source whatsoever, not even Raphaël Dubois who, after all, coined the word thirty years earlier. The odds that Harvey would have come up with the word independently are rather small, so it leaves open the possibility that he wished the word to stick with him at Dubois's expense. But this theory clashes with the image Harvey projected, stating in his autobiographical notes that his mother "taught me strict honesty in all things and a particular dislike of insincerity and hypocrisy." However, there have been cases when young and ambitious investigators sometimes have resorted to short-cuts to promote themselves. No wonder, then, that Dubois was mystified by Harvey's seemingly dubious conduct, as described in chapter 8. Curiously, expanding on the results of this paper in a second firefly paper (Harvey, 1915), Harvey mentioned Dubois while avoiding the use of the "L" word!

In the 1914 paper Harvey added experiments suggesting first that the luciferin is a protein insoluble in water, but in the 1915 paper he retracted that suggestion, only to add "that it will be a vastly more difficult problem to isolate and identify the photogenic substance than might at first be supposed." Despite this setback and probably because his research on topics

Figure 10.2 Ulric Dahlgren. Acc. 90-105 – Science Service, Records, 1920s–1970s, Smithsonian Institution Archives, with permission.

other than bioluminescence was going rather well, Harvey was in 1915 promoted from instructor to assistant professor.

At this phase in Harvey's career, a shadowy Princetonian encroached on his budding image as the reigning bioluminescence specialist. Ulric Dahlgren fits the perception of the proverbial elephant in the room. In Johnson's biography of Harvey, no mention is made of Dahlgren, and Harvey's correspondence is largely silent on him. And yet, Dahlgren was a faculty member at Princeton who produced no fewer than fourteen papers on light organs and bioluminescence between 1915 and 1928. Ulric Dahlgren II (1870–1946) was born in Brooklyn, NY, the son of a naval officer, civil war veteran and mining engineer. He was named after his uncle, Colonel Ulric Dahlgren, a controversial civil war officer who was mortally wounded at the age of twenty-one during the siege of Richmond, Virginia. Dahlgren completed both his bachelor's and master's degrees at Princeton, but he never undertook a PhD program even though "he was generally addressed as Dr. Dahlgren" (Evans, 2015). Although this lack of credential would have been considered a handicap to academic advancement in other elite universities, Dahlgren was appointed instructor at Princeton in 1896 and managed to rise to full professor by 1911, the year Harvey arrived in Princeton. This was a case of academic inbreeding that Harvey would find objectionable by his lofty standards, standards that were not yet Princeton's for academic science.

It was said of Dahlgren that "early in his career he chose the field of microscopic anatomy as his special province. The striking and the unusual phenomena in nature always fascinated him and he vigorously carried on pioneer investigations on such subjects as the histology of light-producing organs and of those which produce electricity in animals" (Evans, 2015). In fact, in keeping with the potential harm of inbreeding, Dahlgren was a unidimensional researcher short on creative power, who was prisoner of his histological technique and modelled his research projects around it; quite the opposite of his colleague Harvey, who asked questions and went about finding the appropriate technical approach to answer them. On the other hand, Dahlgren was a good teacher and a skilled speaker who was adept at promoting himself and his work. He "was dapper, wore a goatee and celluloid wing collars that were in constant danger of bursting into flame, due to the close proximity of the strong cigars he incessantly smoked" (Evans, 2015).

Most of Dahlgren's bioluminescence papers appeared in the *Journal of the Franklin Institute* rather than any of the specialized journals of microscopic anatomy. The papers were devoted to bacteria (Dahlgren, 1915a), protists and Radiolaria (1915b), luminous plants [meaning fungi] (1916a), sponges and coelenterates (1916b), echinoderms (1916c), cephalopods (1916d), worms (1916e), crustaceans (1916f), insects (1917a), elaterid beetles (1917b), lampyrid fireflies (1917c), development and physiology of firefly light organs (1917d), and tunicates and fishes (1917e). Finally, a last paper appeared to show that the alluring light organ of the anglerfish *Ceratias* owed its luminescence to symbiotic bacteria (Dahlgren, 1928). Most of his papers were inordinately occupied with literature reviews, and only a few pages present original histological data with illustrations. The most striking feature was drawings by artists of organisms emitting light, inspired by Dahlgren's own visual observations. Harvey occasionally cited Dahlgren's papers, doing so very icily for a colleague of shared interests. Dubois questioned the accuracy of some of Dahlgren's statements and expressed his scorn of the man in his private papers.

If the intrusion of Dahlgren in Harvey's research territory in 1915 may have given the young man cause for umbrage, it was more than offset by a happy occasion – his wedding. In 1916 he married Ethel Nicholson Browne (1885–1965), a like-minded experimentalist who had followed in Harvey's steps and was granted a PhD at Columbia University in 1913. As a graduate student, she had conducted grafting experiments on *Hydra* that clearly demonstrated embryonic induction and the "organizer principle" (Browne, 1909). It has been persuasively argued that the 1935 Nobel Prize awarded to Hans Spemann for the discovery of the organizer, which he presented fifteen years after Browne's paper, should also have been awarded to her (Lenhoff, 1991). She moved on to her own distinguished career, becoming a prominent embryologist using the sea urchin as her experimental model. The newlyweds embarked on a honeymoon trip to Japan, where Harvey collected and studied the luminous squid *Watasenia*, and where by chance he happened on the ostracod *Cypridina* while visiting the Misaki Marine Laboratory. The latter discovery would have a major impact on his bioluminescence research program.

Perhaps the happiness brought on by his marriage softened Harvey toward greater generosity in granting priority of discovery where deserved, for in an article written from Tokyo while still on his honeymoon he at last acknowledged: "The credit of this discovery [luciferin-luciferase] belongs entirely to Professor Raphaël Dubois, of the University of Lyons" (Harvey, 1916a). In the same article and in another (Harvey, 1916b) he was able not only to confirm Dubois's findings in the click beetle, but also to generalize the luciferin-luciferase scheme to lampyrid fireflies and to luminous bacteria. However, he proved to be overly optimistic in saying "that the problem of bioluminescence has been solved at least in its broad aspects." Questions about the chemical identity of luciferin, and whether the structure of luciferin and luciferase was the same for all luminous organisms still loomed large. Harvey was clear-headed enough to fathom the challenge ahead: the accumulation of large enough quantities of material to be able to extract and purify the key components of the luminous reaction.

Harvey very soon found out how foolish his ebullient message on the prospects of decoding the chemical secret of the luminescent reaction had been. Now, only months later, he found out not only that no luciferin-luciferase could be obtained in three newly examined organisms (*Watasenia*, *Cavernularia*, and *Noctiluca*), but also that in the firefly *Luciola* and the ostracod *Cypridina* the roles were reversed: contrary to conventional wisdom,

Figure 10.3 Photograph of specimens of the ostracod *Cypridina hilgendorfii*, showing their glassy, transparent appearance and the bright fluorescence of their luminous glands. Modified from frontispiece colour photograph in *Bioluminescence in Progress* by Johnson and Haneda (1966). Courtesy of Princeton University Press.

the luciferase acted as if it was the seat of the light emission, and luciferin acted as if to facilitate the reaction (Harvey, 1916c). To reflect this bizarre phenomenon, he labelled the so-called luciferase "photogenin" and the so-called luciferin "photophelein." In so doing, he sank into deeper foolishness, as the many unknowns of his crude extracts could only lead him astray. And he managed to infuriate Dubois, who saw, erroneously, in Harvey's new terminology a dark conspiracy to upstage his original name attributions. Harvey carried on with his pseudo-substances in an expanded paper in which the effects of all manners of physical and chemical factors were tested on the Japanese species he was investigating (Harvey, 1917a).

Harvey returned to Japan in 1917, ostensibly to keep his supply of dried *Cypridina* abreast of the demands of his experimental investigations. The results of these updated *Cypridina* experiments, along with experiments on *Pholas dactylus* material shipped to him by Dubois, finally convinced Harvey of his error. Comparing the results from the two animals, he concluded:

"The difference in our [his and Dubois's] results is, therefore, not to be referred to a difference in method of experiment but to a difference in the substances themselves" (1917b). Indeed, posterity was to show that *Cypridina* luminescence arose from a classical luciferin-luciferase reaction, whereas *Pholas* luminescence stemmed from a photoprotein system of the kind discovered by one of Harvey's academic grandsons, Osamu Shimomura (see chapter 11). To make amends, Harvey restored Dubois's terminology, speaking now of *Cypridina* luciferin and luciferase. He also introduced the term oxyluciferin to designate spent luciferin in the oxidized form, which is not luminous. Harvey found ways to suck out the oxygen of the luminescent reaction product and thus convert oxyluciferin back to luciferin, ready to emit light again upon another oxidation cycle (1918). In his next *Cypridina* paper he went through an exhaustive series of chemical experiments to narrow down the identification of the luciferin and luciferase compounds, but little progress was registered (Harvey, 1919a).

Unable to purify the compounds to his satisfaction, which is an *a priori* for tackling the mechanism of light production, Harvey temporized by testing if carbon dioxide is produced during the luciferin-luciferase reaction (1919b). He found that if carbon dioxide is at all produced, it is much less than its production during respiration. Reasoning that significant heat is always associated with carbon dioxide production, he devised an ingenious and complex apparatus to determine with high sensitivity if any heat is released by the luciferin-luciferase reaction (Harvey, 1919c). No measurable heat was recorded. "The reaction, luciferin → oxyluciferin," Harvey surmised, "therefore involves a relatively slight energy change and should be readily reversible" (1920a). He found that a vast array of reducing agents cause a reversal of the reaction, oxyluciferin → luciferin, by bringing the pH of the solution from the alkaline 9 to near neutral (7.1). Projecting his *in vitro* results to the light organs, Harvey saw the light-on period as a reaction in aerobic conditions, when oxygen is admitted to the luciferin, and the light-off or rest period, as a reductive reaction in anaerobic conditions, when oxygen is removed from luciferin. "What more efficient type of light than this is to be desired," he asked (1920a).

After Harvey was promoted to full professor of physiology in 1919, he marked a pause to take stock of the field and write his first book on bioluminescence, entitled *The Nature of Animal Light* (1920b). The book was commissioned for the *Monographs on Experimental Biology* series edited by a committee headed by Jacques Loeb. Harvey, we saw earlier, was a devotee of the physico-chemical approach to physiology heralded by Loeb, so it is in keeping with this ideology that his book ended up as a monograph in the series. The small book skims through a few aspects of the subject: the identification of luminous animals, their distribution in the animal kingdom, the physics of light emission, the categories of luminescence, and the structure of light organs. Not surprisingly, coverage of the chemistry of light production was favoured. Harvey sounded unduly triumphant when he claimed: "Great advances have been made since the first early guesses that the light was due to phosphorus and was a kind of oxidation. Although the problem cannot be considered as solved, it has been placed on a sound physico-chemical basis. Some material is oxidized. Exactly what this material is and why light accompanies its oxidation are the two more fundamental problems in the field of Bioluminescence" (1920b).

Travelling to Friday Harbor in Washington State, Harvey turned his attention to cnidarians (1921a). He focused on the small hydrozoan jellyfish *Aequorea*, which thrives there. He traced the luminous source to a particulate fraction, but was unable to obtain a luciferin-luciferase reaction. This first foray in Friday Harbor and this jellyfish result paved the way for momentous investigations more than forty years later (a narrative thread that will be pursued later in this book). Harvey also investigated the inhibitory effect of light on ctenophore luminescence and perceptively foresaw that sunlight "must therefore act to prevent the formation of photogenic substance rather than to prevent its oxidation on stimulation. Why sunlight causes the disappearance of photogenic material already formed is a question awaiting solution."

Harvey set no limit on travelling distances in his unrelenting drive to unravel the mystifying mechanisms of light production. Flashlight fishes fascinated him, so he travelled to the Banda Islands in the Dutch Indies "to obtain all the material needed for my investigation without encountering any more difficulty than the general indisposition of the fishermen to work beyond their usual amount, or the advent of rough weather" (1922a). In his study, of which he published a preliminary note the year before (1921b), Harvey asserted that Otto Steche was wrong to assign a secretory function to the flashlight fish light organs: instead, he substituted "the rather startling conclusion to which I have come from chemical investigation, namely, that the light-organ is a mass of tissue designed for the nourishment of luminous bacteria, which produce the light." By squashing the subocular organs on histological slides he was able to obtain a luminous emulsion in which he detected rod-like organisms behaving like bacteria. He could not, however, culture them. He tried unsuccessfully to cross-react the bacterial luciferase extract with *Cypridina* luciferin, or the bacterial luciferin extract with *Cypridina* luciferase; he took the lack of cross-reaction to mean that they do not have a luciferin-luciferase system, on the basis of the dubious assumption that luciferins and luciferases should be similar enough in properties to allow cross-reactions.

The quandary of the cross-reactions induced Harvey to take a hard look at the problem of the specificity of luciferins and luciferases (1922b). He recognized that several causes could be behind the lack of luciferin-luciferase reactions in many species, not the least being the infinitesimal amounts of extractable luciferins, in some cases too small to detect any light. But his broad survey of reactions under optimal conditions among many of the known luminous forms led him to conclude that the *Cypridina* luciferin and luciferase are specific only to ostracods. One cause of the apparent absence of luciferin-luciferase reactions in other experimental models was one that he had no means of grasping at the time: the involvement of photoprotein systems, discussed in the next chapter.

~

By the 1920s Harvey had made his Princeton Physiological Laboratory the most advanced centre of bioluminescence research in the world. Not only was he prolific in his production of highly original work, but he also wrote at intervals authoritative reviews on the progress of the field. One colleague whose work he reviewed in 1924 was Sakyo Kanda (1874–1939), considered one of the pioneers of bioluminescence research in Japan. Born to a long line of Shinto priests but later converted to Christianity (Kanda, 1910), Kanda travelled to the United States in 1907 to study for a master's degree under G. Stanley Hall (1846–1924), the pioneer of American psychology. Kanda was one of the happy few who attended the psychology conference in 1909 at Clark University (Worcester, Massachusetts), where Hall held a professorship and where Sigmund Freud was an invited lecturer (Takasuna, 2005). That same year Kanda presented his master's thesis dealing with religious psychology, and he moved on to the University of Minnesota, where he earned a PhD in zoology in 1915 at the age of forty-one. His thesis on geotropism in snails was influenced by Jacques Loeb's views as much as Harvey's work was.

After his return to Japan in 1915, Kanda "worked as a temporary researcher at various universities. However, it seemed that his personality and attitude were not a good fit for public employment at imperial universities, so he decided to live as an independent scholar in Tokyo" (Takasuna, 2005). For his bioluminescence work on *Cypridina* and fireflies he held a loose affiliation with the Marine Biological Laboratory of Kyushu Imperial University, but his research was financially supported by a wealthy individual named Jihashi Hamano. Kanda, who wrote fluent English, published his papers in the *American Journal of Physiology*. There is no evidence that Harvey had a hand in getting Kanda's work published in this journal. Harvey was a member of the American Physiological Society and he himself published in the journal, but he was never a member of the Editorial Board. However, he could have been called upon to act as referee for Kanda's submitted manuscripts, which revisited the issue of the *Cypridina* chemiluminescent system and sought to settle who of Harvey or Dubois was right about the luminescent mechanism.

Kanda was aware that Harvey's *Cypridina* results conflicted with Dubois's theories based on elaterid beetles and the bivalve *Pholas*, at least in print. But he was in the dark about the private exchanges between Dubois and Harvey. Among Harvey's papers preserved by the American Philosophical Society are letters exchanged with Dubois in the years 1917–19. From the beginning of their correspondence, as seen in a letter dated 14 December 1917, Dubois summed up his quarrel with Harvey by pointing out that they were in agreement about the facts, but differed in their interpretation. In his letter to Harvey of 23 April 1918, Dubois accused his American colleague of confusing the luminous reaction with the trappings of the colour of luminescence due to fluorescent substances in the light organs when Harvey substituted "photophelein" for luciferin, and of using this "confusion" to dismiss his theory. He warned Harvey that "there is no question to replace the *theory of Professeur Dubois* [Dubois's emphasis] with another, which would be very improper and perhaps judged severely." Despite the stern tone of the message, Dubois hoped Harvey and he could reconcile their differences, adding: "It is imperative, in the interest of scientific truth, that we agree on all the points, and that instead of being adversaries we should be loyal allies like your compatriots who currently fight [the First World War] with mine."

In the same letter Dubois expressed his annoyance that Harvey had circumvented him and contacted the Principality of Monaco directly either to visit or to obtain *Pholas* material from them. The war prevented that from happening, but Dubois wrote that he "would have happily placed [*Pholas* extracts] at your disposal in Tamaris, if you would have made your wish known to me, as it would have procured me the opportunity to get better acquainted with you." In the end, the two never met. Harvey stoically bore the brunt of Dubois's frontal attacks. His papers show how diligently Harvey had Dubois's letters translated and how he sounded Dubois out for technical clarification, the latter responding dutifully with his extensive answers as the year 1919 wore on. If they communicated with each other after 1919, no document has surfaced to indicate it.

In his first paper, Kanda stated that he had "attempted to test Harvey's conclusions with *Cypridina hilgendorfii* and [had] found many experimental facts essentially contradictory to those of Harvey and rather in accordance with those of Dubois (Kanda, 1920a)." He attempted to right many perceived misconceptions entertained by Harvey about *Cypridina* and claimed to make several improvements in the chemical procedures to obtain the luciferin and luciferase fractions, starting with the preparation of dried material to increase yield. In so doing he was able to discard Harvey's "phraseology of the light-producing substances" and restore that of Dubois, even to the point of labelling it in French: luciférine and luciférase, in so doing adding salt to Harvey's wound. The false note was Kanda's insistence that no oxidation of luciferin took place in the luminous reaction (Kanda, 1920a,b) and that luciferin is protein-like (1920a,c; 1924).

It was extremely rare that Harvey was served a public dressing down such as Kanda's. In his 1924b review paper, Harvey, did not take it lying down. Although he had already abandoned his "phraseology," Harvey criticized Kanda's methods, which led him to flawed conclusions about luciferin's oxidation and protein nature, and about the chemical fingerprint of luciferase

(1924b). His review also showed how Harvey's conjecture of the steps of the chemiluminescent reaction evolved away from Dubois's. Dubois had envisaged the following sequential steps: co-luciferase + preluciferin = luciferin; luciferase + luciferin = oxyluciferin; and finally oxyluciferin + oxygen = light. Harvey, on the other hand, was now advocating the following sequence: preluciferin → luciferin; luciferin + luciferase ↔ luciferase-luciferin; luciferase-luciferin ↔ oxyluciferin + H_2 + luciferase (with luminescence); and finally $H_2 + O = H_2O$. The innovation highlights here are the reversible reactions and the formation of a luciferin-luciferase complex as part of the process.

In the same year Harvey – forever curiosity-driven – addressed a perplexing phenomenon: the colour differences of luminescence between closely related species (1924a). Some colours are determined by coloured filters in the light organ; the luminous cells emit a light of a certain colour, but the colour is shifted by a filter before radiating out of the light organ. However, this was not what Harvey was after. What, he asked, determines the differences in colour of the luminous material inside the photocytes where no filter is involved? He examined two firefly species and two ostracod species, each pair having one member that emits a bluish light and the other a yellowish light. By cross-reacting the luciferin and luciferase of the two members of the pair, he should be able to verify which reactant was responsible. Anticipating that the colour depends on the product being oxidized, luciferin, he was very surprised to find that it was the luciferase that determined the colour. He concluded that it was not what was oxidized but how it was oxidized (by the luciferase) that determined the colour expression.

Harvey's publications of 1925 reflected his long-standing fascination with the inhibition of bioluminescence by ambient light. He revisited the photoinhibition of ctenophore luminescence, using *Mnemiopsis leidyi* as his experimental model (1925a). This species is readily available in Woods Hole, Massachusetts, in late August and September; one can walk to the wharf across from the Marine Biological Laboratory (MBL) and effortlessly scoop them off the surface waters with a fine-mesh net. Already in 1909 young Harvey had spent his first visit at MBL continuing the research he had undertaken with Alfred Mayer in the Tortugas (Johnson, 1967). Located in Cape Cod, the MBL added a pleasant vacationing atmosphere to the buzz of the research community assembling there every summer. As Johnson explained: "Harvey met most of the great biologists of the time from the eastern United States. In later years he built a cottage near the laboratory and ultimately acquired a delightful summer home on Penzance. From 1909 through the rest of his life he spent a part of every summer in Woods Hole."

Two questions occupied Harvey's mind about photoinhibition in *Mnemiopsis*. At which functional level does the photoinhibition operate, and is oxidation involved in the photoinhibition process? His experiments clearly pinpointed the luminous substances inside the photocytes as the target of the inhibition, but an additional inhibition of the local nervous control could not be ruled out. The latter possibility was based on fresh observations of *Mnemiopsis* luminescence by Arthur R. Moore, then professor of physiology at Rutgers College in New Jersey (Moore, 1924). To the question of oxidation involvement Harvey answered in the negative. In addition, Harvey found that irradiating the ctenophore with near ultra-violet light induced what he called a blue tonic luminescence, which he suspected of being a fluorescence emitted by the oxidation product of the luminous material. This finding prefigured many future reports pointing to the fluorescent properties of luminous cells and to the utility of fluorescence microscopy for detection and characterization of luminous cells.

Harvey also showed that the luminescent secretion of *Cypridina* is inhibited by light, and in this case he demonstrated by clever experiments that luciferin, not luciferase, was the target of the inhibition (1925b). Years later he determined that *Mnemiopsis* luminescence "occurs in complete absence of dissolved oxygen," but also that, for recovery after photoinhibition, luminescence "again returns in the dark in presence of oxygen but not in its absence" (Harvey and Korr, 1938). Harvey concluded that in its luminous form oxygen is firmly bound to the photogenic material – later found to be a photoprotein – and that it takes the inhibitory effect of light to dislodge the oxygen.

Harvey's next focus was of a quantitative nature. He wanted to know how many quanta of light are emitted in relation to the oxygen consumption of the luminous reaction. He estimated that in *Cypridina* a minimum of 50 molecules of oxygen and 100 molecules of luciferin were used up per quantum of light (Harvey, 1927). Harvey (1928) also made measurement that gave him an estimate of the infinitesimal oxygen consumption of single luminous bacteria. These were diversions from the central issue of the chemical mechanism of light production which consumed him, but whose resolution by 1927 was not progressing to his satisfaction. In that same year he struck a friendship with a kindred spirit that diverted him even further from his bioluminescence enterprise. This new friendship with Alfred Loomis was to have a significant impact on Harvey's research career.

Alfred Lee Loomis (1887–1975) was born in New York City, the son of a prominent academic physician. He studied at Yale and at Harvard Law School. He started his career as a corporate lawyer, while at the same time displaying talents for gadgetry and invention, thanks to "a wide-ranging mind and the ability to 'learn all about' a completely new field in a remarkably short time through independent reading" (Alvarez, 1980). The "Great War" gave him the opportunity to exercise these talents in the service of ballistic research. After the war he went into investment banking with his brother-in-law and made a fortune. But, as Luis Alvarez (1980) explained, "Alfred was leading a double life; his days were spent on Wall Street, but his evenings and weekends were devoted to his hilltop laboratory in the huge stone castle in Tuxedo Park." Harvey first met Loomis shortly after the latter had set up the lab in his residential estate. Loomis's lab ran on an alternative basis to university-based research, which was the accepted and dominant model even then. In fact, it was said that the Tuxedo Park Lab was the last relic of the "home model" of scientific enterprise (Alvarez, 1980), the epitome of which was Darwin's Down House.

The "wizard of Tuxedo Park" co-opted mostly physicists to work in his lab, so Harvey, the physiologist, was a rare addition to the roster. Harvey was then involved in two projects: ultrasounds and the centrifuge microscope. Both involved the design and construction of innovative equipment of great potential benefit for physiology and biochemistry. Harvey and Loomis (1928) originally intended ultrasound generators to experimentally affect intracellular organelles without destroying the cells, but their emphasis soon became cell destruction (Harvey and Loomis, 1929) and the doors it opened to obtain cell-free extracts for biochemical analysis in a cleaner way than by mechanical grinding. The benefit to Harvey's bioluminescence enterprise was not lost on him. The development of the centrifuge microscope was hailed as a great technological achievement that earned Loomis and Harvey prizes. As Johnson (1967) explained, "By means of this instrument it was possible to watch and photograph cells, under high magnification, as they revolved at 10,000 revolutions per minute, subjected to strong centrifugal force." The instrument helped Harvey's wife, Ethel, to produce sea urchin half eggs devoid of hereditary material but still capable of developing into embryos, a phenomenon known as artificial parthenogenetic merogony.

Another technological development was produced by Harvey in his own Princeton lab with the help of his student Peter A. Snell. So far, light emissions had only been visually observed, with the lack of objectivity and the imprecision that method entailed. Already in 1928 Harvey and his student Kenneth P. Stevens had succeeded in measuring the brightness of the light emission of the click beetle with a "Macbeth illuminometer" (Harvey and Stevens, 1928), a rather primitive photometer. But now, by developing with Snell a photoelectric cell coupled with a string galvanometer, he was able to record the dynamics of light flashes as short-lasting as those of the firefly with great precision (Harvey and Snell, 1931; Snell, 1931, 1932). The design of the photoelectric cell showed some ingenuity, as it borrowed from the nascent television industry. It used a neon-like tube developed by General Electric and its signal went through an amplifier before hitting the string galvanometer. The latter, originally designed at the beginning of the twentieth century to record fast cardiac pulses, was ideally suited to render graphically fast flashes of light.

Thanks to this groundbreaking apparatus, Harvey proudly included samples of his recordings in his paper. Recordings of the *Cypridina* luciferin-luciferase reaction show a fast rise of the light emission (one tenth of a second), followed by a slower decay (over one second) developing like an exponential curve (Harvey and Snell, 1931). Snell applied the apparatus to the study of the firefly flashes (1931, 1932). His recordings showed that the firefly flash intensity varies greatly but the flash duration is constant at around 140 milliseconds. This suggested to Snell that a variable number of subunits in the light organ are recruited in any flash production. He also found that at a critically low level of oxygen tension flashes are replaced by a continuous glow, and in these conditions a "pseudoflash" is obtained by slightly increasing the oxygen tension again. He concluded that the nervous system controls the access of oxygen to the photocytes through the tracheal end-cell, which acts as the oxygen gateway.

By comparing the physiological results of his studies on various luminous organisms, including the luminescence recordings, Harvey arrived at his first comprehensive views on the evolution of bioluminescence (1932): "In the lowest and simplest luminous forms, the luminous bacteria, I believe the luminous mechanism is a part of the respiratory mechanism and that the development of luminosity in the animal kingdom represents a specialization of one phase of the respiratory process, namely the development of special hydrogen acceptors." From the core chemiluminescent mechanism, Harvey moved to the organismic level and his thoughts drifted to the recognition that bioluminescence may have evolved independently many times:

> There has never been any great evolution in the direction of an entirely luminous group. Perhaps the coelenterates most closely approach this, for luminescence is more widespread there than in any other phylum. Generally, luminescence has appeared sporadically in the living world, with a few luminous species scattered here and there among structurally very close non-luminous relatives. This again means that luminescence can be readily developed in the course of evolution by some slight change in a mechanism already existing within all cells. I believe this mechanism is the respiratory one and that luminescence has resulted from the transformation of some of the hydrogen acceptors of the cell, together with the development of proteins with very actively fluorescent groups, which are excited to luminescence by the energy of the oxidative dehydrogenation.

In 1929, the year his predecessor at the top of the heap, Raphaël Dubois, died, Harvey was elected trustee of the MBL at Woods Hole. In 1930 he joined the editorial board of several scientific journals, and the following year he was named managing editor of a new periodical, the *Journal of Cellular and Comparative Physiology*, which catered to his kind of physiological approach and in which he and his students published many papers. In 1933 he became the Henry Fairfield Osborn Professor of Biology, filling the chair

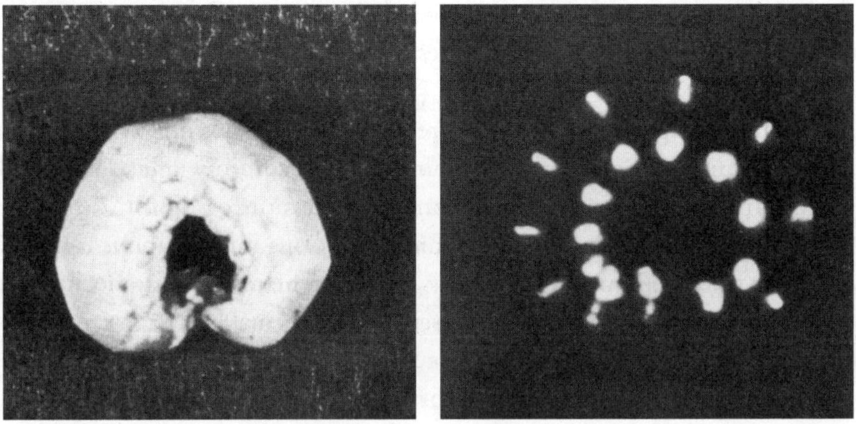

Figure 10.4 A close relative of the railroad worm. A female beetle (*Phengodes*) in a coiled posture, photographed in daytime (left) and by its own light (right). Fig. 49 in Harvey (1940). Courtesy of Princeton University Press.

of his old mentor E.G. Conklin, who was retiring. These distinctions only confirmed the prominent status Harvey had reached in American biology. But now that he had reached his zenith, Harvey reduced his hands-on lab research to a trickle and let his graduate students take over the conduct of the research projects. His own output consisted mainly of review papers and science popularization.

Harvey's second book on bioluminescence (1940) showed characteristics of both literature review and popularization. In *Living Light* Harvey produced an attractive book that stood out for its numerous illustrations and exhaustive literature review. The first chapter, an introduction, was designed to attract a general readership. The second offered a survey of luminous organisms; and the third, which seemed by virtue of its length out of place in a book on living light, covered all types of luminescence other than bioluminescence. The fourth, largely devoted to the work of Harvey's laboratory, discussed the chemistry of light production. The fifth reviewed knowledge related to the physiology of luminous cells or organs in an attempt to understand control mechanisms. And the final chapter was devoted to the physical aspects of the light emission: intensity, colour (spectral distribution), efficiency (light output versus energy investment, for instance). The book well served its purpose of posting where the field stood in 1940 while drawing attention to the hegemony of the Princeton lab in matters of bioluminescence research. Harvey followed it up the following year with a literature review of recent works (1941).

During the American involvement in the Second World War, all ongoing projects on bioluminescence were halted as Harvey and some of his students turned to medically oriented war projects through the Committee on Medical Research of the Office of Scientific Research and Development (Johnson, 1967). However, the tireless American managed to slip in a paper on the railroad worm (*Phryxothrix*) in 1944. The specimens of this spectacular insect were shipped to Princeton from Uruguay. He apparently needed help to locate the light organs, for in a letter dated 17 June 1946 John Buck, already by then an authority on luminous insects, advised Harvey by sending him "a sketch of the anatomy of that region, indicating the approximate position of the little structure I found … If with a binocular you can localize the region more exactly, and indicate the spot on my sketch, it will be a very great help in making certain which is the proper structure." In the paper (Harvey, 1944), looking for the source of the two-colour luminescent display, he described the display thus:

> They showed no light when at rest but if disturbed very slightly by knocking the table gently, or blowing air over them, they responded by shining the red light. When the disturbance was greater the rows of greenish yellow lights appeared and the animal explored its environment with a brilliant display of pyrotechnics. The red light in the head resembled the tip of a glowing cigarette. Sometimes all and sometimes only certain segments with greenish lights would be turned on. Later the greenish lights went out while the red remained on for some time, finally to disappear as the animal became quiet again.

By analogy with a deep-sea squid whose red luminescence Carl Chun believed was due to a red filter, Harvey asked if something of the kind could explain the railroad worm's red emission. Harvey determined that neither the red nor the green emissions were the result of interfering filters, but represented the *bona fide* light of the photocyte's chemiluminescent system.

Harvey's large body of groundbreaking bioluminescence work was specifically rewarded in 1947 with the Rumford Medal of the American Academy of Arts and Sciences. Luminaries such as Thomas Edison and Enrico Fermi kept him company in the roster of prize recipients. Keeping an arm's length from the day-to-day activities of his laboratory, and with reduced teaching loads, Harvey now concentrated more than ever on writing and reflecting. According to his correspondence, he embarked in 1949 on an ambitious book project that led to the publication of his magnum opus, *Bioluminescence*, in 1952. There is evidence that Harvey prepared his documentation very carefully and scrupulously before the writing. He wrote to many colleagues for reprints of their papers he was missing, for permission to use their illustrations, and for specific calls on their expertise. For example, on 16 December 1949 he wrote the Office of Technical Services of the US Department of Commerce for access to Japanese papers on bioluminescence in the wake of the postwar American occupation of Japan. On 6 March 1950 he asked Charles Bonhomme, a recent doctoral laureate from Montpellier in France, for his papers and information on annelid worms. On 25 April 1950 he sent enquiries to Arthur R. Moore at Rutgers University about comb-jelly and jellyfish bioluminescence, especially the role of the nervous system. And on 11 August 1950 he asked Grace E. Pickford of Yale University for a reprint of her monograph on the vampire squid and information on earthworm bioluminescence.

The book's publication in 1952 was greeted with glowing reviews. Harvey's previous books on the subject paled by comparison, none reaching the thoroughness, breath of vision, quality and effectiveness of illustrations, and the full bibliography of what to this day is considered a classic. No other scholarly book on bioluminescence after this achievement has reached these heights or was produced by a single author.

The last significant brainchild of Harvey was a thoughtful musing on the evolution of bioluminescence and of luminous organisms (Harvey, 1956). Harvey did not rate this intellectual task any easier for him than it was long ago for Darwin who, referring to light organs and electric organs in *The Origin of Species*, wrote, "it is impossible to conceive how these wondrous organs have been produced." The root of the problem as Harvey saw it was insufficient scientific data on the role of bioluminescence to the organisms. But even equipped with such data: "there are two problems in the evolution of any unusual ability of animals, (1) the first beginnings, and (2) the subsequent complications which lead to a perfected organ … just as the simple material used by bacteria to luminesce is far removed from the rows of lanterns of a deep-sea fish" (Harvey, 1956). Bacterial mutant strains in which luminescence is no longer expressed or is brighter were known to exist at the time of Harvey's writing, so naturally he regarded mutations as a major potential factor driving the evolution of bioluminescent bacteria or any other single-cell organisms, "but," he said, "it must be admitted that the factors involved in the change of a luminous spot to a complicated lantern are far from clear."

In complex organisms Harvey identified control mechanisms of light organs as important evolutionary adaptations. He cited the case of the "cheek" of subocular light organs of fish. Already in 1931 he had noticed that the cheek organ of the deep-sea fish *Echiostoma* was activated differently than the trunk light organs:

> This cheek organ in *Echiostoma ctenobarba* flashed rhythmically with a bluish luminescence when the freshly caught fish, kept in ice-cold sea water, was handled, while the principal lateral line of light-organs and the numerous minute light-organs embedded in the skin did not light up until adrenaline was injected into the fish. The light of the latter was yellowish, lasted some time, and then went out, while the cheek-organ continued to flash rhythmically during and after the adrenaline injection.

To this case of intraspecific difference of, presumably, nervous control mechanism, Harvey opposed the cheek organs of the flashlight fishes *Photoblepharon* and *Anomalops*, both, unlike *Echiostoma*, involving symbiotic bacteria, but whose shutter mechanisms of the continually emitted bacterial light are dramatically different: "*Photoblepharon* has developed a fold of black skin, like an eyelid, which can be drawn over the organ, obscuring the light. In *Anomalops* the light-organ is hinged at its anterior end and can be turned inward toward the body, thus presenting the black pigmented opaque side outward. Why two genera of closely allied fishes, belonging to the same family, should have developed such diverse methods of controlling the light is one of the mysteries of evolution."

Irrespective of the habitat and the use made of the light organs, Harvey concluded, "The ability to luminesce can be considered an adaptation for living in the dark – in lampyrids a very beautiful adaptation for bringing the sexes together. Among deep-sea animals it would seem as if the complicated lantern is almost too perfect an adaptation. Something less complex would do just as well, if the true purpose of the light is for recognition. A row of luminous spots would function in the same way as a row of lanterns." But, of course, few alternative adaptations were considered in his era, least of which the camouflage of the body silhouette to stay out of sight of predators.

Harvey's last mammoth project (1957) was a compendium of the history of research on all forms of luminescence from Antiquity to the end of the nineteenth century, a work on which we had occasion to comment in previous chapters. As a reviewer aptly remarked, the book "marks one of the very rare occasions when a single individual has been able to comprehend the whole of an important and active field of scientific endeavour and present it in a definitive way. It is a tribute to 'Mr. Bioluminescence' that he has accomplished this task of a lifetime while maintaining broad biological interests in both teaching and research" (Buck, 1958).

Harvey retired in 1956, at which time he recollected his career at Princeton, pointing out, "I have seen Princeton pass from the reputation of a nice country club to one of the highly regarded centers of learning in the U.S. and I am proud to have played even a small part in the change" (Johnson, 1967). Death struck him suddenly on 21 July 1959 in his Woods Hole summer residence. He was seventy-one and the cause of death was heart failure. The high standing of his Princeton laboratory did not die with him, nor did the impetus he gave to bioluminescence research in the United States.

11

The Triumph of the Biochemists

> Harvey died unexpectedly in 1959, but because of his influence the faculty and students around him were inspired to continue the study of bioluminescence and many subsequently went on to make a career in bioluminescence research.
> –Frederick I. Tsuji (2010)

E. Newton Harvey had trained a number of graduate students since the 1920s to assist him in advancing the field of bioluminescence. The majority of their theses were directed to physico-chemical aspects of bioluminescence, the pet subject of their mentor. In the latter half of his career Harvey became frustrated by the slow pace of progress in identifying the chemical structure of luciferins and luciferases. To accelerate that pace, he invested major physical and human resources in his laboratory, but breakthroughs in this regard came late in his career. His students and associates could claim first credit in making this happen, and their drive to lay a firm biochemical foundation for bioluminescence continued after Harvey's death and met with spectacular success.

To put the American situation in context, it is appropriate here to map the development of biochemistry in Europe and North America. The differences between them arose from the need to separate biochemistry from physiology or organic chemistry as an academic discipline in its own right. In Europe departments of physiological chemistry (or biochemistry) were not instituted before the early 1920s, and in four German universities only: Frankfurt, Freiburg, Tubingen, and Leipzig. Otherwise it was considered a subdiscipline of physiology – or, in some cases, organic chemistry – and was taught in institutes of physiology (Krebs and Lipmann, 1988). The word biochemistry (*Biochemie* in German) was coined in 1877 by Felix Hoppe-Seyler (1825–1895), a professor at the University of Strasbourg, on the occasion of the publication of the first issue of the *Zeitschrift für physiologische*

Chemie, a journal which, as a pioneer of biochemistry, he founded. Hoppe-Seyler's appeal to establish biochemical institutes separate from physiology departments was opposed by eminent physiologists such as Eduard Pflüger. This resistance was felt in Great Britain as well as in Germany (Krebs and Lipmann, 1988).

The debate simmered also in the United States, but it was framed around medical schools. As stated by Robert Kohler (1982): "The real issue for most medical schools was where physiological chemistry belonged among the medical sciences: in independent departments, in departments of physiology, or in departments of chemistry? University leaders were not immune to the belief that German ways were best, even if they did not fit American practices." Kohler further explained the different circumstances in America that determined the leadership of American universities to establish separate biochemical programs:

> Ironically, it was the backwardness of American medical colleges in the biomedical sciences that made possible the sudden success of biological chemistry in the 1900s. When medical chemists disappeared, biochemists took their places. The independent department was the result of the reform impulse acting upon an established, independent institution. Biochemistry did not evolve gradually out of physiology, as in Britain; it was not split between organic chemistry and physiology, as in Germany. In the United States, biological chemistry emerged like a butterfly from the cocoon of medical chemistry.

Between 1905 and 1910 the first departments of physiological chemistry took root in US medical schools. The trend soon followed across the border: in 1907 at the University of Toronto and in 1920 at McGill University in Montreal. But in Princeton, where there was – and still is – no medical school, biochemistry as a separate discipline was inaugurated only in 1961, a mere two years after Harvey's death. According to the first chair of biochemistry at Princeton, Arthur L. Pardee (quoted in Leitch, 2015), "Courses in this field had previously been given by the Biology Department for many years. In fact, in the 1920s E. Newton Harvey gave one of the earliest undergraduate biochemistry courses in the country." Pardee further explained that the "need for a more chemically oriented sort of biology became apparent to the University during the 1950s, as the spectacular successes of biochemistry and molecular biology in explaining life's mysteries achieved general acknowledgment."

The long tradition of teaching biochemistry and conducting bioluminescence research from a "physiological chemistry" perspective was so strongly identified with Harvey and the Biology Department that Harvey's successor in Princeton, Frank H. Johnson, shunned the newly created Biochemical Sciences department and remained in the Biology Department. Johnson is only one of Harvey's students who made their mark in the field of bioluminescence and enjoyed successful and in some cases brilliant careers. This chapter focuses primarily on the story of three lineages of Princeton investigators who endeavoured to solve the chemical structures and properties of the *Cypridina*, bacterial/firefly, and jellyfish luminescent systems.

The previous chapter chronicled Harvey's early work on *Cypridina* but glossed over its setbacks. Harvey, who year after year was receiving shipments of dried *Cypridina* material from Japan for his studies, was counting on the ostracod to unlock the mystery of the chemical structure of luciferins, boasting that "the Cypridina material has turned out to be by far the best for biochemical investigation of luminescence" (quoted in Johnson, 1967). One reason for Harvey's optimism was the simplicity of the luciferin-luciferase reaction, requiring only oxygen and uncluttered by co-factor requirements as in other luminescent systems. But Harvey soon grew disenchanted, as Johnson explains:

> From a chemical point of view, however, this material turned out to present some frustrating difficulties, e.g., the luciferin is unstable in solution with oxygen, and it occurs in infinitesimal amounts, recently estimated as averaging about one microgram per organism at best, along with numerous unknown substances inert in luminescence but having similar gross properties of solubility and the like. Thus despite the assistance of able bio- and organic chemists who at one time or another were enlisted as Research Associates (among others, Rupert Anderson,

Aurin Chase, Howard Mason, and Fred Tsuji), efforts to obtain pure luciferin, necessary to learn the chemical structure, were unsuccessful.

Rupert S. Anderson, of whose life and career no trace has been found, pioneered instruments to measure the total light output of luminescent reactions (Anderson, 1933) and improved the purity of *Cypridina* luciferin by his elaborate chemical extraction method (Anderson, 1935); but he fell short of reaching the purity necessary for identification studies. The yellow colour of his semi-purified extract proved to be a trademark of luciferin, and Anderson ventured to suggest that luciferin shared properties with polyhydroxybenzene derivatives.

Anderson's successor in Harvey's lab was Aurin M. Chase. Born in Syracuse, NY, around 1904–05, he was the son of the founder of the Chase Motor Truck Company, which built utility trucks (for farmers mainly) and cars between 1907 and 1919. Chase completed his PhD in biophysics at Columbia University in 1936, and joined Harvey's lab in the late 1930s. For his investigations of *Cypridina* luciferin, Chase enlisted the help of the spectrophotometer, an instrument that measures the absorption spectrum of substances in solution, an expertise that harked back to his graduate years at Columbia. The spectrum of absorbed light wavelengths constitutes the signature of a substance and helps monitor the degree of progress in the purification process. But nagging impurities in the luciferin extracts held him back (Chase, 1948). Chase rose through the ranks of assistant, associate, and full professor in Princeton.

Howard S. Mason (1914–2003) was granted his doctorate at the Massachusetts Institute of Technology in 1939. At Princeton around 1950 he took over the thankless task of prodding the *Cypridina* luciferin into revealing its secret. His stated objective was "to determine the homogeneity of purified *Cypridina* luciferin and to test the suggested structures by means of partition chromatography and its auxiliary techniques" (Mason, 1952). After Chase's spectrophotometer, chromatography was the second modern technology focused on the problem. Mason separated two forms of luciferin, alpha and beta, on a chromatography column, and eluted them as bands on chromatography paper. He could not determine the homogeneity of his luciferin product, but his analysis dismissed all structural models of luciferin suggested to that point. The mysterious structure had eluded him, as it had his predecessors. After Princeton, Mason spent his career at the Oregon Health & Science University. His claim to fame was his co-discovery in 1955 of oxygenases as a class of enzymes, to which luciferases belong.

The fourth investigator called to tackle the problem of *Cypridina* luciferin was Frederick Ishiro Tsuji (1923–2016). Born in Honolulu to parents of Japanese ancestry, Tsuji pursued all his university education on the continent, at Cornell. In his memoirs, he explains how he came to work with Harvey (Tsuji, 2010):

> Following the end of World War II, I was able to resume my education, receiving a PhD degree in biochemistry from Cornell University in 1950. As a student, I learned my biochemistry from James B. Sumner (recipient of the 1946 Nobel Prize in Chemistry for isolating and crystallizing urease and showing that enzymes are proteins). In 1952, I was appointed a postdoctoral research assistant (1952–1955) in the laboratory of E. Newton Harvey, Department of Biology, Guyot Hall, Princeton University. Howard S. Mason (co-discoverer of oxygenase with Osamu Hayaishi) was leaving Harvey's lab and I filled his position.

As he settled into his job, Tsuji learned how Harvey's lab kept being supplied with that exotic ostracod which had become the research lifeline in Princeton:

> The *Cypridina* used for our research were obtained from Dr. Yata Haneda, Director of the Yokosuka City Museum, Yokosuka, Japan, who had hired fishermen to collect and air-dry them in large amounts. Haneda [of whom more in chapter 13] then shipped the dried organisms to Harvey in a bottle with a small amount of anhydrous calcium chloride in a cotton bag as desiccant. When stored under such conditions, the luciferin and luciferase in the organisms remained highly active almost indefinitely.

Tsuji, working with Chase as well as Harvey, refined the technical results obtained by both Chase and Mason. For a long time, investigators had missed the presence of nitrogen in the quasi-purified luciferin. But Chase and Gregg (1949) made "ultra-micro quantitative determinations" suggesting that nitrogen represents 8 percent of the luciferin, although they acknowledged that the nitrogen may be associated with the remaining impurities. Mason (1952) also detected nitrogen in *Cypridina* luciferin and suggested that the luciferin was a chromopolypeptide. Tsuji and his collaborators (1955) added further support, and although this chemical assignment pushed the research onto the right track, they failed to identify tryptophan as one of the suspected constituent amino acids, which turned out later to be an important element of the peptide.

For all these years these investigators tried to crystallize the luciferin, as it would facilitate the completion of the purification and the identification of the structure, but to no avail. In the end, the breakthrough originated not at Princeton, but, ironically, in the country of origin of *Cypridina*, Japan. The young researcher who accomplished the feat was Osamu Shimomura, who later became a central figure of bioluminescence research. Being born in 1928 in Nagasaki meant that he was a teenager when the atomic bomb destroyed his native city. He explained how he avoided the fate of many of his compatriots and landed the lab job that led to his *Cypridina* work (Shimomura, 2008):

> My story begins in 1945, the year the city of Nagasaki was destroyed by an atomic bomb and World War II ended. At that time I was a 16-year old high school student, and I was working at a factory about 15 km northeast of Nagasaki. I watched the B-29 that carried the atomic bomb heading toward Nagasaki, then soon I was exposed to a blinding bright flash and a strong pressure wave that were caused by a gigantic explosion. I was lucky to survive the war. In the mess after the war, however, I could not find any school to attend. I idled for 2 years, and then I learned that the pharmacy school of Nagasaki Medical College, which had been completely destroyed by the atomic bomb, was going to open a temporary campus near my home. I applied to the pharmacy school and was accepted. Although I didn't have any interest in pharmacy, it was the only way that I could have some education.
>
> After graduating from the pharmacy school, I worked as a teaching assistant at the same school, which was reorganized as a part of Nagasaki University. My boss Professor Shungo Yasunaga was a gentle and

> very kind person. In 1955, when I had worked for four years on the job, he arranged for me a paid leave of absence for one year, and he sent me to Nagoya University, to study at the laboratory of Professor Yoshimasa Hirata.

Hirata assigned Shimomura the job of crystallizing *Cypridina* luciferin. This he managed to pull off, with the assistance of Toshio Goto. In the article that resulted (Shimomura et al., 1957), Shimomura made it sound disarmingly simple: "The extract was purified according to a modified [Rupert] Anderson's benzoylation method and by means of partition chromatography on cellulose powder to obtain 'purified' luciferin. Acidification of the purified luciferin solution with hydrochloric acid afforded crystalline luciferin as orange-red needles." But, of course, it was more complex than that. It took him ten months to find the right solvent for precipitation, and the one that worked, high concentrations of hydrochloric acid, was so unlikely a candidate for the job that it is hardly surprising all his predecessors had failed. Ultraviolet absorption spectra showed that hydrogenated *Cypridina* luciferin contains an amino acid with an indole ring, either tryptophan or tryptamine. The complete structure of the luciferin was fully resolved by Kishi et al. (1966) and found to be a small peptide composed of tryptophan, arginine, and isoleucine moieties.

Unravelling the chemiluminescent systems of luminous bacteria and fireflies proved more straightforward, even though they are more complex. The key protagonists of this story were graduate students of Harvey who went on to become themselves leaders in the field and to enjoy brilliant careers. The foundations of modern chemical studies on bacterial bioluminescence were laid out by Frank H. Johnson (1908–1990). Born in Raleigh, North Carolina, he graduated at Princeton in 1930, earned a master's degree from Duke University in 1932, and returned to Princeton to study under Harvey, earning his PhD there in 1936. He remained in Princeton as a professor of the Biology Department for his entire career. Initially he studied the metabolism of luminous bacteria and the effect of physico-chemical forces on their bioluminescence (Johnson and Harvey, 1937, 1938). One of his major contribu-

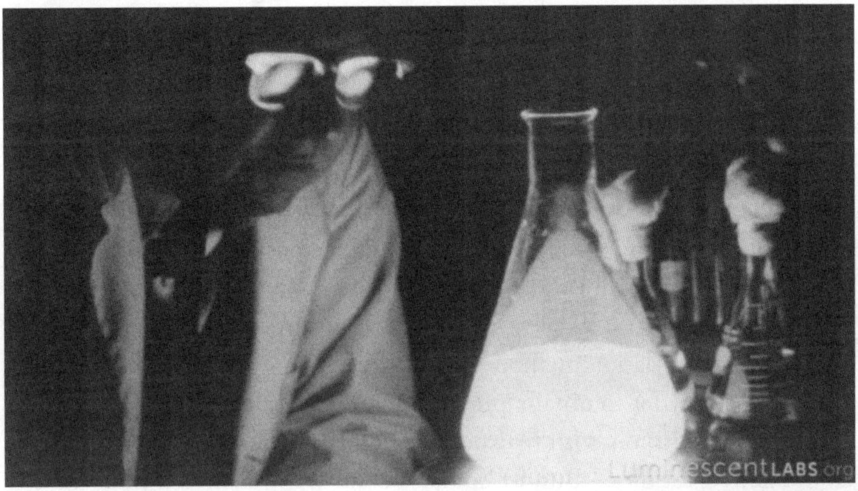

Figure 11.1 Frank H. Johnson, illuminated by a flask of luminescent bacterial culture. From LuminescentLabs.

tions was to develop a bacterial luminescence model to validate the theory of absolute reaction rates.

The theory of absolute reaction rates, or transition rate theory, was developed by theoretical chemist Henry Eyring (1901–1981). Born in a Mexican Mormon colony, Eyring received his PhD at Berkeley in 1927. He was recruited in 1931 by Princeton, where he stayed until 1946, when he moved to the University of Utah. Transition rate theory, which he articulated in 1935, proposed that an equilibrium or transition state is reached in chemical reactions between the initial reactants and the products of the reaction. Chemists believe that the theory was one of the key discoveries of twentieth-century chemistry and that Eyring should have been awarded the Nobel Prize. Some go so far as suggesting that his belonging to the Mormon Church denied him the award. At any rate, the collaboration between Johnson and Eyring in the 1940s ensured that the reaction process of bacterial bioluminescence became the first actual chemical phenomenon to which Eyring's theory was applied (Miller, 1984). Johnson explained why bacterial luminescence recorded by a sensitive detector in real time made it such an ideal experimental model to validate the theory:

> There I think that luminescence had a unique advantage, and we were just lucky because of this. We didn't realize it the first time but the fact is that it's probably the only process in nature where you get an instant measure of reaction velocity. I mean elsewhere, you would have to have a slope of a curve against time to know what the rate is; you measure the rate at one moment and then some moments later. But you know instantly in luminescence because the light intensity is proportional to reaction velocity. So you not only do it instantly but you get it accurately, which you can't always do with other processes by any means. (Miller, 1984)

This was the first clear case of using bioluminescence as a tool in basic research. Many more examples were to follow in the future. Johnson and Eyring (1948) reviewed their findings, and a few years later published a book in which the title included the first usage in the scientific literature of the term "molecular biology" (Johnson et al., 1954). Although this was a fascinating diversion, the problem of determining the chemical structure of the constituents and of their involvement in the chemical reactions of bacterial luminescence fell into the lap of two other students of Harvey: William D. McElroy and J. Woodland Hastings.

Born in Texas, William D. McElroy (1917–1999) moved with his family to California and he graduated from Stanford University in 1939 (Hastings, 2004). He pursued a master's in biology at Reed College in Portland, Oregon, where he met and married a fellow student. In 1941 he crossed the continent to study under Harvey in Princeton. Upon his arrival, he experienced quite a culture shock, as he explained (McElroy, 1976):

> Perhaps the simplest way to describe my experience on going to Princeton in 1941 is to say that I was a country boy heading east. I moved into an environment that was quite foreign to my upbringing. For example, at that time graduate students were expected to live in the graduate college on the Princeton campus, where one had dinner in academic cap and gown. Except for graduation I had never worn this attire and never quite understood the idea behind the Princeton manner. Fortunately I was able to get an exception to the regular dining and living ar-

Figure 11.2 William D. McElroy in 1964. Courtesy of the US National Library of Medicine.

rangements because by that time I was married and therefore was given permission by the Graduate Dean to live in an apartment off the campus.

McElroy's doctoral thesis was actually an offshoot of the Johnson-Eyring project on bacterial luminescence. He looked at the effect of narcotics (barbiturates, chloretone, and so forth) on the chemical reaction of luminous bacteria (McElroy, 1944). He found that depressing the glucose-fuelled oxidative metabolism leads to the inhibition of luminescence. His thesis work was disrupted by the diversion of research resources to the war effort, in which he participated with Harvey. He had this to say of his relationship with his supervisor (McElroy, 1976):

> Learning to do research under the direction of Professor Harvey was a rewarding experience. He was tolerant of false starts due to inexperience and always took the time to point out how to do a particular experiment better. Most of all he was a storehouse of knowledge about

> luminescent organisms. We would spend hours discussing different types of organisms, where they could be found, and what was known about their luminescent characteristics. In later years we spent time together in Jamaica, Woods Hole, and other places doing joint research projects. It is fair to say that my lasting interest in luminous organisms can be traced entirely to Professor Harvey. Almost all scientists working on luminescence in organisms in the United States today were trained either by Professor Harvey or by his students.

McElroy's praises and those of others of Harvey's students are not exaggerated homilies bestowed on a great man's grave. Indeed, Harvey's preserved correspondence with his students and associates testifies to the solicitude he showered on them, the exchange of scientific ideas, and the care he took in furthering their careers or fostering the publication of their research. In McElroy's case his career took off on his appointment in 1947 in the Department of Biology at Johns Hopkins University after a postdoctoral stint at Stanford. Two years later he was also made director of the newly created McCollum-Pratt Institute, whose initial objective was the study of trace metal metabolism but which blossomed in several branches of biochemistry. This aspect of McElroy's accomplishments in Baltimore is well documented by Tulley Long (2009). In time McElroy's industry ensured that Johns Hopkins became as important a research center for bioluminescence as Princeton already was. And this is where efforts to decode the bacterial bioluminescent system were deployed.

To implement his projects, McElroy recruited a graduate student, Bernard Louis Strehler (1925–2001), and a postdoctoral fellow, John Woodland "Woody" Hastings (1927–2014). Strehler was born in Johnstown, Pennsylvania, and he received both his degrees at Johns Hopkins. Hastings, born in Salisbury, Maryland, finished high school in 1944 and spent the rest of the duration of the war in the Navy V-12 unit based on the campus of Swarthmore College (Hastings, 2001a). After graduating from Swarthmore, he writes, "I taught biology in a French lycée and worked on reconstruction in Germany, then went to graduate school at Princeton, with research on bioluminescence mentored by E. Newton Harvey." His thesis work at Princeton provided experimental support for the hypothesis that there are two binding sites on the luciferase of luminous bacteria and fungi: one for

Figure 11.3 J. Woodland Hastings (second from right), with (from left) Yata Haneda (see chapter 13), Fernand Baguet and Peter Herring (both discussed in chapter 16). From the author's archives.

oxygen and the other for luciferin (Hastings, 1952). He found that binding of the two substrates must happen simultaneously for luminescence to occur.

Strehler and Hastings went to work in Baltimore in the early 1950s and the discoveries piled up, especially after Strehler produced workable cell-free extracts of the bacteria. This is how the elated McElroy put it in a nutshell (McElroy, 1976):

> As history has shown, Professor Strehler and his associates were able to demonstrate that this cold-water extract could be stimulated to emit light by adding reduced DPN. In other words, the addition of reduced

> DPN [the coenzyme diphosphopyridine nucleotide] (DPNH) to a crude bacterial extract temporarily restored light. This was the real beginning of the unravelling of the bacterial system. Later Dr. Hastings and I demonstrated that reduced FMN [flavin mononucleotide] was the main substrate for light emission in the bacteria and that the DPNH was needed to reduce the endogenous FMN present in the crude extract. Following these observations, Professors Milton Cormier and Strehler were able to show that a long-chain aliphatic aldehyde was also necessary for bright light emission. This was the second luminescent system that yielded to the isolation and identification of key substrate molecules necessary for light emission. Although Dr. Arda Green and I proposed that the aldehyde was consumed during the light reaction, only in recent years has it been demonstrated conclusively that the aldehyde is indeed used in the light emission.

This requirement of several factors – DPN, FMN and an aldehyde – makes the bacterial luminescent system more complicated than that of *Cypridina*, and a lot more fussing goes on in the cell to achieve the light emission (Strehler, 1955). The basic scheme of the reaction was construed as follows:

$$FMNH_2 + RCHO \text{ (aldehyde chain)} + O_2 + \text{enzyme} = \text{light} + \text{products}$$

The requirement for a long-chain aldehyde was discovered by Milton J. Cormier (Cormier and Strehler, 1953). Cormier was born Milton Hanchey in 1926 in dirt-poor rural Louisiana, and his name changed to Cormier when his stepfather officially adopted him (Cormier, 2007). He studied at the University of Louisiana in Lafayette, with a two-year interruption serving in the US Navy late in the Second World War. After a master's degree at the University of Texas in Austin, he worked as a technician at the Oak Ridge National Laboratory in Oak Ridge, Tennessee, assisting Strehler, who had taken a position there after leaving McElroy's lab. Cormier advanced from technician to PhD candidate, completing his thesis under Strehler's colleague in Oak Ridge, John R. Totter. He stayed in Oak Ridge for postdoctoral work before accepting an appointment as assistant professor of biochemistry at the University of Georgia in Athens, where he advanced in rank and remained for his entire career. We will have occasion later to return to his bioluminescence career in Athens.

Concurrently with the bacterial work, McElroy had embarked on projects to unravel the chemical luminescent system of the firefly. He needed no reminder that fireflies should be at the top of his to-do list for the simple reason that Johns Hopkins is in Baltimore, which in Hastings words "is the firefly epicenter on the east coast, and they were everywhere on campus (Hastings, 2001a)." Collecting myriads of the insect needed for the research, Hastings added, "involved passing out nets to local children and trying not to be duped into paying them for more than their catch (going rate: a penny apiece)." Shortly after his arrival in Baltimore McElroy obtained surprising results with firefly extracts: he found that "if one adds a small amount of adenosine triphosphate (ATP) to this crude extract a brilliant flash of light appears immediately and lasts for a considerable time depending on the ATP concentration (McElroy, 1947)." However, because ATP is known to be involved in energy metabolism, he assumed that the firefly luciferin, by binding ATP and releasing phosphates, "may act in both phosphate and electron transporting systems." He recanted this view later, and in the meantime he and Bernard Strehler discovered that magnesium was required in addition to ATP for the light reaction to take place (McElroy and Strehler, 1949). Strehler also purified the firefly luciferin, which showed a yellow-green fluorescence very similar to the colour of the firefly flash (Strehler and McElroy, 1949), but the study provided no clue as to the chemical identity of the substance.

McElroy (1976) could not help remarking on a strange historical twist: Raphaël Dubois's luciferin in his "firefly" extract was in fact ATP, although the Frenchman could not know that. "In retrospect," McElroy wrote, "it is interesting to note that DuBois was the first to show a physiological function for ATP, even though it would be 40 years (through the work of Lohman and Fiske and Subbarow) before we would know that such a compound existed." As for luciferin itself, its chemical identity was disclosed in 1961 in the Johns Hopkins lab; it was the first luciferin molecule to become known, six years ahead of *Cypridina* luciferin. It was a small molecule containing a cysteine residue and a benzothiazole aromatic ring (White et al., 1961). The purification and crystallization of firefly luciferase was performed six years earlier (Green and McElroy, 1956). The suggested reaction scheme was as follows: luciferin reacts with ATP and luciferase, in the presence of magnesium, to form a luciferase-luciferin-AMP (adenosine monophosphate) complex and the release of pyrophosphate. Next the complex reacts with oxygen and the oxidized complex is the light emitter, with the concomitant formation of water (McElroy and Seliger, 1966).

The amazing discovery that the total light output of the firefly luminescent reaction is directly proportional to the quantity of ATP and luciferin present, and that a single light quantum is emitted for each luciferin molecule consumed (Seliger and McElroy, 1959), opened the door for an exquisitely sensitive assay of ATP as a commercial application. The firefly bioluminescence ATP assay kit, as it became known, was introduced in the 1970s and was soon a common fixture in clinical laboratories of hospitals to help diagnose metabolic diseases. The kit included all the chemical ingredients for the firefly luminescent reaction, and a luminometer, an apparatus to detect light in a vial inserted in a light-tight chamber, read luminescence values calibrated to represent ATP concentrations. This was the launching pad for many more applications from basic bioluminescence research that saw the light of day in the years to come.

~

The third major breakthrough in identifying the chemical structure of bioluminescent reactants concerns a jellyfish and harks back to the Princeton lineage. Impressed by Osamu Shimomura's success in purifying and crystallizing *Cypridina* luciferin, Frank Johnson invited Shimomura to come and work with him at Princeton as a research associate (Shimomura, 2008; Tsuji, 2010). The young Japanese arrived at Princeton in 1960 right after his PhD was granted, ostensibly at first to continue working on the *Cypridina* system. He purified and characterized the *Cypridina* luciferase (Shimomura et al., 1961), but the achievement was spoiled by Frederick Tsuji, who had left Princeton and had established his own lab at the University of Pittsburgh. Tsuji and his assistant had also purified the luciferase, and their paper sat awkwardly next to Shimomura's in the same journal (Tsuji and Sowincki, 1961). A scientific race could not have squeezed closer to the finishing line!

Figure 11.4 Osamu Shimomura. Courtesy of Nagoya University (Japan).

Soon after Shimomura settled down in Princeton, Johnson asked him, in his words, "if I would be interested in studying the bioluminescence of the [hydromedusan] jellyfish *Aequorea*. I was strongly impressed by his description of the brilliant luminescence and the abundance of the jellyfish at Friday Harbor in the state of Washington. I agreed to study the jellyfish (Shimomura, 2008)." We may recall from the previous chapter that Harvey had studied the jellyfish in 1921, and he may have conveyed his enthusiasm for the beast to Johnson in later years. During the first bioluminescence conference ever, held in the spring of 1954 at the Asilomar Resort near Pacific Grove, California, Johnson heard discussions "regarding the unique aspects of the *Aequorea* bioluminescence system," which caused him to change the course of his research program (Tsuji, 2010). So Johnson and Shimomura drove all five thousand kilometers from Princeton to Friday Harbor in the summer of 1961. Shimomura (2008) recalls:

> At the University of Washington laboratory there, we carefully scooped up the jellyfish one by one using a shallow dip net. The light organs of *Aequorea aequorea* … are located along the edge of the umbrella, which we called a ring. The ring could be cut off with a pair of scissors, eliminating most of the unnecessary body part.

At the time, it was a common belief that the light of all bioluminescent organisms was produced by the reaction of luciferin and luciferase. Therefore, we tried to extract luciferin and luciferase from the rings of the jellyfish. We tried every method we could think of, but all our efforts failed. After only a few days of work, we ran out of ideas. I was convinced that the cause of our failure was the luciferin-luciferase hypothesis that dominated our mind. I suggested to Dr. Johnson that we forget the idea of extracting luciferin and luciferase and, instead, try to extract a luminescent substance whatever it might be. However, I was unable to convince him. Because of the disagreement on experimental method, I started to work alone at one side of a table, while, on the other side, Dr. Johnson and his assistant continued their efforts to extract a luciferin. It was an awkward, uncomfortable situation.

The situation was unusual also at another level. In those days of scientific exchange involving Japanese nationals, the hosts in Europe and North America were accustomed to facing young apprentice scientists too eager to please, too submissive to authority for their taste. But they resigned themselves to accepting that the cultural ambience in Japan made these traits hard to shake off. From this author's recollections, Shimomura was exceptional in this regard not only for his imposing height – he is about six feet tall – but also for his strong personality spiced with a dab of stubbornness. That Shimomura stood up to Johnson in the disagreement, therefore, is significant; and by pursuing his own idea so forcefully he made a stunning discovery. However, initially he was getting nowhere, as he explained (Shimomura, 2008):

> I tried very hard, but nothing worked. I spent the next several days soul-searching, trying to find out something missing in my experiments and in my thought. I thought day and night. I often took a rowboat out to the middle of the bay to avoid interference by people. One afternoon, an idea suddenly struck me on the boat. It was a very simple idea: "Luminescence reaction probably involves a protein. If so, luminescence might be reversibly inhibited at a certain pH."
>
> I immediately went back to the lab and tested the luminescence of light organs at various pHs. I clearly saw luminescence at pH 7, 6 and

5, but not at pH 4. I ground the light organs in a pH 4 buffer, and then filtered the mixture. The cell-free filtrate was nearly dark. But it regained luminescence when it was neutralized with sodium bicarbonate. The experiment showed that I could extract the luminescence substance, at least in principle …

But a big surprise came the next moment. When I threw the extract into a sink, the inside of the sink lit up with a bright blue flash. The overflow of an aquarium was flowing into the sink, so I figured out that seawater had caused the luminescence. Because the composition of seawater is known, I easily found out that Ca2+ activated the luminescence. The discovery of Ca2+ as the activator suggested that the luminescence material could be extracted utilizing the Ca-chelator EDTA, and we devised an extraction method of the luminescent substance.

Shimomura, Johnson, and a technician, Yo Saiga, had a busy summer in Friday Harbor processing ten thousand jellyfish for photoprotein extracts. On their return to Princeton, the photoprotein, which they called aequorin, was purified and its selective activation by infinitesimal amounts of calcium was confirmed (Shimomura et al., 1962). This most simple, two-component luminescent system was heralded as a new biochemical system as opposed to the luciferin-luciferase system. Soon it was exploited by cell physiologists in their research. Calcium is critically important in the control of many cellular activities, and aequorin afforded a technique whereby calcium levels can be monitored inside the cell in real time. Once aequorin became available, Ellis Ridgway and Christopher Ashley at the University of Oregon in Eugene wasted no time. They microinjected aequorin into single muscle cells and they recorded the time course of a light emission, thereby reflecting an increase in intracellular calcium when the muscle cell contracted (Ridgway and Ashley, 1967). It would be hard to imagine a more sophisticated and elegant way of following the dynamics of physiological events inside cells.

In a footnote in their seminal 1962 paper, Shimomura and his co-authors drew attention to another protein extracted along with aequorin: "A protein giving solutions that look slightly greenish in sunlight though only yellowish under tungsten lights, and exhibiting a very bright, greenish fluorescence in the ultraviolet of a Mineralite, has also been isolated from squeezates. No indications of a luminescent reaction of this substance could be detected." This was a significant observation because Shimomura was well aware that aequorin in a test tube gives out a blue light but the living jellyfish emits a greenish light. As Johnson, Shimomura, and co-authors explained in a paper published later the same year, "it is reasonable to believe that the greenish quality of the light emitted from living photocytes is due to the absorption by the green protein, of blue luminescence, followed by fluorescence at longer wavelengths" (Johnson et al., 1962). James G. Morin and Woody Hastings (1971) called it the green fluorescent protein (GFP) and the new name stuck. They showed how an energy transfer mechanism shifts the colour of the photoprotein to the colour of the GFP.

If the acronym GFP rings a bell in the reader's mind, the cause is probably the media attention the green fluorescent protein has attracted as a tool that has revolutionized molecular biology and medical research. As discoverer of the protein, Shimomura shared the Nobel Prize for Chemistry in 2008 and helped push bioluminescence research into the limelight. The triumph of the biochemical approach could not have been more complete, and E. Newton Harvey, who would have considered Shimomura his academic grandson, could not have been prouder had he lived to see this apotheosis.

12

Through a Glass, Brightly – William Beebe's Bathysphere

A large transparent jellyfish bumped against the glass,
its stomach filled with a glowing mass of luminous food.
–William Beebe (1934)

While E. Newton Harvey brought the entrepreneurial spirit to bear on his bioluminescence research and made Princeton the headquarters of Bioluminescence Inc., it can be said that William Beebe showed the same quality in his drive to get acquainted with deep-sea animals and their light displays. But Beebe was cut from quite a different cloth. Unlike the Princeton professor, who embraced the physico-chemical approach and the laboratory bench, Beebe seemed a throwback to the heyday of the nineteenth-century European model of naturalists, who always fret to roam the natural world freely and shun the armchair and laboratory life. He also showed no patience with the career demands of academia. He ran counter to the scientific culture that was then blossoming in America, and yet, he reaped the fruits of his labour with dividends of fame as a naturalist and writer.

His life and career are now well documented thanks to Carol Grant Gould's biography, *The Remarkable Life of William Beebe* (2004), which is remarkable itself for its insights and thoroughness. This chapter relies heavily on her biography as well as on an account by Brad Matsen of the story behind the bathysphere adventure (Matsen, 2005).

Charles William Beebe (1877–1962) was born in Brooklyn, NY, the son of a paper mill businessman. Soon after his birth the family moved across the Hudson River to East Orange, NJ, where Will, as he became known, could give free range to his budding passion for the lush natural world around him, away from the stifling urban asphalt. Hunting for and collecting all sorts of specimens – rocks, plants, and animals, especially birds and insects – was his most cherished childhood bliss. He became a frequent visitor of the natural wonders preserved in the American Museum of Natural History in New York City, which had first opened its doors the year of his birth. In his teenage years Will learnt to shoot and stuff dead animals, mostly birds. In 1896 he went to Columbia University, where he was soon noticed by a distinguished professor there, Henry Fairfield Osborn (1857–1935), who also happened to be president of the American Museum of Natural History. Osborn, the son of a railroad tycoon, was a Princeton graduate and Harvey's Chair of Biology at Princeton bore Osborn's name.

In 1899 Osborn helped Beebe, who had failed to finish his bachelor's degree, to get a job as assistant curator of birds (ornithology) at the newly created Bronx Zoo of the New York Zoological Society. Will would often benefit from Osborn's protection and from the benevolence of other wealthy members of the New York Zoological Society for the pursuance of his career and expeditions. The New York Zoo was still under construction when Beebe was appointed, and he quickly made his mark as an ornithologist and organizer on its grounds. But his *Wanderlust* soon awoke, as his self-appointed mission to hunt for specimens and enrich the zoo's menagerie and the American Museum of Natural History's collections overrode the demands of his desk job. Despite the reservations of his boss, Beebe was always able to count on his well-nurtured connections with Osborn and with well-heeled millionaires to overcome such obstacles and to raise funds.

In 1902 he married Mary Blair Rice, a Southern socialite, at a time when he was trying his hand at writing books on birds. The response of the reading public to his published efforts was such that he kept pushing his pen effortlessly and produced an extraordinary number of highly successful books in the course of his life, blending perceptive scientific observations with an engaging style that appealed to a wide audience. For many years to come, Beebe and his wife roughed it in exotic, mostly tropical locations – the Florida Keys, Mexico, Trinidad, British Guiana – where they made ecological and behavioural observations and acquired numerous specimens. In addition, Beebe was commissioned by a rich patron to track all species of pheasants around the world, a one-year trip that resulted in a highly acclaimed monograph. But in 1913 Beebe's walk on clouds was abruptly interrupted by the end of his marriage, his wife having filed for divorce. The best way he found to manage his grief was to engage in more expeditions and create a tropical research station in British Guiana.

Beebe's shift to the study of marine life occurred relatively late in his life, in 1923 when he was forty-six. It came out of his worship of Darwin and his fascination for the islands Darwin had envisioned as an ongoing laboratory of evolutionary processes, the Galapagos. One of his wealthy contributors, Harrison Williams, who had made his fortune in the utilities business, was persuaded by Beebe to foot the "outrageous" bill for an expedition to the Galapagos, stressing "that, given the current state of scientific knowledge and technology and a competent staff, he would be able to amass the hard data that Darwin, on his brief, solitary visit, had been unable to collect" (Gould, 2004). So Beebe chartered a large steam yacht, the *Noma*, which explored the fauna of the archipelago, but very little of the deep sea. The book Beebe published from that experience, *Galapagos – World's End*, was a best-seller for months in 1924. The tantalizing glimpse of deep-sea life around the Galapagos and the fascination Sir John Murray's and Johan Hjort's book, *The Depths of the Ocean*, aroused in him, pushed Beebe to mount an expedition during which the study of the deep sea was to figure prominently. This materialized as the *Arcturus* expedition.

Beebe recruited helpers to produce "a design for an oceanographic research vessel that would be able to trawl and dredge at the greatest depths then possible, and would provide the stability for microscopic and dissection work that had been lacking in the *Noma*." A 2,400-ton steam yacht owned by one of his benefactors was refitted to specification and baptized *Arcturus*. It sailed for over six months in 1926 along a familiar trajectory in the Caribbean Sea and through the Panama Canal to the Eastern Pacific and the Galapagos. At two locations Beebe sampled a column of ocean water extensively, from the surface to the bottom, instead of sampling at multiple stations over vast distances as was the rule with the classic oceanographic expeditions of the past. This was an approach that he had pioneered in the tropical jungle, whereby a "cylinder" of jungle environment from undersoil to canopy, from low plants to tall trees, was subjected to an exacting census. Here Beebe may have anticipated the field of community ecology.

In his book chronicling the expedition (Beebe, 1926), Beebe waxed lyrical when he commented that "for some reason good fortune was with us and again and again deep sea fish and other organisms lived from two minutes to as many hours, – and swam and breathed and sent forth barrages from their luminescent batteries – the strength of which sometimes lighted up the whole dark-room where I studied them." The hauls undertaken in the ocean column off Cocos Island yielded several descriptions of luminescent displays, some of which were original. Of the luminescence of a shrimp and its possible role as a defensive smokescreen, Beebe wrote:

> I saw a dull glow from what I took to be some one-celled organism, perhaps a dying Noctiluca. To my astonishment it increased in size, and, bringing near the illumined face of my watch, I saw the source of the fiery flow was the prawn itself. The light now took the form of a liquid pouring out into the water, and soon the entire contents of the aquarium was aglow, while, swimming about in it, the prawn could be seen as a black, inchoate mass. Suddenly the significance of this occurred to me – this red crustacean was playing the same trick as the squid, but adapted to the darkness of six hundred fathoms. The squid had its cloud of smoke by day, the prawn its pillar of fire by night.

The eyes and light organs of the hatchetfish also intrigued Beebe, and he was genuinely perplexed by the apparent functional disconnect between them:

> Argyropelecus … however, is the first of many deep-sea puzzles because, while the lower sides are lined with large luminous organs, the light from which is thrown downward rather than sideways, yet the eyes, which are very large and bulging, are directed straight upward. Why this fish should be denied the ability to enjoy its own pyrotechnics is not apparent. If the downward sheet of light acts as a lure to attract its prey there still seems considerable need for anatomical alteration, for the mouth, like the eyes, is turned almost straight upward. Twice I secured living specimens and three times I was able to distinguish the illumination.

Today our understanding is that the eyes are turned upward because the hatchetfish's prey lies above in the water column, and the light organs shine downwards as a camouflage strategy to hide from the hatchetfish's predators lying below. Another source of perplexity, for anyone adhering to the view that luminescence is used as a searchlight, was the discovery of an eyeless fish with light organs:

> Another confusing condition of affairs came to light (in every sense of the word), when I found a brightly illumined blind fish. The lights may have persisted from the time when its eyes were better developed, but a more probable explanation is that the rays act as a lure for small edible creatures, and the fish, through sensations other than sight, is able to detect their presence and to seize them. Until we actually know the cause, however, we can only speculate, and allow it to bring such absurd similes to mind as a blind Diogenes stumbling along with a lighted lantern in his hand.

In the same vein, Beebe was mesmerized by a strange-looking crab with similar visual deficits:

> Rendered conspicuous in this colorful throng by the neutral tones of its monk-grey garb was a small globular crab which seemed at first glance as immutable, as lifeless, as a bit of Archean rock from the ocean floor, but which, upon closer inspection proved to be one of the most remarkable crustaceans captured by the expedition … It has evidently long been an inhabitant of the abyss, for the eyes are small and degenerate and the antennae are exceedingly long and tactile. And finally and most unexpected, situated at the base of these antennae and opening just in front of the mouth cavity are the ducts from paired luminous organs. When released by the opening of the magical circular door which is formed by the first joint of the antennae (a segment lost in most crabs) the luminescent substance glows like a tiny lantern, and may well serve to attract a host of small creatures who are promptly devoured.

Deep-sea anglerfishes were already assumed to use the luminous bulb of their tentacle as bait, but Beebe found a "diabolical" species that goes a step further: "In Diabolidium not only the tip of the tentacle, but all the larger teeth were dimly outlined with luminescence – apparently a mucus like that given off from the numerous pores of the skin. I tried to estimate roughly the relative proportions of the mouth and the rest of the body and in two species found it quite four-fifths of the entire animal." He described the fish as nothing less than an "eating machine."

After Beebe's second wedding in 1927, to writer Elswyth Thane, the couple honeymooned in Bermuda. Beebe quickly saw the advantages of creating a marine station in the British colony. Thanks to the steep slope dropping off the volcanic island where they stood, his team would only have to venture out a few miles from a shore station to reach ocean depths of up to two miles. Harking back to his approach on the *Arcturus*, Beebe planned to "study a carefully delineated eight-mile-square area with unprecedented thoroughness" and to "classify every creature they came across, and collect specimens wherever possible for the zoo, the American Museum, the aquarium, and subject-hungry researchers" (Gould, 2004). Starting in 1928, these became known as the Bermuda Oceanographic Expeditions.

Beebe soon felt that his ambitious project was thwarted by the frustration of not really knowing what was brewing in the deep. No matter how sophisticated the trawling and dredging equipment had become by the late 1920s, there was always the nagging suspicion that the catches were not representative of what was there, what with the ability of agile creatures to dodge the net or slip out of it. Moreover, many of the caught specimens were dead or dying, or heavily damaged by the rough and tumble of the hauling and by decompression. Of course, Beebe was only echoing the preoccupation of the many oceanographers who preceded him. There is no substitute to being physically present if one is to accurately assess the deep-sea fauna in its living interactions, including bioluminescence. But at the time, the deep ocean seemed as inaccessible to human presence as outer space.

The remarkable thing about Beebe is that he did not let this lurking hurdle dampen his drive and imagination. He had already used diving helmets, heavy and clumsy as they were, to explore shallow waters and especially coral reefs. The blueprints he came up with in 1927 and 1928 expanded on the helmet concept, by creating a steel cylinder presumably able to withstand the pressures found one mile deep and to house a person inside it (Matsen, 2005). It would be lowered at depth by a steel cable off a ship winch. A young man read Beebe's plan in the newspaper and saw right away that the geometry of a cylinder was unfit to withstand the pressures of the deep. Frederick Otis Barton (1899–1992) was born in New York, the son of a successful textile merchant who died when he was only six. He grew up in an environment of privilege, studying at the best schools (Groton and Harvard), but "he was known as a loner and a daydreamer" (Matsen, 2005). As a teenager Barton had experimented with homemade helmets in the waters of Cape Cod's Vineyard Sound where his family had a summer home, and from then on, the thought of exploring the oceanic depths never left him.

After his Harvard graduation Otis Barton toured the world and enrolled at Columbia University, where his fantasies about diving contraptions continued to simmer. About the time period when Beebe was toying with such ideas, Barton had conceived of a steel sphere as the ideal shape and material for the safe distribution of the burden of pressure at great depths. To give respectable credibility to his brainchild and thereby make it palatable for a sales pitch to Beebe, he enlisted the help of the firm Cox & Stevens, renowned worldwide for its designs of ships with wooden, iron, or steel hulls. Needless to say, with no experience in designing submersibles, the firm hesitated but finally agreed to give it a try. Overcoming the obstacles erected around Beebe's inner circle to fend off crackpots and other nuisances, Barton succeded in booking a meeting with the famous man. Barton's design appealed to Beebe and the submersible project was on.

Barton forged ahead with the project with Cox & Stevens, bracing for one challenge after another: the hitherto unattempted casting of a steel sphere of that size (nearly two meters in diameter) and thickness (up to five centimeters), the necessity to opt for the newly emerging technology of quartz glass for the portholes, the best way to manage oxygen storage and carbon dioxide disposal inside the sphere, the strength demands of the cable descending and pulling up the 5,500-kilogram sphere over a vertical stretch of more than 800 meters, and the need to adjoin an umbilical cord of telephone and electrical lines for communication between the ship and the sphere and for power supply (Matsen, 2005). After some trial and error, all

Figure 12.1 William Beebe (left), Otis Barton (right), and the bathysphere in between. Fig. 98 in *Half Mile Down* (Beebe, 1934).

these design challenges were met successfully, but Cox & Stevens had to contract out to many highly reputable firms to achieve this, firms that went out on a limb to push their technologies beyond their previous limits.

Meanwhile Beebe was tending to the scientific side of the project. First, he kept trawling at various depths the very vertical cylinder of Bermuda waters where the bathysphere, as he baptized the submersible, was to descend, so that he had as good a census and visual recall of the species as possible to readily identify whatever drifted by the portholes. Second, he wanted to photograph the floating creatures from the porthole, a serious challenge given the short and unpredictable time frames the fleeting moments of encounter would allow snapping a picture. In the end he and Barton lacked the time to take pictures because they were too busy with observations and looking after the instruments in the bathysphere.

The partnership of Beebe and Barton was always uncomfortable, at times awkward and very tense. This stemmed from their differences of personality and motivation. A source of unease for Beebe was that he was accustomed to being in complete charge of a project, and here was Barton, sinking his fortune into the mounting costs of constructing the bathysphere and laying claim to ownership of the contraption. Barton used this leverage to ensure that he occupied one of the two spaces in every descent of the bathysphere. Barton was caught up in the quest for celebrity that animated many a young New Yorker of the Jazz Age. As Matsen (2005) put it: "Otis Barton had always dreamed of adventure, but his current fantasy of exploring the oceanic abyss was also about becoming famous. He loved the vision of himself as a celebrated explorer featured in newspaper stories." And he wanted to make big money to offset the horrendous costs of the project. Beebe balked at these shallow motives. He had become famous not out of hunger for celebrity, but by striving for scientific achievement and writing the story of his expeditions in a style attuned to a wide readership. Although some zoologists considered him an amateur, "Beebe would spend his life waging a battle to be recognized as a scientist first, a writer second" (Gould, 2004). Letters from E. Newton Harvey to Beebe in 1939 testify to the respect and admiration some dyed-in-the-wool scientists bestowed on him.

Everything conspired to have the first dive in the spring of 1930: "Barton brought with him in Bermuda the great sphere, thirty-five hundred feet of steel cable, a full 800 meters of the solid rubber electric cable, and the multitude of necessary extras. I was able to provide the seven-ton *Arcturus* winch and sheaves, the [ship] *Gladisfen* for towing out to sea, and my staff for cooperation in the actual operation" (Beebe, 1934). On 11 June Barton and Beebe, both six feet tall, squeezed themselves into the cramped space of the bathysphere and started a descent 740 meters deep. Beebe used a searchlight to see through the porthole glass, but more often than not "I chose to have it cut off, for I wanted more than anything to see all that I could of the luminescence of the living creatures." He noticed that the deeper they dove the fewer colours showed in the downwelling light except for the deepest hues of blue.

It was during this seventh descent that a bonanza of luminescence spotting started to happen. At 120 meters deep only a few flashes from lanternfishes were seen, and the most populous fishes in these zones, the bristle mouths (*Cyclothone*), never displayed luminescence from their numerous and small light organs. Curiously, it was at a depth of 300 meters that significant events occurred. Beebe "saw a series of luminous, colored dots moving along slowly," which emanated from "a school of silver hatchetfish,

Argyropelecus, from a half to two inches in length and gleaming like tinsel" (Beebe, 1934). At their deepest location a dragonfish, *Stomias*, gave out a continuous light.

Their 1930 diving season over, Barton and Beebe returned to New York the toast of all the newspapers and newsreels, which "always mentioned their discovery of creatures and the weird light in the abyss, but the meat was in the danger" (Matsen, 2005). "Diving in the Bathysphere," Matsen continued, "was an act of courage or insanity that ranked Beebe and Barton among the outrageous daredevils who were capturing the imaginations of a nation laboring under the crushing burdens of the Depression." Barton should have been ecstatic, but the media insisted on pinning Beebe to the limelight instead, probably because he already was a recognizable celebrity. Barton, who was less at ease socially, was cast as Beebe's sidekick.

Beebe had to wait two years before the bathysphere yielded more dividends in luminescence currency. On 22 September 1932, at the depth of 500 meters, no more daylight filtered down. In that environment "as black as Hades," Beebe saw "a school of brilliantly illuminated lanternfish with pale green lights [who] swam past within three feet [one meter] of my window, their lights being exceedingly bright." A little deeper he saw several sabre-toothed viperfishes (*Chauliodus*), whose "eyes shone with a dull glow, and their bodies were covered with a multitude of tiny lights." At a depth of over 600 meters a fascinating spectacle appeared:

> When the darkness closed down on the path of the [search]light again, I saw we were in the midst of a large number of shrimps, and almost at once two large fish dashed into the midst of them, rolling them over and over, all these creatures and their actions silhouetted only in their own light. One at least of the fish had an isolated light, blue and pale reddish, which kept following it about, and I realized that this was a barbel light, whipping about as the fish turned.

All the way down to 670 meters Beebe kept seeing luminescence, even from large, hitherto unknown fishes. There was a great richness of actively swimming, bioluminescent organisms that conventional trawling techniques had missed and could not have anticipated. After another hiatus of two years, during which technical and financial problems kept the submersible out of operation, the bathysphere was finally back in service thanks to a grant from the National Geographic Society. A groundbreaking dive, the thirtieth, occurred on 11 August 1934. Beebe called it "a descent into perpetual night." At 360 meters "a strong glow shot forth, covering a space of perhaps eight inches. Not even the wildest guess would help with such an occurrence. Then the law of compensation sent, close to the window, a clear-cut, three-inch, black anglerfish with a pale, lemon-colored light on a slender tentacle." At 580 meters he saw an almost round fish with big eyes and long, high vertical fins: "Along the sides of the body were five unbelievably beautiful lines of light ... Each line was composed of a series of large, pale yellow lights, and every one of these was surrounded by a semicircle of very small, but intensely purple photophores ... In my memory it will live throughout the rest of my life as one of the loveliest things I have ever seen."

The last dive, going 800 meters down, was on 15 August 1934. Beebe saw large red shrimps darting about with an "outpouring fluid of flame," obviously a copious luminescent secretion. Other shrimps emitted a pale greenish light from photophores. Pale green luminescence was also seen from the "headlights" of lanternfishes (*Diaphus*). These are but a tiny sample of the luminescence sightings Beebe described at the lower depths, many from unidentified sources. Fish were dominant, but a variety of invertebrates also emitted all kinds of light displays. Beebe (1937) provided statistical estimates of deep-sea fish numbers based on his extensive trawling surveys in the Bermuda waters. His expeditions had captured over one third of the deep-sea fish fauna known at the time, distributed among ten systematic orders, forty-six families, sixty-five genera, and two hundred and twenty species. Beebe himself captured and described thirty-one new species. The top eight families in number of individuals – bristle mouths (Gonostomatidae), lanternfishes (Myctophidae), hatchetfishes and pearlsides (Sternoptychidae), bigscales (Melamphaidae), viperfishes and black dragonfishes (Stomiidae), barracudinas (Paralepididae), anglerfishes (Ceratioidei) and sawtooth eels (Serrivomeridae) – accounted for 66 percent of all the luminous deep-sea fish species, but luminescent individuals formed an astounding 96.5 percent of all captured individuals belonging to these eight families.

If the general impression felt by one of Beebe's assistants, John Tee-Van (in Beebe 1934), is to be trusted, the observations during the bathysphere's dives reflected these statistics fairly. He wrote that from 700 feet down "lights were constantly visible, sometimes single and shining continuously or flashing on and off, sometimes in groups that moved along without changing their relationships to one another, which indicated that they belonged to a single fish or other animal. At other times the lights moved about independently of each other showing that they were lights on different fish in a school." Against a backdrop of pitch darkness the seascape outside the bathysphere windows reminded Tee-Van of "a journey through the heavens on some yet-to-be invented machine at unheard-of speeds – a constellation suddenly appearing and disappearing with the rapidity of meteors arriving in our atmosphere."

A most remarkable finding was that lanternfishes, the most abundant after bristle mouths, swam in schools and their ventral (belly) photophores emitted a continuous light. Beebe and Mary Vander Pyl (1944) later added that the "lower battery," when going full blast, casts a solid sheet of light downward, so strong that the individual organs could not be detected. Five separate times when I got [lantern]fish in a large, darkened aquarium, I saw good-sized copepods and other organisms come close, within range of the ventral light, then turn and swim still closer to the fish, whereupon the myctophid twisted around and seized several of the small beings." Notwithstanding the assignment here of a role for the belly photophores in prey capture, this observation was to weigh heavily on biologists of future generations when they conceived of, and tested experimentally, the theory of counter-illumination.

Beebe astutely realized that lanternfish bioluminescence may have several more functions. The distribution of lateral photophores was the object of his scrutiny (Beebe and Vander Pyl, 1944):

> Perhaps the best distinction between various species of lanternfish is the arrangement of the lateral light organs, and in the darkroom in absolute darkness, I could tell at a glance what and how many of each species were represented in a new catch, solely from their luminous hieroglyphics. When several fish were swimming about, these side portholes were almost always alight, and it seems reasonable that they may serve as recognition signs, enabling members of a school to keep together, and to show stray individuals the way to safety.

Beebe lavished particular attention on the enigmatic tail lights (caudal luminous organs) of lanternfish that made them so well tooled for repertoires of luminescent behaviour:

> The light scales of the tail are apparently of considerable importance. Ordinarily when the whole fish is glowing with the pale greenish light of luminescence, these caudal lights are seldom seen. A clue to their use is found in the fact they show a sexual difference, the males having them on the upper side of the peduncle and the females on the lower side … Although it is very evident that the caudal flashes have some sexual significance, yet another very important function seems that of obliteration … When the ventral lights die out they do so gradually, so that the eye holds the image of the fish for a time after their disappearance, but the eye is so blinded by the sudden flare of the tail lights that when they are as instantly quenched, there follow several seconds when our retina can make no use of the faint, diffused, remaining light, but becomes quite blinded. A better method of defense and escape would be difficult to imagine.

After the last dive of the bathysphere Barton and Beebe parted, never to meet again (Matsen, 2005). Barton embarked on a lowbrow career of adventure filmmaker that earned him Beebe's contempt, especially when he exploited Beebe's good name and past partnership for crass purposes. But Beebe had little time to worry about Barton's next moves as he worked feverishly to write a book on the bathysphere adventure. The book, *Half Mile Down*, appeared in stores just in time for the shoppers of Christmas 1934. It was on the bestseller list for several weeks. As Matsen explained, it "was a good thing for Beebe, who was down to his financial reserves, as the Depression was at its nadir and he still owed money for expedition expenses."

~~~

There is a postscript to this story that intersects with E. Newton Harvey. Harvey got to know Beebe through the latter's friendship with Edwin Conklin, Harvey's old mentor and head of the Biology Department at Princeton.
~~~

Conklin, an expert on developmental biology and evolution, obtained from Beebe specimens of the lancelet (Amphioxus), a form that was then believed to be a missing link between invertebrates and vertebrates and which was found in Bermuda waters (Matsen, 2005). It is not clear exactly when Harvey himself became acquainted with Beebe, but their common interest in bioluminescence must have been a catalyst. Both men must have felt crestfallen that luminous deep-sea fishes could not be photographed from the bathysphere. Harvey thought hard about the problem and outside the box, that is, the bathysphere. In a letter dated 8 July 1939 Harvey wrote to Beebe:

> Dear Will:
> I have just returned from Bermuda and was very sorry to have missed you for they told me at the Laboratory that you had returned on the boat which passed mine coming down. My trip to Bermuda was to try and take some pictures of deep-sea fish. Not possessing the spirit of adventure which you have, I contented myself with sending down a camera in a pressure chamber equipped with automatic mechanisms for turning on both camera and light when a certain depth had been reached. Five successful descents of the camera were made, one to a mile and a half [2.4 kilometers], and the pressure chamber actually touched bottom (by mistake). I sent the film off to be developed and will hope for the best.

The results, which he described in a short paper published by *Science* magazine (Harvey, 1939), were completely disappointing. Later he explained that in spite of "the taking of 17,000 individual frames, not a single fish appeared in any picture. The method is somewhat like pointing a camera at the sky hoping to obtain a picture of a bird, but it is rather surprising that no fish was recorded. The element of complete chance was eliminated by hanging a lure in the water before the camera, a model of a deep sea fish with luminous spots painted on its sides with luminous paint. Perhaps the deep sea fish easily recognized the deception, or perhaps the light which suddenly flashed on and off, or the noise of motors and clicking of relays scared the fish away" (Harvey, 1952).

The technological breakthrough happened when Beebe's and Harvey's careers were well over. A master's student at the Massachusetts Institute of

Technology (MIT), Lloyd Robert Breslau, designed and built a "luminescence camera," with advice from his thesis supervisor, Harold Edgerton (1903–1990). Edgerton was known as "Papa Flash" for his pioneering work using stroboscopes and short duration electronic flash to photograph fast events (Vandiver and Kennedy, 2005). Breslau summarized his project in disarmingly simple words: "A submarine camera named the luminescence camera was designed and constructed for the purpose of investigating marine bioluminescence. The camera and strobe light unit contained an electronic control system which utilized a photomultiplier tube as the primary detector. Upon 'seeing' a bioluminescent flash the camera is triggered and a picture of the organism causing the bioluminescence is obtained. This camera has been used extensively at sea and has produced many pictures" (Breslau, 1959). Indeed, the camera, lowered to depths of several thousand meters from the deck of Jacques-Yves Cousteau's *Calypso*, among others, brought back stunning pictures of siphonophores, comb-jellies, krill, jellyfish and squid, but not fishes.

In 1936–37 Beebe and his team conducted more oceanographic expeditions in the Eastern Pacific on board the *Zaca*, a schooner with an auxiliary diesel engine owned by the San Francisco–based multimillionaire Templeton Crocker (Gould, 2004). Crocker's difficult personality and his proneness to depression made life on board hell for Beebe. This may have precipitated the swan song for Beebe's marine adventures, for soon afterward he went back to his studies of tropical (jungle) ecology. He continued working until the late 1950s, but in early 1960 he suffered minor strokes, which slowed him down considerably (Gould, 2004). An aggressive pneumonia caused his death on 4 June 1962.

PART FIVE

OFF CENTRE STAGE

13
The Peculiar Career of Yata Haneda

Japan is like a treasure box of luminous organisms.
–E. Newton Harvey (quoted in Johnson, 1967)

Twenty years after E. Newton Harvey first set foot on Japanese soil and observed the luminescence of the ostracod *Cypridina*, a young Japanese embarked on a bioluminescence research odyssey that led to one of the greatest contributions in the field. His career eventually intersected with that of the Princetonian dean of bioluminescence research.

Yata Haneda was born in Ogaki, Gifu Prefecture, about 40 km northwest of Nagoya, in 1907. Perhaps he was predestined to develop a passion for all organisms luminous, for it is said that the famous haiku poet Matsuo Bashō (1644–1694), who wrote: "Alas! the firefly seen by daylight – Is nothing but a red-necked insect" (Miyamori, 1932), chose Ogaki as the final destination of his long journey in his work *The Narrow Road to the Deep North*. Haneda was born into, and grew up in a Japan that was gratifying its imperialistic impulses, following up its conquest of Taiwan with the occupation of Southern Manchuria (Kwantung) and the annexation of Korea by 1910. As an adult he was to witness more imperialistic forays. According to an article of memoirs by Haneda preserved at the Harano Agricultural Museum in Kagoshima, his father was a physician and head of his village in Ogaki. As the eldest son, Yata was pressured to enter the medical profession by his father. But it seems that he was crippled by crises of self-confidence and confusion about his future which delayed his higher education. In 1931, at the age of twenty-four, Yata experienced Japan's colonial life in Taiwan, where he entered the Emperor Dali School of Agriculture. The following year he finally yielded to paternal wishes and enrolled in a preparatory year at the Jikei University School of Medicine in Tokyo. This institution was founded in 1881 by Kanehiro Takaki (1849–1920), who distinguished himself by eradicating the disease beriberi in the Japanese Imperial Navy through dietary changes.

According to Nobuyoshi Ohba, one of his most prominent disciples, Yata graduated from Jikei in 1936, when he was already twenty-nine. During the Jikei years Yata "served as an assistant, a lecturer, and an assistant professor in the Department of Hygiene and Bacteriology" (Buck et al., 1995). One of his Jikei professors, Yoshiwo Yasaki, who welcomed him to his bacteriological laboratory in his third and fourth university years, particularly inspired him. Yasaki had studied diseased shrimps infected by luminous bacteria, and his awakened interest in luminous bacteria led him to demonstrate in 1928 that the luminescence of the knight fish was caused by a harmless symbiosis with luminous bacteria (see chapter 9). Yata's exposure to the work of his professor initiated his lifelong fascination with bioluminescence and anchored his determination to pursue research on the subject.

Haneda was fascinated with bacterial symbiosis in both luminescent fishes and squids. From his interest in the bacterial luminescence of squids, no scientific contribution emerged in the early years. He was aware, of course, of the work of his Japanese predecessor, Teijiro Kishitani, of whom little is known except that he was associated with Hiroshima University. Kishitani's research on luminous symbiotic bacteria of squids was supported by the Tokugawa Biological Institute, a presursor of the philanthropic model epitomized today by the Howard Hughes Medical Institute. Its instigator was Marquis Yoshichika Tokugawa, of whom more later. Kishitanis's work on squids became the material for his doctoral thesis (Kishitani, 1932) under the supervision of Hirotaro Hattori, who had been Marquis Tokugawa's teacher in Tokyo and whom the marquis had hand-picked to head his Biological Institute. Hattori had also tutored Crown Prince Hirohito in biology and had taught him Darwinism and the use of the microscope (John K. Corner, 1990).

Kishitani found that the myopsid squids *Euprymna* (Kishitani, 1928a), *Sepiola* (1928b) and *Loligo* (1928c) possess glandular light organs with an exit opening and filled with culturable bacteria, whereas oegopsid squids (*Watanesia*, *Abralia*, *Enoproteuthis*, *Chiroteuthis*) possess closed light organs (no exit opening) and contain rodlets suggestive of bacteria, although no bacteria could be cultured (1932). He also found that each myosid squid

Figure 13.1 Yata Haneda, probably in the late 1930s. Courtesy of Harano Agricultural Museum, Kagoshima, Japan.

harbours its own distinctive species of luminous bacteria, and finally concluded that the bacterial colonies of the light organs are not inherited from generation to generation, but are acquired anew from surrounding waters through the pores giving access to the light organs.

Starting in 1934 Yata studied the symbiosis of luminous bacteria with fish in Yasaki's laboratory. His research produced two papers by which Yata received his baptism of scientific authorship (Yasaki and Haneda, 1935, 1936). In the 1935 paper, the pair described ten species of rat-tails – macrourids, related to codfishes – all of which share luminous organs of the same type: an abdominal photogenic mass filled with luminous bacteria, bordered on the dorsal side by a reflector and on the lower side by pigment cells, and connected to the rectum by a duct. Twenty years later Haneda recalled that he "was able to obtain several strains of luminous bacteria from each species of the Gadidae and Macrouridae … All the bacteria had the same general biological characteristics but they varied in their optimum temperature, being higher in some cases but never varying greatly. I think all these luminous bacteria are of the same group" (Haneda, 1955). Similar findings applied to another macrourid, the softtail grenadier *Malacocephalus laevis* (Haneda, 1938). The next year they added a new fish to the list of luminous species. The fish became known as the glowbelly (*Acropoma japonicum*) and for good reasons. Yoshiwo Yasaki and Yata Haneda (1936) observed that "the entire surface of the abdominal area is constantly lit of a pale blue." The light organ resembles that of rat-tails, but the reflector and abdominal lens are arranged such that the bacterial light emission is transmitted broadly through translucent abdominal muscles, resulting in a diffuse light over the entire ventral surface.

Then in 1939 Haneda discovered a zoological oddity, an earthworm exposed to seawater, which also happened to be luminescent (Haneda and Kumagai, 1939). The earthworm, *Pontodrilus matsushimensis*, is found in many intertidal flats of Japan. As Haneda recalled, "The luminosity of this worm was discovered by the late Dr. Kanda [see chapter 10] and myself when we saw them in the wet sand at the tidal line near Yokohama. This worm is also not luminous under normal conditions, but it will discharge yellow luminous mucus from mouth and anus on strong stimulation or injury."

After such auspicious beginnings, Haneda was eager to maintain the momentum and push aside any thought of practising medicine. While keeping his position in Yasaki's lab, he seized a unique opportunity to branch out and get acquainted with more luminous organisms by taking a staff position at the Palau Tropical Biological Station. Palau is an isolated archipelago of the western Pacific, 900 km east of the Philippines' Mindanao Island. Japan had occupied German-ruled Micronesian Islands during the First World War, and in 1922 Japan was granted a mandate over Palau by the League of Nations. Soon the Japanese immigrant population, especially fishermen, outgrew the natives, and the Palau Branch Government oversaw the building of many colonial houses and a thriving economy (Peattie, 1988). The Biological Station was built in 1934 in the capital, Koror (Izumi, 1992). So although Haneda may have landed in an exotic landscape in 1937, he felt at home. His association with Palau lasted from 1937 to 1942.

The first paper by Haneda to come out of his research in Palau concerned luminous fungi (Haneda, 1939a). He was much taken in by the display of these organisms, as he later recalled (Haneda, 1955):

> It is a most remarkable fact that many species of luminous fungi appear at night in the rainy season in the forests or jungles of the tropics. The decayed wood that had grown luminous fungi was collected and

brought back to Japan from tropical countries, where the fungi continued to glow in my laboratory in Tokyo during the summer season and I was able to observe in detail their ecology, the intensity of light, and color of their light.

Up to now this book has ignored the luminescence of fungi, which "is certainly not among the most conspicuous manifestations of bioluminescence" (Wassink and Kuwabara, 1966). But now the time has come to remedy that oversight. Harvey (1957) pointed out that fungi as sources of luminescence came to the fore only in the first half of the nineteenth century, when naturalists became aware that the luminescence of dead wood originates in the vegetative part of fungi (mycelium) embedded in the putrefying wood. Generally the mycelium is the only luminescent part, but there are some species, mostly in the tropics, where the fruiting body or toadstool is also luminescent. Haneda was instrumental in adding many species in which the gills (or lamellae) of the fruiting bodies are luminous (Haneda, 1939a, 1942). He even measured the emission spectrum of their bioluminescence, which showed a peak around 525 nm, that is, perceptibly in the yellowish green range.

So far the clam *Pholas dactylus* was thought to be the only luminescent species of the boring clams. But in Palau, Haneda discovered another species, *Rocellaria grandis* (1939b). The distribution of the luminescent zones on the body is strikingly similar to that of *Pholas*. Haneda also discovered new species of luminous fish. The shallow-water ponyfishes (Leiognathidae) do not appear to be luminous as no external light organ is apparent. But Haneda (1940) dug deeper and found a bacterial light organ arranged as a ring around the oesophagus. The light radiating from the organ is refracted through semi-transparent muscles that act as a lens to illuminate the thorax and abdomen evenly with a bluish white light. Haneda also found that the luminescence increased in intensity when the fish was excited or when the fish uttered its weird "snore" sound.

During a tour of Southeast Asia, Haneda observed synchronously flashing fireflies in New Guinea. Fireflies of the American continent can synchronize their flashing in low shrub or grass areas, but that synchronicity does not compare with the otherworldly display of the Asian male fireflies in trees. Others had reported on these tree light shows in the past and in Haneda's day by Thomas Morrison (1929) and Hugh Smith (1935) in Siam (today's Thailand). Naturalists were at a loss, and at odds with each other, to explain the spectacle, and Smith, the latest observer, had rejected courtship as an explanation, purporting that females were not involved.

Haneda entered the fray through the observations he wrote as part of the travelogue of his trip, which included anthropological notes on New Guinea aborigines (Haneda, 1941). He intended to devote a full scientific article on the synchronous fireflies once he was back home and he could go through the firefly specimens he had collected during the trip. But, as he explained, the specimens "were burned in Tokyo during the war. Later specimens I collected were lost in Singapore, so unfortunately I am unable to report on them (Haneda, 1955)." However, Haneda later recollected in English what he had written in Japanese in 1941:

> Although my specimens from these countries were lost, I can never forget the amazing spectacle of synchronous flashing of fireflies in New Guinea. I happened to see it in March, 1940, at the Rabaul Botanical Garden, Rabaul, New Britain. On the leaves of a big silk tree countless numbers of fireflies were alighting and flickering rhythmically, causing the whole tree to appear as if it were breathing. This species, with black wings, was 7 mm long. Its flicker is distinct, because when the light disappears, it does so instantaneously and completely.

There in New Guinea Haneda made breakthrough observations that went a long way toward explaining satisfactorily what the synchronous flashing was all about. For this reason, it is worthwhile to quote his statement at length:

> 1 The silk tree was a big one, and the fireflies alighted forming three groups, one on the upper part, one on the middle part, and one on the lower part. The flicker was transmitted rhythmically, the upper group extinguishing its light first, followed by the middle group, and last the lower group. Sometimes the rhythmical flashing was transmitted from the lower part to the upper. The flashes were repeated at the amazing speed of seventy per minute. The phenomenon continued every day for one week while I was there, lasting from sunset to dawn,

notwithstanding the rain. It resembled the description reported by Smith (1935), except that he noted the phenomena occurring when the moon was half full.

2 When a strong electric light was directed on them for a few seconds the synchronous flashing became irregular. After thirty seconds new synchronisms arose from some other groups in the tree and extended over the whole tree.

3 Fireflies on the tree were male and female in equal number. This fact differs from the observations reported by Morrison (1929) and Smith (1935). I observed about 100 males and females, each in separate cages in the darkness. Only the males continued to flash synchronously. On the contrary the females showed irregular flashing.

4 Not only the luminous organs of the male and female, but the colors also differ. The difference is discernible with the naked eye. The color of the light of the males is yellow, while that of the females is bluish green. Their light looked like scatterings of yellow and bluish green powder when the tree was shaken.

5 Even after dawn, with the sun shining brightly, the fireflies remained on the leaves of the tree.

6 This species of firefly selected thin-leafed trees. Sometimes these fireflies flashed synchronously as they flew through the air.

7 Many copulating fireflies were found on the grass under the tree. At that time the males were not emitting light. In view of this observation, it seems the synchronous flashing is a behavior pattern by which the males invite the females to a group. I do not believe there is any permanent leader in the group that acts as a continual pacemaker for the synchronous flashing. I think that when some individual or group emits light, it has a stimulating effect that causes the light to spread throughout the whole group as a wave.

This synchronous firefly (*Pteroptyx effulgens*) is known to be mildly toxic when eaten. Incidentally, Haneda's ex-student Nobuyoshi Ohba discovered recently, along with Victor B. Meyer-Rochow, that at least twenty insect species living in the Papua New Guinea flashing trees mimic some of the appearance and behaviour of the firefly (Ohba and Meyer-Rochow, 2012). They suggest that these insects use the mimicry and their residency in the tree to ward off potential predators, which have learned to avoid preying on the distasteful fireflies.

On a return trip to New Guinea, Haneda was fortunate to watch the flashlight fish *Anomalops* and its "large half-moon shaped luminous organ below the eye." Here is how he recalled his observations based on a paper published in Japanese (Haneda, 1943):

> In June 1942, I saw schools of *Anomalops* at the surface of the water in the harbor of Manukuwari on the northwest coast of New Guinea. At that time I collected three small specimens and was able to observe their luminescence. When this fish is swimming under natural conditions, the luminous surface appears and disappears intermittently. If the fish is caught and put into a glass jar, its luminous display becomes irregular, if the water is in any way unsuitable. A fish in a dark place which is suddenly illuminated by switching on an electric light will cease to display luminosity in one or both organs. In daylight the fish will not display its luminosity, but if the place in which it is kept is suddenly darkened, its luminosity is immediately displayed and appears as a bluish green light.

Japan was at war, and no matter how absorbed Haneda was in his pursuit of luminous organisms, the reality of war was knocking at his door. According to his memoirs some of Haneda's classmates had already been killed in combat. In the fall of 1942 the Science Council of Japan appointed him Director of the Raffles Museum in Singapore. Singapore, then a British crown colony, had surrendered to the Japanese military forces of Lieutenant-General Tomoyuki Yamashita on 15 February 1942 (Turnbull, 2009). Yamashita was "then called the Tiger of Malaysia and [was] hanged for his war crimes after the war as the only Japanese war criminal whose sentence was either not commuted by General Douglas McArthur or who was not given the privilege of a more honourable execution by a firing squad" (Arditti, 1989).

A certain Marquis Yoshichika Tokugawa (1886–1976), who was related to the emperor and "took advantage of his wealth, name, and connections to become a major force in Japanese politics" (Pitelka, 2011), accompanied

Yamashita's campaign through the Malay peninsula and helped restore order in Singapore insofar as the preservation of scientific treasures were concerned, especially the Botanical Gardens and the Raffles Museum (John K. Corner, 2013). A forestry expert, Tokugawa was, in the words of John K. Corner (2013), "first and foremost, a man of science and his code was probably not that of the military. This was perhaps to allow him to achieve the conservation of all things scientific." Emperor Hirohito, himself a practising marine biologist, was no doubt won over to the preservation of scientific specimens and documents in Singapore and backed Tokugawa's appointment as president of the Museum and Botanic Gardens, effective 1 September 1942 (John K. Corner, 2013). Tokugawa, in turn, went back to Japan and brought Kwan Koriba (1882–1957), a distinguished botanist who had just retired from the University of Kyoto, along with Yata Haneda, back to Singapore to take the directorships of the Botanical Gardens and the Raffles Museum, respectively.

What led to Haneda's appointment is unclear. His memoir suggests that superior officers of the Japanese occupation army in New Guinea were impressed by him at the time of his scientific visit there in 1942 and that good words about him trickled back to Tokyo. One way in which Haneda may conceivably have secured the good graces of the army was in advising them on the use of dry crustaceans (*Cypridina*) in military campaigns. Haneda had told Osamu Shimomura (2006) that "the military had planned to use the material as a source of low-intensity light during the war in New Guinea and other places in the southern Pacific. One of the intended uses was to mark the backs of soldiers at night with the glowing substance, allowing soldiers to identify and follow one another silently through the dark jungles." Apparently, Japanese soldiers also rubbed dried cypridinid ostracods in the palm of their moistened hands at night to read maps while avoiding detection by their enemy (Chase, 1967).

At any rate, another factor in Haneda's appointment was probably Marquis Tokugawa himself, who was already acquainted with Haneda; according to acknowledgments in one of Haneda's papers, he had supported Haneda's research in Palau through the Tokugawa Biological Institute, which was created in 1918 (Pitelka, 2011). So for Haneda it was simply a matter of pondering his acceptance of the Singapore appointment with his mentor, Yoshimo Yasaki, and his department head at the Jikei Medical School. Thus, the medical doctor with strong natural history leanings and no military background came to occupy an administrative office of high responsibility overseas. He and Koriba were given officer ranks in the Japanese army, apparently an inescapable condition of their appointments.

Between the British surrender and the arrival of Haneda, a few of the British scientists serving in the colony were kept on the job by Marquis Tokugawa to ensure continuity in the Gardens and Museum as the occupation army worked to stop looting and rigorously restore order. One of them was John (E.J.H.) Corner (1906–1996), a distinguished botanist and mycologist, and assistant director of the Singapore Botanic Gardens when the Japanese invaded. Corner wrote a memoir of his years in Singapore during the Japanese occupation (John E.J.H. Corner, 1981), and his son John K. wrote a very personal, informative and touching biography of his father (John K. Corner, 2013). The British botanist was to play an important role in Haneda's life and career. In 1986 Haneda recalled how their paths crossed:

> In late 1942 I arrived at Syonan [Japanese renaming of Singapore] and was assigned to the Syonan Museum. I found there three English scientists, Corner, [William] Birtwistle and [Eric] Holttum. It was Marquis Tokugawa's idea to let them pursue their studies, as their cooperation would be needed for saving cultural properties at the Gardens and the Museum from plundering. I became a sort of their supervisor. Because I had no superior officer [to] whom I was accountable, I was free until the end of the war to do whatever was needed for dealing with them and administering the Museum. These three scientists cooperated with me for preserving assets from damage of war, recovering those left behind by evacuated people, storing them in the church next to the Museum, sorting the important books from others, and categorising them. Birtwistle [an ichthyologist] helped my study of luminous fish, Corner my study of luminous mushroom with no interference. (Communicated by John K. Corner)

And indeed there was a complicity between them that led to enjoyable scientific excursions. Haneda would walk together with Corner in the Gardens after dark to collect luminous insects and fungi (Arditti, 1989). According to John (E.J.H.) Corner (1950), he and Haneda discovered new

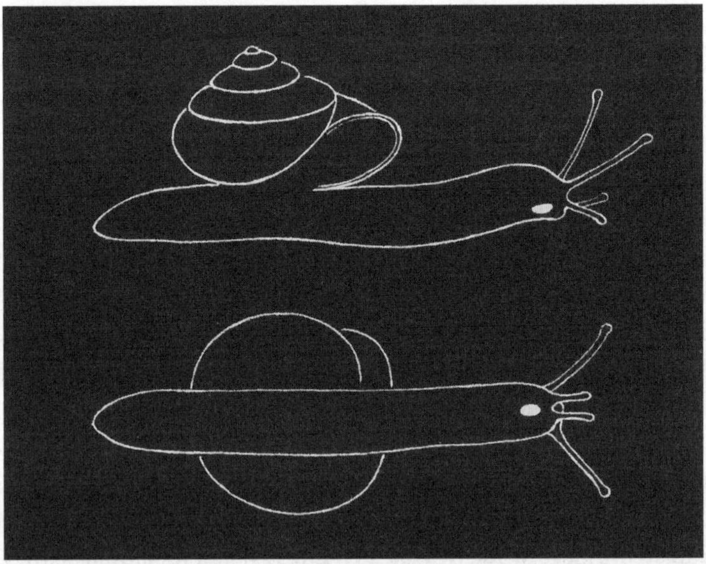

Figure 13.2 Diagram of lateral (top) and ventral (bottom) views of the snail *Dyakia striata*, highlighting the luminous organ near the tentacles. From Harvey (1952) after Haneda (1946).

luminous fungi (agarics) in Singapore, and Haneda had prepared a manuscript on them in 1944, but the manuscript was lost. It became incumbent on Corner to write the paper after the war. About the species *Dictyopanus luminescens*, he wrote: "Dr. Haneda showed me the bright green phosphorescence of the whole fruit-body … (1950)." He said of the other species, *Mycena rorida*, that it "is remarkable among luminescent fungi because only the fresh damp spores, as seen round the base of the stem on the stick or leaf, are luminous."

Such excursions led also to the discovery of the only known case of luminescence in a land snail at that time (*Dyakia* = *Quantula striata*). Haneda recalled the night of September 1943: "when Mr. Kumazawa, entomologist, was collecting luminous larvae of fireflies on the lawn of the Goodwood Park Hotel, Scotts Road, Singapore, he saw a weak light from a small land snail and informed me of the possibility of luminescence in land snails. The next evening we went to the place and were astonished to observe a true luminescence in this animal … The light appears inside the anterior region of the foot and cannot be seen when the animal is irritated and has withdrawn within its shell. When expanded the bluish white light passes through the translucent muscles of the head and of the foot and flickers like a firefly" (Haneda, 1955). The paper reporting the discovery only appeared after the war (Haneda, 1946).

Sometimes Haneda would host Corner and Birtwistle for breakfast and they talked shop. Both Koriba and Haneda showed solicitude for their British colleagues. John K. Corner wrote of an incident in late 1944: "my father had two bouts of flu and was bed-ridden for several days. Dr. Ogkawa, a medical friend of Dr. Haneda, gave him injections of camphor as a last resort for a failing heart" (John K. Corner, 2013). In January 1945 the British trio (Corner, Birtwistle, and Holttum) was picked up by the Japanese military police and interned for five days to remind them whose rule they had to buckle under. "When they were freed," John K. Corner wrote, "the three were collected by Professor Koriba and Dr. Haneda in their cars. My father recalled that they were taken straight to Professor Koriba's house for a bath and then to a banquet specially prepared. Their internment had grieved Koriba and Haneda."

Four days after Japan's surrender was announced on 15 August 1945, "all the Japanese records and also many of Dr. Haneda's records and papers were burned by the Japanese (John K. Corner, 2013)." This was a serious blow to Haneda, because precious scientific records and manuscripts were erased along with the politically sensitive documents. What followed is dramatically recounted by John K. Corner (2013):

> Koriba, Haneda and the other Japanese scientists were interned by the British on 11 September 1945. When my father heard this, he tried to arrange the release of Koriba and Haneda to work in the day time, if not to stay at the Gardens at night, but he was told that they preferred to remain with their own people. My father never knew if that was the truth, but guessed that they expected that he and Birtwistle would quickly leave for England thus leaving them in the hands of unsympathetic foreigners…
>
> Early in October, my father submitted, via Lieutenant Colonel [Gilbert] Archey [a New Zealand officer], personal reports on both Koriba and Haneda which he, with important help from Birtwistle,

had written. These were destined for the British Military Administration. According to Ms. Yukie Kanie, Dr. Haneda's daughter, these reports certainly saved her father from being tried as a war criminal.

Now, why should Haneda, a mild-mannered administrator and scientist, be charged with war crimes? Did the British Military Administration possess information that incriminated Haneda by association? Haneda had struck a friendship with Marquis Tokugawa, and many letters, which are preserved at the Tokugawa Memorial Foundation in Shibuya-ku, Tokyo, were exchanged between the two in the course of the war. But their relationship went back to the Palau years (1937–42) when Tokugawa's Biological Institute had supported Haneda's research. The marquis escaped inquiry by the British because he "had already been recalled to Japan in the middle of 1944. Why was he recalled? Was it realised that the war was lost and a relative of the Emperor was to return to his homeland" (John K. Corner, 2013)? After all, as supreme consulting advisor to the Japanese colonial administration of Singapore, he had his say in wartime planning (Pitelka, 2011). When I requested access to Haneda's letters, I was told that the letters "were not made public because of their sensitive content on the state of the war at the time" (communication between Rie Koyama-Hayashi and the Tokugawa Memorial Foundation). Access was not granted. However, Marquis Tokugawa's diary was examined by Japanese scholars, who found "an inexplicable gap in the diary between 11 February 1942 and 19 May 1942, which covers the period of the Chinese purge [in Singapore]" (Allen, 2003). This raises suspicions that can hardly be articulated, let alone verified. John (E.J.H.) Corner had no such qualms, and in the widely read magazine *Nature* shortly after the war he emphasized that "in the interest of science, one must distinguish carefully between the 'Japanese' of popular conception and the Japanese men of science who in Malaya, at least, endeavoured to serve science with impartiality" (John [E.J.H.] Corner, 1946).

Haneda was released from the internment camp early in 1946 and by 11 February, according to his memoirs, for the first time in four years, he stepped on Japanese soil at the Otake harbour in Hiroshima Prefecture. He took the night train back home and reunited with his wife. He pondered his future in a bedraggled country: was he going to make a living as a physician or seek a civil service job around Tokyo? In the end he served as Head of Health Education for the Yokosuka City Board of Education between 1947 and 1954, while retaining a loose affiliation with his old *alma mater*, the Jikei Medical School.

Yokosuka City has an interesting history, dating back to the middle of the nineteenth century (Tompkins, 1981). When pressure was brought to bear by Commodore Matthew Perry on the Tokugawa Shogunate Government to end its isolation from world trade in 1853–54, the government turned not to the Americans, but to the French Government to help them build a shipyard capable of producing large ships for export. In the 1860s the French engineer Léonce Verny spotted the sleepy little village of Yokosuka in the Miura peninsula, at the entrance of the Bay of Tokyo, which to Verny resembled geographically the navy port of Toulon in France. The shipyard was built there in the late 1860s and in 1886 was reorganized as the Yokosuka Navy Yard. What was originally intended as a merchant port developed into the main instrument of naval deployment for the aggressive imperialistic aims of Japan in the twentieth century. In the course of the Second World War the shipyard turned out aircraft carriers as well as battleships, and served also as a gun factory and arsenal depot.

After the war the naval base was occupied by the US Marines, and the US Fleet Activities settled there. The Fleet Activities Commander between 1946 and 1950, Benton W. Decker, made every effort to help put Yokosuka's economy, political institutions, and infrastructure back in operation. Old Navy buildings were converted into hospitals and schools. It is likely that the ensuing investment in education opened up opportunities for Haneda in Yokosuka, where he remained for the rest of his career. His pre-war affiliation with a medical school department of health and bacteriology made him a good fit for the job he embarked on. However, it left him little time to indulge his passion for bioluminescence field work.

For this reason Haneda published a paper on a luminous insect, the star worm (*irat intan* or *irat bintang* in Indonesian), which he had observed in Singapore and other South Asian locations in the 1940s (Haneda, 1950a): "This animal has a pair of luminous dots on the second and twelfth segments, three luminous dots on each segment from the third to eleventh, and one dot on the last segment. The insect is very beautiful, emitting a bluish green light from each dot. The body is larval in form and wingless, similar to *Phengodes* or the railroad worm of South America." To his surprise Haneda discovered that the larva-like worm (*Diplocladon hasselti*, family Rhagophthalmidae) is actually the mature female and that the adult male resembles a winged firefly but without light organs.

Haneda also recycled material from his tropical explorations in the late 1930s and early 1940s by converting the papers published in Japanese journals for English-speaking readers (Haneda, 1950b, 1951), but he did little original work. One such work was a paper on a fish of the deep-sea hatchetfish family in which he showed that the colour of the light coming out of the photophores was determined by a colour filter inside the light organs (Haneda, 1952). "I have examined living *Polyipnus stereope* in the dark," he wrote, "and have found the luminescence, as emitted, to be a greenish blue, whereas the color of the filter is violet."

During his years at the Department of Health Education in Yokosuka, Haneda struggled not only to find time for research, but also to get his papers published. The occupying powers came to his rescue. The Scientific and Technical Division of the Economic and Scientific Section of the General Headquarters of the Supreme Commander for the Allied Powers – a bureaucratic mouthful – was instituted in 1947 to help "the Japanese get back on their feet scientifically and back into the mainstream of world scientific and technological activity after nearly a decade during which they were almost totally cut off from the world scientific community" (Dees, 1997). A member of the division staff, the entomologist Donald G. Pletsch, was directly instrumental in financing Haneda's early publications in the journal *Pacific Science*. Another source of funding came from the Tokugawa Biological Institute, the brainchild of Marquis Tokugawa mentioned earlier in this chapter. Haneda maintained links with the Tokugawa family after the war.

Now residing on the Miura Peninsula, Haneda decided to take a survey of luminous organisms there. He described dinoflagellates, sea pens, jellyfish, comb-jellies, copepods, ostracods, insects, squids, pyrosomes, and fish (Haneda, 1953). The next year he reported on a small intertidal snail,

Planaxis, now known as *Angiola* (1954). "When the animal is under natural conditions on a rock with its foot well extended," Haneda (1955) wrote, "no light can be seen. However, the light appears on strong stimulation. If many specimens are placed in a bottle and well shaken in the dark, some of them become luminous and twinkle. The light continues one or two minutes and then gradually disappears. If the body of the snail is irritated, the light reappears, and if placed in fresh water, the light continues for a longer time." A large patch of the dorsal mantle contains many clusters of luminous cells.

Haneda's definite escape from the postwar scientific wilderness happened when he was invited to attend the first-ever conference on bioluminescence, which was held in the spring of 1954 at Pacific Grove on the Monterey Peninsula of California. The venue provided Haneda with an unequalled opportunity to make his great body of work known to the international scientific community (Haneda, 1955). In particular, he was able to meet face to face with E. Newton Harvey for the first time and forge strong ties with him. However, Harvey's correspondence shows that he had already in December 1949 initiated steps to obtain the early publications of Haneda in Japanese or German through the offices of Donald G. Pletsch, who had been instrumental in financing Haneda's early publications in the journal *Pacific Science*, as mentioned earlier. This led to Haneda himself sending Harvey in February 1951 a handwritten translation of his 1946 paper on the Singapore land snail, to be included in the book he was writing, *Bioluminescence*. Shortly after that, Haneda started shipping dried cypridinids to Princeton; in April 1952 he received five hundred dollars from Harvey for expenses incurred in paying collectors, chartering boats, and other miscellanies. Harvey's solicitude extended to the promotion of exchanges between Haneda and other influential Americans. In a letter to Haneda dated 9 February 1953, Harvey wrote:

> This letter will introduce to you Dr. Harold J. Coolidge, the Executive Director of the Pacific Science Board, who will shortly make a trip to Japan. I gave him your home address and also the Jikeikai Medical College in Tokyo and he will look you up.
>
> It is very possible that the Pacific Science Board will sponsor a study of bioluminescence in the South Pacific and I would be pleased if you

> could give him information regarding scientists in Japan interested in that subject. I also told him that I hoped he would be able to employ you to help with the scientific research. You will find him a very fine person and I hope his visit will stimulate the study of luminous organisms in Japan and particularly to help you with your own work.

After the bioluminescence conference of 1954, Haneda moved on from California to various locations in the United States where he indulged his knack for detecting luminous organisms in the outdoors. He sent to Harvey live luminous earthworms from Key West, Florida, so that Harvey could observe their luminescence in his lab. In a letter of 31 May 1954, Haneda, by now back in Japan, wrote to Harvey:

> It was my utmost pleasure that I was able to talk to you at Pacific Grove and Princeton and I herein express my sincere gratitude to you and Mrs. Harvey who kindly invited me to your home full of various Japanese souvenirs [from past travels by Harvey and his wife]. Besides, the famous Parker's fountainpen and pencil which you presented to me is highly appreciated and I will keep it with the utmost care as a memorial gift. Above all I thank you so much for your good offices in giving me the chance of visiting the United States which was so splendid and wonderful.

That same year Haneda became the first director of the newly created Yokosuka City Museum. His tenure as Head of Health Education may have triggered a train of thought on the best way of exposing the citizenry to science, but according to a personal communication of Professor Mikiko Ishii, Kanagawa University, to John (K.) Corner, the idea of a museum in which research is conducted while specimens and documents are displayed was fostered by memories of the conduct of research at the Raffles Museum with John (E.J.H.) Corner. Haneda enlisted the help of Masayoshi Nagano, mayor of Yokosuka, to get the project off the ground. Bioluminescence as a field of study benefitted from the new Museum in that exhibits of luminous organisms raised the awareness of the Japanese public to the phenomenon. And in 1956 Haneda also created a scientific periodical, the *Science Reports of Yokosuka City Museum*, which disseminated around the world scientific studies of bioluminescence by Japanese biologists, albeit largely in the Japanese language.

~

A chance discovery in 1958 propelled Haneda again to the forefront of bioluminescence research as he reached the age of fifty-one. Luciferin-luciferase reactions, which so far had been documented in pholads, fireflies, ostracods (*Cypridina*), the Bermuda fire worm, shrimps, the freshwater limpet-like *Latia neritoides*, a dinoflagellate and bacteria, were believed to be highly specific. Numerous attempts over decades to cross-react the luciferin of one organism with the luciferase of another, or vice-versa, had failed unless the two organisms were closely related – members of the same genus or family. Haneda had started collaboration with Frank Johnson, who had taken over the bioluminescence lab from Harvey at Princeton. A new luminous fish discovered by Haneda in Japanese waters, the pigmy sweeper *Parapriacanthus beryciformis* (now *ransonneti*), was found to produce a luminous secretion that was not bacterial and did produce a luciferin-luciferase reaction – a first for fishes (Haneda and Johnson, 1958a).

A letter of Haneda to Harvey, dated 19 November 1958, expressed his perplexity at finding that the luciferin of a luminescent cardinalfish, *Apogon marginatus*, cross-reacts with the *Cypridina* system, a preliminary report of which appeared the same year (Haneda, Johnson, and Sie, 1958b). Here was a clear case of cross-reaction between a fish and a crustacean, an interphyletic relationship. Harvey quizzed Haneda about the possibility that the fish had eaten cypridinids, thereby making the *Cypridina* luciferin available for the luminescent reaction of the fish. But in a letter of 7 January 1959 Haneda replied that "according to my experience so far I have failed to find out any Cypridina, or its traces in the digestive tract of Apogon … Besides, I cannot believe that Apogon could eat Cypridina also from the fact that the Apogon is an oceanic animal, living in 40 fathoms deep of water, while Cypridina is a coastal and would live in far shallower waters than the former."

Matters rested there even when the full paper on the interphyletic luciferin-luciferase reaction appeared two years later (Johnson, Haneda, and Sie, 1960). When subjected to a battery of chemical and physical tests, the luciferin of *Apogon* displayed similar properties with *Cypridina* luciferin. The luciferases, however, showed divergences in their properties. "Although the similarities in the luminescent systems of a fish and crustacean could represent a rare, evolutionary coincidence," they concluded, "they as likely indicate that more of a thread of unity exists in the comparative biochemistry of luminescence among diverse types of organisms than has been hitherto supposed." But Haneda was in for a surprise when he studied the anatomy of the pigmy sweeper and the cardinalfish (Haneda and Johnson, 1962). Shining an ultraviolet lamp on the guts of the fish betrayed the strong greenish-yellow fluorescence typical of *Cypridina* luciferin in the pyloric caeca – finger-like extensions of the stomach – and in the anterior portion of the intestine. He also found that the thoracic luminous organ opens into the pyloric caeca via a thin duct and the anal light organ similarly connects to the rectum. Luciferase is only present in the light organs, so the gut cannot light up even with the luciferin stored there. And finally, Haneda found cypridinids in the stomach of some of the fish.

At this point everything was in place to validate the hypothesis that these fishes obtain their luciferin through their diet of cypridinids, that they are, in the words of Thérèse Wilson and Woody Hastings (2013), plagiarists and thieves. But the evidence did not discard the possibility that the fish produce their own luciferin in addition to acquiring it from an outside source. The best way to eliminate this alternative, as Haneda, Johnson, and Shimomura (1966) explained, is "to cultivate such fishes experimentally for a considerable period of time under conditions where *Cypridina* is not available as food to see if these fishes cease to emit light." Fate had it that another fish, which possesses skin photophores wholly unconnected with the digestive system, was instrumental in solving the riddle. Milton Cormier and his team at the University of Georgia, inspired by Haneda's and Johnson's work, extracted from the Atlantic midshipman fish, *Porichthys porossisimus*, a luciferin identical to that of *Cypridina* (Cormier, Crane Jr, and Nakano, 1967). But on the West Coast, the local species (*P. notatus*) is luminescent from Mexico to San Francisco, with an isolated population around Washington State – where *Cypridina* is absent – which is not luminescent yet possesses photophores. Frederick Tsuji, Anthony Barnes and James Case (1972) took advantage of this peculiarity to demonstrate that when non-luminescent fish are fed cypridinids, they become bioluminescent. It was later found that the ingested *Cypridina* luciferin is directly incorporated into the photophores and does not induce *de novo* synthesis of a fish luciferin (Thompson et al., 1988), thus providing overwhelming evidence that bioluminescence in the midshipman fish is entirely dependent on the food chain.

Haneda's pioneering observations planted the seed of a scientific discovery of great ecological and evolutionary significance. First, it ushered in the awareness that the role of marine trophic chains is not just to provide access to food for sustenance and energy, but also to acquire new biological functions for the feeding organism, such as bioluminescence. Once the luciferin of the prey is incorporated into the predator, it is a matter of adaptative evolution for the predator to develop light organs and express its own brand of luciferase in these organs. And second, it helps explain why so few chemical luminescent systems evolved in comparison to the great number and diversity of bioluminescent species. Indeed, if luciferins are available in the food chain, why bother investing energy in fabricating it? The biodiversity of luminous organisms across many phyla is maintained by selection for traits such as light organs and an oxidase to make bioluminescence work for the animal.

Haneda continued to collaborate with the Princeton lab in the 1960s, but he also teamed up with Tsuji in Pittsburgh (see chapter 11). Their collaboration was predictable; Tsuji was American, but his having both parents of Japanese descent ensured that he spoke decent Japanese (as I know from personal acquaintance). Haneda, on the other hand, although he improved over the years, always struggled with the English language. The upshot was that Haneda felt comfortable working with Tsuji who, like the Princeton group, gave him access to top American scientific journals for disseminating his results. Their collaboration started in 1965 when Haneda hosted a bioluminescence conference under the auspices of the United States–Japan Cooperative Science Program. It was held in Hakone National Park, about 70 km from Yokosuka and a great tourist destination offering majestic views of Mount Fuji. Among the proceedings of the conference, edited by Haneda and Johnson, was a paper by Tsuji and Haneda (1966) in which they reported differences in chemical properties between the luciferases of the cardinalfish and *Cypridina*. This was followed by a study of the luminous cardinalfishes of the Philippines (Haneda, Tsuji, and Sugiyama, 1969) and a study of the luminescent system of a lanternfish, which they reported, erroneously as it turned out, to cross-react with the *Cypridina* luciferin-luciferase system (Tsuji and Haneda, 1971).

Haneda relinquished the directorship of the Yokosuka City Museum in 1974, but he continued there as a researcher. One area of research that he particularly nurtured late in his career was the study of Japanese fireflies. His fascination with firefly flash synchrony climaxed with a paper on the fireflies of New Guinea (Haneda, 1966), but he yearned for a full inquiry of flash communication as exploited by Japanese fireflies, spurred in this direction by the work of the American John Buck (see next chapter), who visited him in Japan. Haneda found in Nobuyoshi Ohba the person he needed to pursue his dream. Ohba became curator at the museum in the 1970s. He helped establish a firefly rearing program and a world inventory of firefly species, including forty-six Japanese species. But his major research contribution was his exhaustive study of flash communication in Japanese fireflies, which was reviewed for the English-speaking audience (Ohba, 2004). Ohba, in his turn, is now retired.

Haneda travelled extensively throughout the United States and Europe in the 1960s and 1970s, and developed a network of colleagues who shared his dedication to bioluminescence research. The last meeting on bioluminescence he participated in was in Dunedin, New Zealand, in 1983, on the occasion of the fifteenth Pacific Science Congress. Two years later he published a book on luminous organisms for a general Japanese readership (Haneda, 1985). He died on 30 January 1995 in Yokosuka, at the age of eighty-eight, when the field of bioluminescence had moved on from the physiological and biochemical approaches to embrace the molecular age.

14
Circling the Luminaries

> Knowing the fascinating chemistry and biochemistry was all very well, but it rarely told us how a firefly flashes and a glow-worm glows.
> –Anthony K. Campbell (2008)

Among the satellites gravitating around the core of Harvey's students and collaborators, three investigators, aside from Yata Haneda, must be acknowledged: John Bonner Buck (1912–2005), Joseph Arthur Colin Nicol (1915–2004), and Jean-Marie Bassot (1933–2007). Buck was a pillar of field research on firefly bioluminescence and its physiological control, Nicol an important investigator of physiological control of bioluminescence in marine animals, and Bassot a key player in our understanding of the dynamics of light sources inside luminous cells. All three worked largely outside conventional academic institutions, being employees of government-sponsored research institutions during their bioluminescence careers; none of them developed large laboratory teams of students and technicians. That they were not engaged in biochemical research in no way deterred from the importance of their roles in the development of the field in the years spanning four decades starting in the mid-1930s. After all, they asked questions of unassailable importance: how bioluminescence fits into the grand scheme of the natural world, what the organization of light organs and photocytes can tell us about the control of light emission, and, basically, how light organs work and evolve.

John Buck was born in Hartford, Connecticut, and in his teenage years his family moved to Baltimore, a hotbed of firefly night life. The best source of information on Buck, which this chapter relies on, is by James Case and Frank Hanson (2004), who noted that in 1933, while "still an undergraduate, John independently undertook a summer vacation study on the flashing behavior of *Photinus pyralis*, which abounds in Baltimore every July." The inspiration for this piece of research was an undergraduate class taught by Johns Hopkins professor Samuel O. Mast (1871–1947), who himself had published on firefly behaviour (Mast, 1912). In due course, Mast filled the dual role of Buck's doctoral thesis supervisor and father-in-law; his daughter Elizabeth was to play the dual role of Buck's wife and research partner.

Buck's first scientific paper came out with a bang. It appeared in the prestigious journal *Science* and directly addressed the problem of firefly flash synchrony (Buck, 1935). In the preceding chapter I have described at length Yata Haneda's observations and views on the subject of mass synchrony, but Buck went beyond the phenomenology and tried an experimental approach to addressing the mechanism of synchrony. He later recounted how he came to this approach (1937b):

> The fact that the female invariably responds to the flash of the male after an interval of approximately 2 seconds suggested that the male might respond to an artificial light if it were flashed, like that of the female, 2 seconds after his flash. This experiment was performed by resting the end of a five-cell flashlight on the ground so that the light was directed into the grass at an angle, and flashing it about 2 seconds after each flash of a male flying nearby. The male turned and flew toward the flashlight in precisely the same manner that males respond to the flash of a female. The method was astonishingly successful, and it enabled the writer to exert an almost magical control over the male firefly. Individuals were "attracted" straight to the flashlight from distances of over 50 feet without difficulty, and fifteen or twenty males were often "attracted" to the flashlight at once.

This pragmatic experimental approach was to stand Buck in good stead for the future as he enhanced the sophistication of the method. But meanwhile he was busy working on his thesis, out of which two papers were published. In the first (1937a) he asked whether the daily rhythm of flashing, whereby fireflies initiate flashing at dusk and terminate it when the evening darkness sets in, is activated by ambient light levels and temperature, or is determined by an internal clock. His elaborate sets of experiments decided in favour of the existence of an internal clock and a circadian rhythm, although temperature had an effect on flash frequency as well. In the second paper (1937b) he determined in the field that the time interval between the male flash and the female flash response – two seconds in the case of *Photinus pyralis* – was the only meaningful mechanism of communication between the sexes, a species-specific frequency code for mating, so to speak.

Buck was granted a fellowship for postdoctoral studies at the California Institute of Technology, followed by a professorship at the Rochester Institute of Technology in New York State. Only when he joined the Maryland-based National Institute of Health, in 1945, did he fully resume his firefly studies. Soon he produced "his masterful work on light organ structure and physiology (Buck, 1948) which was certainly by far the most complete documentation at the time and still is a valuable resource" (Case and Hanson, 2004). In it Buck did more than simply review the literature on the organization and physiology of firefly light organs; he also offered valuable insights. For instance, he distinguished six types of light organs according to the arrangement of their tracheal – and therefore oxygen – supply. He carefully evaluated various hypotheses regarding luminescence control, dismissing many and delivering a hung verdict for some. Indeed, some problems were left for the future to resolve. "Most of these problems centre around the question of whether the photogenic cell is stimulated by direct nerve action, presumably in an environment always adequate in oxygen, or whether the stimulus which sets off luminescence is the passage of oxygen into the cell" (Buck, 1948).

The 1948 paper had a long gestation, starting in 1946 when it was the topic of the earliest surviving correspondence between Buck and E. Newton Harvey. Buck was circulating the manuscript around for comments or criticisms, and Harvey was one of the recipients. Harvey, in his turn, sought advice from Buck about luminous insects of all stripes, and this accelerated as Harvey was writing his greatest achievement, the book *Bioluminescence*. At the time Buck described himself as a cytologist in his letters, to emphasize the keen interest he had developed in the cellular make-up of insect light organs. He jettisoned this self-depiction when he became head of the Laboratory of Physical Biology at the National Institute of Health in the early 1950s.

The issue of luminescence control stayed with Buck for many years to come. He couched his reflections on the subject in a paper which embraced

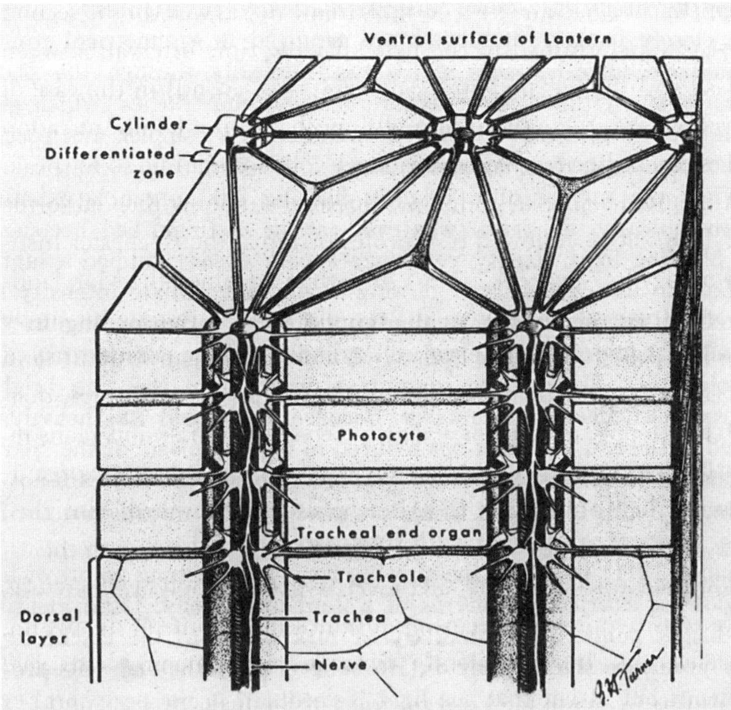

Figure 14.1 Internal organization of type 6 firefly lantern, showing the tracheal arrangement in relation to the photocytes. Fig. 1 in Buck (1966). Courtesy of Princeton University Press.

a comparative approach (Buck, 1955). He pointed out some general trends in the ways the light organs can be controlled:

> The simpler organisms such as bacteria, protozoa, coelenterates, ctenophores, and polychaetes seem either to luminesce continuously or to light up only in response to external stimuli, and it is only in groups with rather well-developed nervous systems that photogeny becomes subject to the precise sort of regulation so well exemplified in the mating signals of fireflies. Secondly, it is clear that the ability to emit sharply delimited flashes of light does not depend on a highly developed nervous system but on intracellular light production. Thus, for example, in *Chaetopterus* and *Cypridina*, animals with well-developed nervous

> systems, light production is extracellular and the luminosity decays slowly, whereas the primitively organized *Noctiluca*, *Renilla*, and *Mnemiopsis* produce flashes not inferior in temporal control to those of many fireflies. Extracellular luminescence is characterized, in addition, by rapid "fatigue" due to exhaustion of the luminous excretion, whereas intracellular photogeny may sometimes persist through many hundred successive flashes.

Buck also stressed the concept of the functional unit. By this he meant the smallest individual sources of light inside the light organ. A single flash may represent a sum of numerous, but not all, unit activities of varying intensity and duration, and in the next flash a different set of units may be recruited for activation in an ever-shifting map of microsources. As a result, "the time-intensity curve for the whole organ represents an integration of many separate asynchronous events rather than the simultaneous firing of all the photocytes in the organ" (Buck, 1955). The functional unit may also be the target of an individual nerve ending controlling it, but Buck did not venture too far in this direction. He did, a few years later, when he embarked on a project of neurophysiological recordings of firefly flashes.

To discuss this phase in Buck's career one must introduce James F. Case (1926–2013), who became inextricably linked with Buck in a lifelong collaborative partnership. Case left an unpublished memoir of his life on which the following is based. Jim Case was born in Bristow, Oklahoma, and his schooling was interrupted for a year by malaria. He grew up tinkering with radios and other devices, and showing interest in things biological. After graduating in zoology from the University of Kansas, in 1951 he enrolled in the post-graduate program of Johns Hopkins University. Interestingly, Case shared living quarters in Baltimore with Bernard Strehler (who made an appearance in chapter 11), although Case was still a few years away from warming to bioluminescence research. He had this to say about Strehler: "I shared a tiny apartment with Bernie Strehler, a near-genius biochemistry graduate student, who spent odd hours romping in his bed with his girl-friend, a lovely biochemistry grad student."

After Case successfully defended his PhD thesis "on the role of the pituitary in development," he was drafted in the Army Medical Corps as the Korean War unfolded, and was assigned "to study the effects of war gases on cockroach nerve cells and synapses." It was probably during this period that he became familiar with, and met John Buck, as both were based in Maryland and circulated among the small world of entomologists at the Agricultural Research Center in Beltsville, Maryland. This signalled the incipient stages of Case's research career as an insect physiologist. After discharge from the army he took an assistant professor job at the Zoology Department of the University of Iowa, where he continued to work on insect neurophysiology. There he supervised students – Katherine N. Smalley, Albert D. Carlson, and Frank E. Hanson, Jr – who were to leave their mark in the field of bioluminescence. Ever the tinkerer, Case started constructing "bioluminescence detectors," probably crude predecessors of the RCA photomultiplier tubes he would later use in his collaboration with John Buck.

Case and Buck started working together on firefly neurophysiology in the late 1950s (Case and Hanson, 2004). Both were in the habit of migrating each summer at the Marine Biological Laboratory in Woods Hole, so they set up a joint research lab in the basement of the "Old Main," as the most ancient building of the Woods Hole campus was called. To examine the role of the nervous system in firefly flash production, a subject of debate still contended at the time, intact fireflies or "lantern" preparations were subjected to electrical stimulation via electrodes placed at various locations on the experimental preparations, and stimulus parameters were controlled by a stimulator instrument. The luminescent responses were "followed with an RCA 931-A photomultiplier leading directly to one channel of a dual beam oscilloscope and photographed simultaneously with a second trace carrying a stimulus marker" (Buck and Case, 1961). Except for the RCA low-light detector, the stimulating and recording instrumentation came from the Grass Instruments Company, based in Quincy, Massachusetts, which was to furnish many a bioluminescence lab in the future. The company was run by Albert Grass, an MIT graduate in electrical engineering, and his wife, Ellen, both extraordinarily solicitous of their scientist customers (Zottoli, 2001).

Their first paper (Buck and Case, 1961) showed that the firefly lanterns obeyed the same rules as other neuroeffector organs such as muscles: response latency compatible with synaptic events of neural elements effecting intermediary and facilitated responses. In the second paper (Case and Buck, 1963) brain excitation was found to be associated with normal spontaneous flashing. With fine platinum-iridium electrode wires placed in the vicinity

Figure 14.2 James F. Case in New Guinea with his electronic instruments for recording firefly flash activity. Photo archives of James F. Case.

of lantern nerves, they were able to record volleys of action potentials directly associated with flash events. The third and final paper of the series (Buck et al., 1963) was a physiological dissection of a three-step control chain. A very strong electrical stimulus elicited a response delay of only one millisecond and was interpreted as representing direct excitation of the photocytes. Medium strength stimuli produced delays of 18 milliseconds necessary for the intervention of tracheal end organs; and mild electrical stimulation producing response delays of 70 milliseconds exposed the intervention of nerves and tracheal end organs upstream from the photocytes. All in all, it brought a satisfactory resolution accommodating a role for both nerves and tracheal end organs.

John and Elizabeth Buck rekindled their deep-seated interest in Asian synchronous fireflies by paying them a visit in Thailand in October 1966. Already in that year (Buck and Buck, 1966) the Bucks had suggested that "the firefly trees serve as a quasi-permanent rendezvous for mass mating and that synchrony increases their efficiency as beacons." But observations so far had all been visual, with the potential biases vision entails. Skeptics in the past had even explained away the display as the result of the observer's twitching eyelids, and other optical illusions were also alleged to be the culprits (Buck and Buck, 1968). This time the Bucks wanted to come out clean by travelling to Thailand with a partner that operates under no illusions – sophisticated recording equipment. John Buck's student and assistant, Frank Hanson Jr, built them a portable luminescence recording apparatus specially designed for field work, consisting of an RCA photomultiplier and an "electrocardiograph" chart recorder with battery-driven transistorized electronics. In addition, they backed up their visual observations by filming the tree displays with a Bolex 16-mm camera and high-speed film sensitive to the low-lights of dusk, an instrument provided by the National Geographic Society (Buck and Buck, 1968).

Thus equipped, the Bucks set up camp in "the classic locality for synchronous displays," along the Chao Phraya River south of Bangkok, where the fireflies congregate in mangrove trees. To approach the fireflies at close range, the Bucks slipped a boat at high tide up near the trees where the fireflies dwell. They were greeted by spectacular flash displays: "Each time we saw this hurrying, soundless, hypnotic, enduring performance it impressed us anew as uniquely different from any behavior we had ever seen" (Buck and Buck, 1968). Their photometric and cinematographic data showed that the male fireflies (*Pteroptyx malaccae*), when aggregated in trees, flash every half a second and that the synchrony between them is so precise that the lag is never more than 20 milliseconds. They concluded that the synchrony is regulated by feedback in the central nervous system from previous flash cycles, by analogy with the human synchronized steps in military parades. The study was considered of great significance by their peers, and it produced a long article in the magazine *Science*.

Now John Buck was totally hooked and eager to study other synchronous fireflies, this time in Papua New Guinea (see chapter 15), where Haneda had made original observations almost thirty years earlier. The four accomplices, the Bucks, Jim Case, and Frank Hanson Jr, produced an impressive paper based on their work in New Guinea (Hanson et al., 1971). In contrast to the Thailand trip, they now conducted experiments with pulsed or pacer artificial light to test how the male fireflies respond with their own flash rhythm.

The fact that firefly flash periodicity was set in motion by the pacer artificial light suggested to them that "the pacer signal resets the flash-timing oscillator in the brain, thus providing a mechanism for synchronization."

As for the biological relevance of mass flashing in trees, theories were split between two opposing camps. The Bucks articulated the controversy by opposing James Lloyd's view (Lloyd, 1973a), couched in cost-benefit argumentation, to their own (Buck and Buck, 1978). Lloyd emphasized the benefit of synchrony to the individual males who compete with nonspecific males in the tree for reproductive advantage; to him the male is calling the shot. But the Bucks saw it differently, stressing that "the female selects her mate on the basis of the intensity of his signal relative to those of other males visible to her simultaneously." Adhering to the classical Darwinian bias for survival of the species, not the individual, they argued that "flash synchronization per se, being a group behavior, cannot serve as the competitive agent for promoting the reproductive fitness of an individual participating male, no matter how small the synchronizing group." Mass congregation in trees, in their scheme, is a group adaptation to the particular physiology of the South Asian fireflies, whose brains can synchronize automatically the rhythmic flashing of conspecific males, and to the local jungle, in the sense that male-female bioluminescent communication would be problematic in dense jungle canopy, whereas congregation in trees enhances conspicuousness. The controversy remains unresolved. As Sara Lewis (2016) explained: "We don't yet have the necessary evidence to distinguish among these ideas, so exactly why fireflies synchronize remains an intriguing mystery."

In other spheres John Buck continued to add reflections on the role of bioluminescence at large. In his chapter in the book *Light and Life* edited by William McElroy and Bentley Glass, Buck speculated on two issues of interest for marine bioluminescence (Buck, 1961). The first was the striking anatomical similarity between eyes and photophores in higher invertebrates (squids and shrimps). Buck was beguiled by it and keen to envisage common developmental origins for the two structures; but surprisingly, the most parsimonious explanation – that optical management of the light may impose converging physical constraints on both photophores and eyes – is hardly considered. The second issue, "whether bioluminescence can account for the well-developed eyes of certain bathypelagic or abyssal forms," had been considered "one of the most baffling problems associated with the biology of deep-sea animals (Welsh and Chace Jr, 1937)." Buck made much of the "eye-photophore functional association," noting that at depths where there is so little natural light that eyes are expected to be degenerated, shrimps possessing photophores actually possess the largest eyes. Those which lack photophores are blind at these depths. Buck also promoted "the idea that bioluminescence must develop before eyed forms can venture permanently into stygian depths."

Seventeen years later, John Buck returned to speculative thoughts (Buck, 1978), mulling over the twenty-some catalogued biological roles attributed to bioluminescence, and over the evolutionary origin and evolutionary trends of bioluminescence. Buck aptly explained the complexities attending anyone who attempts to propose a role for a bioluminescent display: "For example, a luminous display by organism A may benefit it by attracting food organism B, but simultaneously it makes A more vulnerable to predator C. But A's luminescence helps it see and avoid C and may also attract other individuals of species A (perhaps to mutual advantage in courtship opportunity or to disadvantage as competitors for food or mate) and these new specimens of A also dilute the predatory threat from C." The possibilities seem endless and strike a cautionary note against the temptation of reducing the function of a luminous display as if the organism is disconnected from its ecosystem. Similarly, Buck was at a loss to detect any clearly traceable evolutionary trend in the historical development of bioluminescence among living organisms, and the worn saying that bioluminescence evolved independently many times in the course of evolution seemed still the only sensible one.

The Bucks continued to publish original contributions on fireflies and synchrony in their later years, with the periodic involvement of Jim Case and Frank Hanson Jr. Buck died in 2005 from non-Hodgkin's lymphoma, at the age of ninety-two.

An almost exact contemporary of John Buck, J.A. Colin Nicol, brought to marine bioluminescence research the broad scope of interest that Buck showered on fireflies. Born in Toronto on 5 December 1915, he was the firstborn of George Nicol, an immigrant from Scotland, and Josephine

Petrie, herself born in Sherbrooke, Quebec, of an English father and French-Canadian mother (Maddock, 2006). The family soon moved to various locations in New England and New Jersey, where Colin, as he was known and an anagram on his last name, suffered bullying in schools because of his French accent. The upside of his American childhood was that school-organized visits to the New York Zoological Gardens and Aquarium kindled in Colin an early interest in natural history (Maddock, 2006). When he turned ten his family returned to Canada, first to Toronto, then to Montreal, in the Lachine suburb. There he observed birds, trapped mammals, and became a fixture of local natural history societies. Here is how his brother Donald described him during these years (quoted in Maddock, 2006):

> Colin was a quiet, reserved, studious individual, never sociable in the sense of engaging in sporting activities or of going along with the crowd, but he always had a few close friends with related scientific interests. As long as I can remember he had a study with a desk, at which he spent long hours both day and night. Colin collected things, stamps, butterflies, animals, birds, reptiles and, of course, books. Once I remember walking into the kitchen where my mother was preparing a dish of mushrooms from a bushel or so that Colin had brought in. They were colourful: reds, yellows, browns, whites with flecks of black. Not as trusting as my mother, I asked Colin if he knew what he was doing. His disdainful answer was that of course he knew what he was doing! The Nicols survived this and other eating experiences so one would have to conclude that he was right.

Colin enrolled in zoology at McGill University in 1934, graduating in 1938. He completed a master's degree at the University of Western Ontario in 1940 with a thesis on the development of a fish's air bladder. He met his wife, Helen Cameron, while studying in London, Ontario. His enrolment for a PhD at the University of Minnesota was beset by multiple problems, and he gave it up to join the Canadian Forces in war-torn Europe, but not before wedding Helen Cameron in 1941. He spent the wartime in the Royal Canadian Corps of Signals, the scientific result of which was a publication on the value of the homing sense of carrier pigeons for warfare (Nicol, 1945).

Figure 14.3 J.A. Colin Nicol. Courtesy of his grandson, Richard Shelton.

Retired from the army, Nicol received a grant from the Canadian government and a British Council Fellowship to pursue a doctorate at Oxford under the guidance of J.Z. Young, discoverer of the squid giant axon, which itself led to the discovery of the basic mechanism of neuronal excitability and a Nobel Prize. Nicol completed his dissertation on the giant axons of polychaete worms in only two years, pressed by being "32 years of age with a wife to support (Maddock, 2006)." He took an assistant professorship at the University of British Columbia in Vancouver, but the teaching load was heavy and the salary meagre. Within two years he was back in the UK, taking a position as an experimental zoologist at the Plymouth Laboratory of the old and prestigious Marine Biological Association (MBA) of the United Kingdom. The committee that presided over the creation of that institution in 1884 included three figures depicted in chapter 4 in relation with the *Challenger* expedition: H.N. Moseley, G.J. Allman, and John Murray. The Plymouth Laboratory was inaugurated in 1888 (Southward and Roberts, 1987).

When Nicol joined the laboratory, repairs to bombing damages incurred in the Second World War had recently been completed, as well as lab renovations. Other reminders of the war were apparent. "Life was rather spartan for the Nicol family for several years," Maddock (2006) explained. "Colin's salary was low and there was still postwar rationing." But Nicol wasted no time scouting for projects, and by 1951 he had settled for the study of a field ripe for investigation: the nervous control of bioluminescence in marine animals. His first object of investigation was the luminous tube-worm, *Chaetopterus*, readily available by dredging in waters near Plymouth (Nicol, 1952a). The luminescence of this species had been studied before (see chapter 9), but the issue of control had not been adequately addressed. Nicol, like Jim Case at about the same time, recorded light emissions with an RCA photomultiplier tube.

On the basis of various regimes of tactile and electrical stimulation, of nerve sectioning, and the effects of anaesthetics, Nicol concluded that the secretion of luminous mucus is controlled by the nerve cord, and that propagation of luminescence excitation along the body is mediated by interneuronal facilitation (Nicol, 1952b,c). He found no evidence of peripheral (neuro-effector) facilitation. He also suggested, on weak evidence, that acetylcholine may act as an excitatory neurotransmitter for the luminescence effector. Evidence of neuro-effector facilitation, and of the roles of acetylcholine as an excitatory neurotransmitter and gamma-aminobutyric acid (GABA) as inhibitory neurotransmitter, were provided only thirty years later (Anctil, 1981). Nicol then had no access to an electron microscope; it was electron microscopy that led to the hypothesis that musculo-epithelial cells, by their contraction, pressed on the luminous cells and squeezed out the latter's mucous (Anctil, 1979). Thus, contrary to what Nicol thought, nervous control of luminescent mucus secretion in the tube-worm seems indirect, via muscular epithelial cells.

Nicol went on to examine minor aspects of the luminescence of the tube-worm, such as the effect of the chemical environment (Nicol, 1954a) and response fatigue (1954b). But his attention was already drawn to worms with intracellular luminescence: scale-worms which, like the tube-worm, had intrigued naturalists for the previous century, as earlier chapters have chronicled. Prior to Nicol's work, it had been established that the luminous cells (photocytes) of the scale epithelium do not secrete a mucus like the tube-worm photocytes (Bonhomme, 1940). Charles Bonhomme, a French histologist based at the Medical School of Montpellier, had also observed that when he removed a scale (elytra) from the worm, a patch of the scale's surface emitted flashes of light; he was convinced that the excitation of the nerve plexus inside the scale caused the luminescence. And indeed, Nicol (1953, 1954c) confirmed through a battery of neurophysiological tests that flashing was the normal display for these worms and that it was under nervous control, involving the central nerve cord and downstream to the elytral ganglion inside the scale. He also noted how the rhythmic flashes showed first facilitation, then summation, and finally fatigue. He correctly deduced that the flashes result from the simultaneous excitation of all the photocytes, and that flash facilitation occurred both at neuro-photocyte junctions and inside the photocytes.

Next Nicol ventured away from worms to take on the sea pansy *Renilla*. What brought this about goes back to January 1953, according to the correspondence of E. Newton Harvey. Nicol, it seems, felt boxed in professionally on British soil for want of new research opportunities and contacts with his peers. He had made overtures to scientists at Scripps Institution of Oceanography (SIO) in La Jolla, who advised him to contact the Pacific Science Board of the National Research Council for funding. This he did, and in his submitted proposal, dated 22 January 1953, he laid out his plans:

> This will give me the opportunity of extending my researches to some forms which can be secured readily on the Pacific coast, but which can be collected only with difficulty, if at all, elsewhere, namely luminescent fish, sea fans, and the sea pansy *Renilla*. This information is being collected with a long view of preparing a monograph dealing with the physiology of luminescence in animals.
>
> Further advantages which I hope to derive from my visit to the United States are these. I plan to call at Princeton University to see Dr. E.N. Harvey, the foremost research worker in the field of vital luminescence. I shall also have the opportunity of deriving first hand experience with the Pacific fauna, of meeting American workers in marine zoology, and of seeing an American marine laboratory in operation. I

> would like to return across Canada, and call at several Canadian universities and marine stations, in order to renew my acquaintance with my Canadian colleagues.

He finally secured a Guggenheim fellowship for his US visit. Nicol had not informed Harvey of his intentions, but Harvey learned about them from the executive director of the Pacific Science Board, who queried him about Nicol's proposal. In a reply dated 2 February 1953, Harvey endorsed Nicol's proposal, adding: "He would be a good man to have on the bioluminescence program." The program alluded to was the Conference on Luminescence, scheduled for the spring of 1954 and sponsored by the National Research Council. Nicol was formally invited to the conference, where he gave a talk on the physiological control of luminescence (Nicol, 1955a). And he fulfilled his work plan as laid out: publications based on his research in California and Washington State appeared on *Renilla* (Nicol, 1955b,c), jellyfish (Davenport and Nicol, 1955), sea pens (Davenport and Nicol, 1956), and the midshipman fish *Porichthys* (Nicol, 1957).

Unbeknownst to Nicol, John Buck was engaged in a study of bioluminescence in *Renilla* (Buck, 1953) while Nicol was pleading for travel funds. However, Buck's work was basically repeating Parker's findings of thirty-five years earlier (see chapter 9), so Nicol felt free to start anew, with Buck's blessing when the two met in 1954. But Nicol's own results (1955b,c) did not add significantly to those of George Parker except for the advantage of luminescence detection by a photomultiplier tube and its recording on an oscilloscope that allowed more sophisticated measurements and interpretations. Although George Parker discovered the inhibition of bioluminescence by light before him, Nicol conducted ingenious experiments that allowed him to conclude that light inhibition operates at the neuro-photocyte junction or inside the photocytes.

Of more interest was his collaboration with Demorest Davenport (1911–2004). Davenport was a Harvard-trained entomologist who, after serving in the Second World War, joined the Santa Barbara College of the University of California system, where Jim Case was hired once it had gained university status as we saw earlier. Davenport was no stranger to bioluminescence, having published a paper on a new luminescent milliped of the genus *Motyxia* found in the Sequioa National Forest in Northern California (Davenport et al., 1952). Working on a campus by the seaside led Davenport to shift his research interest from insects to marine biology, and Nicol's visit, however short, cemented a fruitful collaboration.

On the personal side, Nicol and Davenport were a good match. Both tended to be quick-tempered and judgemental. Davenport, as Jim Case noted in his memoir, got on the nerves of many people with his sanctimonious and condescending aspersions on people's perceived shortcomings. Similarly, Nicol "had formal and somewhat old-fashioned manners and was inclined to be brusque. This could be alarming for students and for those who did not know him well" (Maddock, 2006). Civility and tolerance fell by the wayside: "He [Nicol] had no time for bureaucracy and was impatient with those who were incompetent or did not take science seriously."

Nicol and Davenport investigated hydromedusan jellyfish bioluminescence at the Friday Harbor Laboratories of the University of Washington (Davenport and Nicol, 1955). They found that the luminescence was intracellular and localized to "photocyte masses" along the margin of the bell. Upon stimulation, light appeared as flashes showing facilitation and fatigue like in other neurally controlled luminescent systems. However, luminescence remained local; no propagation along the margin occurred. The following year they produced a paper on the luminescence of sea pens, which belong to the same family as the sea pansy and share very similar colonial organizations. Unsurprisingly, their results were similar to those reported in *Renilla* (Davenport and Nicol, 1956). They concluded their paper by stating: "The variety of coordinated muscular activities and luminescence are controlled by a diffuse nerve net. There exists a problem of great interest: to determine how these complex activities, involving at least two different effector systems, are co-ordinated and regulated by the simple plexiform nervous system found in these animals." The problem they identified was addressed in Jim Case's lab almost twenty years later, by Case's graduate student Peter A.V. Anderson (Anderson and Case, 1975).

Anderson devised custom suction electrodes and electrophysiological recording apparatus to pick up the tiny action potential signals (spikes) reverberating from the minute neurons of the nerve nets of the sea pansy. He found a signalling pathway associated with the protective retraction of the feeding polyps, and another associated with the colonial nerve net that controls bioluminescence and contraction of the colonial mass as part of the sea pansy's escape response.

The last paper emanating from Nicol's Californian visit concerned the midshipman fish *Porichthys* (Nicol, 1957). Nicol wanted to challenge the Greenes' assertion that luminescence control in this fish is exclusively hormonal (Greene and Greene, 1924). He found that the skin photophores are innervated by nerves associated with the sympathetic nervous system and that the transmitter is probably adrenaline. He did not rule out dual nervous and hormonal control, as adrenaline also acts as a hormone in fish. He also investigated stomiid fishes caught during short oceanographic cruises on board research vessels in the North Atlantic, cruises which betrayed his acute seasickness. He was intrigued by the large light organs below the eyes (Nicol, 1960a), as was Haneda (1955), who had this to say:

> Recently I observed luminescence of the cheek organ of a deep sea luminous fish, *Astronesthes ijimai* Tanaka (1908) [Stomiidae], which was collected in a shrimp trawl net in Suruga Bay. This fish has a cheek organ and two rows of minute photophores along the ventral and lateral walls. The structure of the cheek organ is very similar to that of *Anomalops* [flashligh fish]. Although luminescence is continuous, the luminous surface appears and disappears at will by rotating the cheek organ as in the case of *Anomalops*. Comparative studies of the structure and substance of both luminous organs should prove interesting.

Nicol answered Haneda's call by examining the histology of the cheek (subocular) organ of several stomiid fishes and he discovered that a muscle is attached to the light organ and rotates it to mask the luminescence from view. The muscle and light organ, he found, are innervated independently, suggesting that the light emission is subjected to a dual nervous control. Nicol put his observations in the context of speculation about the roles of the different light organs:

> The large subocular organs may be used as torches, for illuminating the surrounding water. The serially arranged photophores may permit the animals to recognize each other. The result may be mutual repulsion,

thus keeping the fish spread out in hunting territories delimited by the intensity of the light and the distance at which it can be seen; or mutual attraction when the animals differ in sex.

An important historical contribution of Nicol's to the field of bioluminescence was his systematic recording of the luminescence of oceanic (pelagic) organisms with photomultiplier tubes and recording equipment on board ships over several years (Nicol, 1958a,b). Nothing of the sort or on this scale was ever undertaken before. He surveyed dinoflagellates, radiolarians, jellyfish, siphonophores, comb-jellies, crustaceans, tunicates (pyrosomes), and fishes. He measured the intensity and spectral (colour) composition of their light emissions, as well as obtaining permanent records of the dynamics of the emissions over time. The brightest performers on the intensity scale, Nicol pointed out, were siphonophores, jellyfish (scyphomedusae), comb-jellies, pyrosomes, and a deep-sea fish, the tubeshoulder *Searsia*, whose luminescent secretion had never been observed before. This survey signalled the incipient stages of the deep future interest in reflectors in light organs, eyes, and body surfaces, and their roles in visual ecology.

Nicol wrote several review articles on the physiological control of bioluminescence, starting from the published form of his talk at the 1954 Conference on Luminescence (Nicol, 1955c). Another followed in 1960, in which a synthesis of the new findings was achieved thanks to the quantitative methods of the investigations he pioneered using photo-electric recording methods (1960b). That same year, he published a large treatise on marine biology, which earned him critical acclaim; a chapter was devoted to bioluminescence (1960c). A review restricted to fish bioluminescence (1967) and an essay on the relations between vision and bioluminescence (1978) completed his contributions to the field. The 1978 essay deserves our attention.

When Nicol was measuring the intensity and colour spectrum of light emissions, questions lurked in the back of his mind. What distance can these emissions travel in dark waters and still be perceived by other organisms? Does the vision of deep-sea organisms need to be attuned to the colour of the light emissions, including the animal's own? To Nicol it all revolved around the notion that "many animals respond in various ways to luminescence and depend on 'living lights' for essential behavioural activities" (Nicol, 1978). Vision and luminescence, Nicol pointed out, can be interlocked as pawns in deep-sea ecological strategies. Fishes, for example, may have favoured ventrally directed luminescence to conceal their lights from same-depth predators, but the emergence of predators with eyes directed upward may have pressured potential prey to use their ventral bioluminescence to conceal their silhouette by counterillumination. In addition, Nicol pointed out the rich literature on the various and sometimes strange adaptations of the eyes of deep-sea animals, all tending to enhance sensitivity to light at the expense of visual acuity, and to assist in the detection of bioluminescence. However, he said little about vision in terrestrial luminescent animals. Recently morphological evidence was presented that the compound eye of the New Zealand glow-worm is adapted for enhanced light sensitivity to suit the lowlight cave environment in which it lives (Meyer-Rochow, 2016).

After seventeen years in Plymouth Nicol was ready for new challenges and scouted marine facilities in the US where he could find employment. In 1966 he was appointed visiting professor at the University of Texas Marine Science Institute (Maddock, 2006), and two years later he became professor of zoology at the University of Oregon and director of its Marine Institute. But, after being accustomed to minimal bureaucratic humbug in Plymouth, he soon became disenchanted with administrative duties. He settled back in Texas, with a permanent appointment as professor of zoology at the University of Texas. He never returned to bioluminescence work, concentrating until the end of his career, in 1981, on, among other topics, reflectors and other features of fish eyes (Nicol and Somiya, 1989). He spent his retirement years in a secluded property in Cornwall in the UK, where he died in 2004 at eighty-nine.

~~~

If Nicol's body of work was the first major contribution to the field from the British Isles since the heyday of the pioneering oceanographic expeditions, the output of Jean-Marie Bassot (1933–2007) must stand as the greatest French contribution since Raphaël Dubois. The following account of his life and work is based on an obituary by a practitioner of bioluminescence research (Campbell, 2008) and my own personal recollections. Anthony Campbell said of Bassot that his "curiosity and strength of intellect shines
~~~

so bright that it can be dazzling!" I cannot agree more, and can only add that for Bassot science was first and foremost an aesthetic experience.

Born in Paris, Bassot spent part of his childhood during the Nazi occupation – a traumatic time for him. But he grew to be a tall young man full of confidence, very soon hooked on natural history and especially marine biology. Shortly after turning twenty-one he travelled to the Indian Ocean and boarded the *Calypso*, Jacques-Yves Cousteau's famous research vessel, in Madagascar to make underwater observations as part of his marine biology training. His mentor, Pierre Drach (1906–1998), was a distinguished marine biologist at the Faculté des Sciences de Paris, in turn administrator of French seaside biological stations and the Centre National de la Recherche Scientifique (CNRS). A strong advocate of oceanography, Drach had already taken up "scuba" diving in 1947, when it was in its infancy. He instilled in his students his enthusiasm for that activity and his conviction of its importance as a tool for marine biologists. Bassot wasted no time embracing his mentor's passion for diving. It served him well in his observations of live displays of bioluminescence under water.

But for the moment it was the histology, histochemistry, and the ultrastructure of luminous organs rather than the observation of light emissions, that constituted Bassot's doctoral dissertation topic under Drach. In the typical French academic tradition of the period, it took him ten years to earn his Doctorat d'Etat, granted in 1967. Along the way he published preliminary accounts of his thesis work on deep-sea fishes (Bassot 1958, 1959a, 1960a, 1963), on the clam *Pholas* (1959b, 1966a), krill (1960b,c), and scaleworms (1964, 1966b,c). A synthesis of these works in the English language appeared on the occasion of the Conference on Bioluminescence held in Japan in September 1965, discussed earlier in chapter 13, in which Bassot took part (Bassot, 1966d).

Bassot's originality already shone in these early contributions. Not only did he gain unparalleled insights into the organization of luminous organs and the cytological make-up of photocytes, but it allowed him to suggest evolutionary pathways for these organs. In particular, he followed the secretory cycles of photocytes in deep-sea stomiid fishes, and he obtained arresting images of the accessory components of the organs which allowed him to see in them reflectors, interference filters, or light guides. In general, the high quality of his techniques highlighted the architectural beauty of these organs. One strange discovery led to a lifelong fascination with scaleworms. In the photocytes of scaleworms he spotted in electron microscopic images large "paracrystalline" bodies made of mosaic-like arrays of tubular membranes showing continuity with the endoplasmic reticulum and the nuclear membrane of the cell (Bassot, 1964). No such extreme differentiation of a cell's endoplasmic reticulum had ever been described before. He was later to propose a hypothesis on the role of these exotic structures in the light emission (1966b).

Ever on the look-out for new bioluminescent encounters, Bassot used the occasion of the 1965 conference to travel in search of Asian fireflies, advised in this by another conference participant, John Buck, with whom he developed a lasting friendship. He travelled to Malaysia, where he observed firefly synchrony (Bassot and Polunin, 1967). In Thailand he watched ponyfishes under water and shipped by air two live specimens to colleagues at the Institut Pasteur in Paris, who studied the symbiotic luminous bacteria of these fishes (Boisvert, Chatelain, and Bassot, 1967). They found that these bacteria, which in culture gave out a bluish green light, were distinct from any other and were named *Photobacterium leiognathi*. Four years later Woody Hastings repeated this study without mentioning the original French paper (Hastings and Mitchell, 1971).

Bassot next moved to Cambodia, where he discovered a new luminous snail, *Hemiplecta weinkauffiana* (Bassot and Martoja, 1968), only the second such finding since Haneda's original 1946 report on *Dyakia* (*Quantula*) *striata* in Singapore. The light organ, found only in the foot of juvenile specimens, is essentially composed of giant photocytes, up to half a millimeter long, which luminesce on and off at intervals of two seconds. The strangest thing happens when the snail reaches sexual maturity; phagocytes invade the organ and remove all the photocytes one by one until it becomes a mere cyst in the foot. But later Jonathan Copeland reported that in *Dyakia* luminescence persists in adults, although flash intensity is lower and duration shorter in adults than in juveniles (Copeland and Daston, 1993). Moreover, whereas flashes are simple in adults, in juveniles they are multimodal. Flashing seems to play a role in intraspecific communication by facilitating congregation of individuals (Copeland and Daston, 1989).

Life moved on when Bassot returned to France. He was on the staff of the Institut Océanographique in Paris, which, we may recall from chapter

Figure 14.4 Jean-Marie Bassot in Quebec City, 1986. From the author's archives.

4, was founded by Prince Albert I of Monaco in 1911. He had married a beautiful and bright woman, Clothilde, and the couple now had three children. He had made a name for himself in the field of bioluminescence and in 1969 he received an invitation from John Buck to participate in the *Alpha Helix* Bioluminescence Expedition. He jumped at the chance of renewing acquaintance with Asian luminous organisms. In New Guinea his "unquenched curiosity," as Campbell (2008) put it, took him in many directions: observations on the luminescence of various species, and even a general study of biological colonization on new islands created by volcanic activity. He would "sit under water in the river to watch the flashlight fish *Anomalops* and *Photoblepharon* for hours on end" (Campbell, 2008).

The next stop in Bassot's intellectual journey with scale-worms resulted from a collaborative project with the laboratory of Max Pavans de Ceccaty (1927–2009). Born in Tunisia, Pavans de Ceccaty studied at the Université de Montpellier where he completed a doctoral dissertation on sponges. He rose through the professoral ranks in Montpellier, later moving to the Université Claude-Bernard in Lyon, where he was running a laboratory of the

CNRS. University science in France was two-tiered: many of the professors had heavy teaching loads and competed for meagre research funds from the university while the lucky few on campus held state positions as *directeur de laboratoire* of the CNRS, which entailed less or no teaching and better access to state research funds. Pavans's lab dealt with various aspects of the microscopic anatomy of lower invertebrates, especially the nervous system at the electron microscope level. In the summer months Bassot was based at the Station biologique de Roscoff in Brittany, where he had access to scale-worms dwelling on the rocky shores, and where Pavans de Ceccaty and his student André Bilbaut came to pitch in.

In their first paper together in 1972, they obtained better electron microscopic images of the scale-worm photocytes – compared to those of Bassot in the 1960s – which showed that the paracrystalline granules, now called photosomes, connect with a special kind of endoplasmic reticulum along the cell membrane to form a junction through which the excitation of the luminescent source may occur and spread (Pavans de Ceccaty et al., 1972). In 1973 a *laboratoire de bioluminescence* was created for Bassot at the CNRS complex in Gif-sur-Yvette, about twenty kilometers southwest of Paris. This is where, since 1946, laboratories were devoted fully to fundamental research beyond the constraints of university academic life. It took time for Bassot to set up a functional lab; as a consequence there is a hiatus in his scientific productivity in the ensuing years. A student, Marie-Thérèse Nicolas, joined his lab, and everything was aligned for Bassot to enjoy a great scientific career.

In a series of papers published in French in 1977, Bassot and his collaborators returned in force to the perplexing problem of luminescence control in the luminous elytra of scale-worms. Worth mentioning here are those of Bassot and Bilbaut (1977a,b) in which they show: (1) that bioluminescent microsources (photosomes) move transiently inside the photocytes from one flash to the next, and (2) that a fluorescence develops and increases in the elytra as more and more flashes are emitted. These findings testified to the remarkable dynamics of light sources inside the photocytes and to the production of a key substance, which happens to be fluorescent, in the process of light emission. Ten years later Bassot revisited this system for the Anglo-Saxon readership with more sophisticated tools.

Bassot (1987) addressed the dynamics of the microsources by recording the developing fluorescence of an elytrum (scale) through a microscope with an image intensifier providing sequential time frames of facilitating flashes. He was able to show that, "within a given cell, individual photosomes can be either coupled and respond to stimulation or uncoupled and quiescent, that the coupled state has a basic lifetime of about 1 second which can be lengthened by reinforcement, and that this state must be established in a matter of milliseconds as a result of the stimulation. In preparing an increased response to a forthcoming stimulation, coupling acts as a short-term memory." In an accompanying paper (Bassot and Nicolas, 1987), Bassot was able to visualize the dynamics of photosome coupling at the resolution of the electron microscope thanks to a new technique whereby flashing elytra are deep frozen within fifteen milliseconds and the chemical fixative is substituted in the tissue as it defrosts. The technique provided for unparalleled high-quality images and time-lapse details of the goings-on inside the photocytes. Thus Bassot discovered a new way for cells to conduct excitation intracellularly through the ever-modulated coupling of endoplasmic reticulum membranes with plasma membranes and photosomes.

In those years Bassot touched on other aspects of bioluminescence. With his collaborators he examined the luminescent system of a siphonophore, a pelagic and colonial hydrozoan found at the Station zoologique de Villefranche-sur-Mer on the Mediterranean coast (Bassot et al., 1978). This system showed similarities with the scale-worm elytra in that luminescence was localized in epithelial cells and spread by conduction of excitation through the epithelium, thanks to gap junctions between the cells. Bassot also enlisted the help of Osamu Shimomura in Princeton (see chapter 11) to uncover the chemical nature of the scale-worm luminescent system (Nicolas et al., 1982). It turned out that a membrane-bound photoprotein, called polynoidin, is the light emitter. This was not a surprising result, as the photosomes are replete with lattices of membranes. Oxygen, but not calcium, is required through the production of superoxide or hydroxyl radicals by the oxidation of reduced riboflavin, the source of the fluorescence during light emission *in vivo*. Curiously, earlier work by the American Albert A. Herrera and by Bassot's collaborator, André Bilbaut, showed that the photocyte action potentials require calcium to trigger flash activity

(Herrera 1979, Bilbaut 1980). To reconcile this apparent contradiction, Nicolas et al. (1982) suggested that only the step of riboflavin oxidation is calcium-dependent.

Bassot became engrossed in the problem of the role of calcium in the facilitation of the flash responses of the scale epithelium. He was looking for an elegant way of explaining it and he thought he had found it. He took the opportunity of a Bioluminescence Symposium organized by Woody Hastings and held in Maui, Hawaii, in February 1993, to present his views. But he was genuinely stunned when Paul Brehm's talk, which preceded his, reported the exact calcium mechanism he had in mind, in the hydroid *Obelia*. Brehm had done his PhD at UCLA under Jim Morin on the luminescent system of brittle stars and had gone on to specialize in neurobiology, especially in synaptic mechanisms involving calcium. While he held an associate professorship at Tufts University in Massachusetts, Brehm became interested in the epithelial luminescent system of *Obelia*, which involves a calcium-dependent photoprotein, as a convenient experimental model to visualize by the light emission what goes on during calcium signalling.

Brehm and his team had made a surprising discovery: in order to excite *Obelia* photocytes to luminesce, calcium must enter neighbouring non-luminescent cells and then cross to the photocytes through the channels of intercellular gap junctions (Dunlap et al., 1987). Next, they found that the facilitation setting in after the initial trigger is explained by local calcium entry directly across the plasma membrane of the photocytes (Brehm et al., 1989). What Brehm presented in Maui, which left Bassot utterly crestfallen, was a clever model according to which not one, but two mechanisms of calcium-dependent facilitation co-exist in the photocytes. Obelin will only emit light if it binds three calcium ions. "Changes in flash intensity during successive action potentials," Brehm's team explained (Naranjo et al., 1994), "result from calcium bound persistently to unexpended obelin, effectively lowering the number of calcium ions required for subsequent activation. Accordingly, facilitation or decrement results from the time-dependent availability of singly and doubly bound obelin." Bassot had come up with a similar scheme for scale-worm photocytes.

Woody Hastings, whose early biochemical work we followed in chapter 11, called on Bassot's expertise in fast-freeze technology to solve the riddle of the microsources (scintillons) in the dinoflagellate *Gonyaulax*. Hastings and Beatrice Sweeney extracted the luminescent system of *Gonyaulax* in 1956 and discovered the daily rhythm of its bioluminescence in 1958. It is time again to pick up the thread of Hastings's career. His outstanding biochemical and molecular research on bacterial and dinoflagellate luminescence earned him positions first at Northwestern University, and then at the University of Illinois. In 1966 he was appointed professor of biology at Harvard University, where he remained for the rest of his distinguished career. Besides his work on bacterial bioluminescence, Hastings and his students established the dinoflagellate *Gonyaulax* as an important experimental model for understanding the molecular mechanisms of biological clocks (Hastings, 2001).

The problem of the source of bioluminescence in dinoflagellates goes back to the pioneering work of Roger Eckert in the 1960s. Roger Otto Eckert (1934–1986) was a young professor at the University of Syracuse when he undertook to study the excitation mechanism for luminescence in *Noctiluca* (Eckert, 1965). This was no easy trick, as he applied microelectrodes to a unicellular organism to record the action potential preceding by three milliseconds the onset of the bioluminescent flash. Because flash intensity was independent of the all-or-none action potential that triggers the flash, Eckert assumed that microflash activities in the cell add up to the overall macroflash response of the whole cell. With the help of George T. Reynolds (1917–2005), a Princeton-trained physicist who worked on the Manhattan Project, made Princeton his home after the war and in the sixties developed image intensifier technologies to deal with recording spatio-temporal displays of bioluminescence (Reynolds 1978), Eckert was able to locate fluorescent and luminescent microsources of 0.4 micrometer which appeared responsible for the microflashes (Eckert and Reynolds, 1967).

However, the physical nature of these microsources remained unknown, and the relationship of the microsources in *Noctiluca* to the scintillons identified by Hastings's team in *Gonyaulax* by biochemical extraction procedures (DeSa et al., 1963) was unclear. Bassot and Marie-Thérèse Nicolas entered the fray by applying to *Gonyaulax* the fast-freeze technique Bassot had employed so successfully with scale-worms, and by labelling thin sections of the cells with a specific antibody against *Gonyaulax* luciferase (Nicolas et al., 1987). With electron microscopy they found dense bodies about the size of Eckert's microsources and corresponding to the scintillons, in which the luciferase for light emission is located. The bodies protrude into the cell's vacuole, and they suggested that "bioluminescent flashes can be elicited by a proton influx resulting from a triggering action potential propagated along the vacuolar membrane."

Elated by these results, Nicolas and Bassot extended their reach to luminous bacteria. Unlike eukaryotic cells, bacteria lack membrane-bound compartments, so Hastings (1978), also on the basis of luciferase properties, had assumed that bacterial luciferase is bound to the cell wall membrane. But Nicolas et al. (1989) found instead that the luciferase was located in the cytoplasm and, more importantly, in small "patches" which appear to be present in luminescence-competent bacteria but not in dark mutants. Thus, thanks to the high resolution of their technique, Bassot's lab helped unravel a long-lasting, seemingly intractable mystery: what microstructures inside the luminous cells participate in light emission and how their dynamics achieve the luminescent outcome. But Bassot's active career was soon to face a wall.

From my earliest epistolary exchanges with Jean-Marie in 1968 until our last meeting in Hawaii in 1993, I came to know a man who was dealt many blows of a personal and professional nature, leaving many scars for his intimates to see. The end of his marriage in the early 1980s was one of them, with its baggage of loneliness and depression. He suffered a severe case of Guillain-Barré Syndrome, which left him paralyzed and bed-ridden for quite a while and endangered his life. In response he sought solace in meditation and yoga. On the professional side, his sensitive nature made him feel, all too often and deeply, low blows by colleagues, administrators, or politicians. As Campbell (2008) put it:

> He did not suffer in silence the strictures and stupidity, as he saw it, of the "system." He hated corruption and hypocrisy or, worse still, scientific dishonesty. His promising career as a young TV naturalist was cut short when he decided to run a whole programme on the analogy between animal parasites and the advertising industry. His TV producer called him in next morning and informed him, "Don't phone us, we'll phone you!"

As the anecdote portends, his perceived eccentricities closed door after door to him. I witnessed in Roscoff the professional isolation in which peers and administrators held him, and the "family feud" that pervaded the scientific milieu there. Around 1987–88 Bassot lost his bioluminescence laboratory in Gif-sur-Yvette and was transferred to a microscopy service of the CNRS in Paris. The final blow was the cavalier way in which the directorship of CNRS closed his Paris laboratory late in 1991 and forced him into an early semi-retirement, with the ripple effect that his highly performing associate, Marie-Thérèse Nicolas, suddenly had to find another lab to resume her career.

Throughout his ordeals, Bassot displayed the Gallic charm and the qualities that endeared him. As examples close to home, I recall his long evening walks with my young children when he would casually catch fireflies for them and patiently explain how and why they flash; and he would seize a chance event, like a cicada fallen from a tree, to entertain them further with insect stories. There was also the time when he exercised his iconoclastic tendencies; in a family restaurant where he lost patience with the numbered menu – anathema to the French food culture – and proceeded to teach my children the art of bending forks!

Bassot retired in the property he had inherited from his forebears, called Le Marteau, about 200 km south of Paris, surrounded by his brood of children and grandchildren living nearby. He died on 10 October 2007, at the age of seventy-four. As Campbell (2008) remarked, "perhaps Jean-Marie's greatest legacy is his greatest puzzle of all – why are we animals so curious?"

15
A Bioluminescence Expedition

> As I wrote the proposal I found it difficult not to appear in the role of a promoter because – I suppose like any dream – the thing began to grow in my imagination.
> –John Bonner Buck (1967)

The R/V *Alpha Helix* Bioluminescence Expedition carried out in 1969 was a unique event in the history of this realm of inquiry. For the first time, bioluminescence researchers gathered not to participate in a symposium or conference, but to converge in one corner of the earth rich in bioluminescent species and engage together in the glory and bliss of their favourite pastime: tinkering with luminous organisms. The planning and execution of the expedition proved to be a harrowing experience at times, however, full of twists and turns that reflect the complexities of such endeavours. The following account is based on the examination of the John B. Buck Papers deposited at the Scripps Institution of Oceanography (SIO) Archives in La Jolla, California.

In August 1967 Theodore H. Bullock (1915–2005), a distinguished neuroscientist and pioneer of the field of comparative neurobiology, wrote from his desk at the University of California in San Diego a letter to his friend John Buck enjoining him to "jot down a couple of sentences" for one leg of an expedition with SIO's research vessel *Alpha Helix*. Bullock was acting as chairman of the National Advisory Board overseeing the *Alpha Helix* expedition program. The program was the brainchild of Per F. Scholander (1905–1980), a Swedish-born marine physiologist who was then director of SIO's Physiological Research Laboratory. According to a report prepared by the National Advisory Board for the R/V *Alpha Helix* in May 1971, the goal of the program was to create "a unique facility for experimental studies of organisms associated with the oceanic environment. Its focal point is a laboratory ship with supporting home-based laboratories and technical facilities which would provide experimental biologists, both here and abroad, with the opportunity to apply their wits and various skills to the many tasks which can be handled only by a well-equipped laboratory on the site of the natural habitat."

The newly constructed ship, named after the helical structure of DNA, was delivered to SIO early in 1966 and the program of cruises of the 133-foot floating laboratory was immediately inaugurated with a half-year cruise to the Great Barrier Reef. The National Science Foundation (NSF) provided an annual grant for the program, which was divided into two parts: one for ship operations and the other for the project leaders to meet the costs of doing research.

The *Alpha Helix* was scheduled to sail to Papua New Guinea between February and November 1969. The expedition was to accommodate three parties of scientists in successive legs: phases A, B, and C. Buck was approached for phase A. As Buck was tossing ideas around in his mind about projected lines of research for such an expedition, he began to develop cold feet owing to his total inexperience in leading a party of scientists, having so far done his field research alone with his wife, Elizabeth. In a letter to Bullock dated 4 November 1967, he wanted "to make it clear – and this is not being coy – that it would suit me much better personally to be able to stowaway in a corner of someone else's expedition than to lead it." In the tentative proposal accompanying this letter, Buck explained what the expedition should be about:

> The principal objective would be to make possible the association and collaborative research over several weeks, of experts in various aspects of light-production by animals and plants: biophysics, biochemistry, physiology, cytology, behavior and general biology. A second consideration would be that the expedition gives opportunity to study organisms and phenomena that are uniquely interesting in some respects and have not already been or cannot be [studied] in shore-based laboratories in the United States.

The proposal also listed a cast of colleagues he wanted on board and the projects they would be associated with. Buck and Jean-Marie Bassot, who was spending a year in Buck's lab in Bethesda, Maryland, would tackle the physiology of firefly synchrony. In addition to Buck, the biology of firefly synchrony would be handed to Yata Haneda and James E. Lloyd, the latter an up-and-coming entomologist from the University of Florida with expertise on firefly ecology and behaviour. Lloyd had attracted the attention of scientists and the general public alike with his discovery of the firefly "femmes fatales" – female fireflies that bait and devour males of another species by mimicking the flash pattern of the females of that species (Lloyd, 1965). The biochemistry of luminous fish was devolved to Frederick Tsuji and Woody Hastings who, after leaving William McElroy's lab at Johns Hopkins in 1953, had taken a job as an assistant professor at Northwestern University near Chicago, where he initiated studies on the biochemical regulation of bacterial bioluminescence. He soon embarked on studies of dinoflagellates with Beatrice M. Sweeney (1914–1989), a phytologist then based at SIO. Sweeney developed a technique to generate mass cultures of the dinoflagellate *Gonyaulax*, which allowed the preparation of extracts demonstrating an oxygen-dependent luciferin-luciferase reaction (Hastings and Sweeney, 1956). Together they found that the bioluminescence of *Gonyaulax* displayed a daily rhythm that persisted in constant light or darkness, a signature of biological clocks (Hastings and Sweeney, 1958). Buck assigned both Hastings and Sweeney to study dinoflagellate luminescence.

The behaviour of luminous fishes, squids and crustaceans would suit the expertise of Bassot, Haneda, Buck, and Jim Case, the latter having moved in 1963 from Iowa to the University of California, Santa Barbara (UCSB). The biochemistry of luminous fishes and the study of their luminous bacteria was assigned to Tsuji and Hastings, and finally the cytology and ultrastructure of light organs would be handled by Bassot and Buck. As the planning evolved, these assignments underwent quite a few rounds of musical chairs. Among other caveats Buck adhered to the school of thought that regards biochemists as poor naturalists. In a letter of 22 April 1968 to Bullock he wrote: "I feel a little uneasy about their ability to improvise (that is, about biochemists' intrinsic potential to switch into another problem if material or equipment obstacles appear) and would be interested in your personal experience with biochemists on other expeditions."

In addition, Buck's proposal required the vessel to be retrofitted with some logistical demands pertaining to bioluminescence work: making the portholes of the laboratories light-tight, installing a fume hood for biochemical work and chemical fixation of light organs, providing trawling equipment for subsurface and midwater depth, and – oddly reminiscent of William Beebe's era – having available a small diving bell with window for observation photography, strobe flash, and telephone connections with surface.

By the end of January 1968, the National Advisory Board of the *Alpha Helix* had approved Buck's bioluminescence proposal, to take place between 15 February and 1 May 1969 in Australian New Guinea. Of the prospective participants in the expedition, only Haneda had experience in Papua New Guinea, so in March 1968 he advised Buck on what to look for and expect. Then a scouting party, which included among others Buck and Robert Haines, the captain of the *Alpha Helix*, spent the second half of May 1968 travelling to Australia and the Territory of Papua and New Guinea. They met with Australian officials and scientists who would supervise local assistance, and from Port Moresby they looked around for potential anchorage sites for the research vessel and for shore facilities to conduct field research. They recommended a bay area near the town of Madang for anchorage of the *Alpha Helix*, a spot Buck nicknamed "world's end." A scenic lowland forest about forty-five minutes by car from Madang, Maiwara, was chosen for building a field house that would act as a base and make it convenient for scientists to stay overnight.

By October 1968 the news about funding turned grim. Federal legislators had reduced appropriations to all science-granting agencies and NSF decided to cut funds across the board by 15 percent. This meant scrounging for alternative sources of monies to cover part of the scientists' travel expenses. In a letter to Buck dated 10 October 1968, Woody Hastings questioned the wisdom of carrying on with the expedition plans at a time when "it is far more important to keep my research group here intact than to undertake field research this year." Six days later Buck replied by defending the philosophy of the expedition concept:

> I think the idea itself is valid and exciting and I think it is at least arguable that more can be accomplished per individual in a concentrated, cooperative full-time 10 week effort than in a year of isolated work full of academic interruptions: but I have to admit that I got enmeshed in the business originally not because of any deep philosophical motivation

> or, still less, because of thirst for administrative or executive position, but from a quite simple-minded desire to use the Alpha Helix to carry forward research in which I am deeply interested.

As a result of these upheavals, the expedition for 1969 was maintained but postponed to June. By the end of the year the dates had been pushed further, between October and December – switched from phase A to phase C. As the New Year kicked in, Buck was wallowing in all manner of expedition micromanagement, especially the local arrangements on shore at Madang and the allocations of meagre funds to various logistical ledgers. The strain of the job soon showed, and in a letter dated 18 March 1969 to the coordinator of the 1969 expeditions, Robert R. Hinchcliffe, Buck contended that "from the Chief Scientists' standpoint there are some inexcusable organizational flaws in the Alpha Helix program." He cited several examples of decisions at the top when "it apparently did not occur to anyone that the Chief Scientists ought to be consulted." In summing up his frustration, Buck tried to maintain a tone of optimism: "if a truculent tone has crept in here and there, put it down to the year of stewing, frustration and slaving that has gone by, rather than to anything personal. I know that you guys have sweated too and that what you have thought and done was in good faith and represented your best judgement of the moment. I hope we'll all live to look back on it and chortle at some future Alpha Helix reunion."

Another blow to Buck's morale hit in mid-August, less than two months before the start of Phase C, when he was informed that new budget constraints might force the *Alpha Helix* to stay at SIO and not join Buck's leg of the expedition. Then, seeming to play yo-yo with Buck's nerves, the SIO reversed its decision. SIO Director William A. Nierenberg, it seems, averted the latest crisis by injecting in-house funds. "I gather that special thanks are due to you," Buck wrote to Nierenberg on 25 August, "for pulling the rabbit out of the Alpha Helix hat at the eleventh hour. Needless to say I'm grateful to you, both for salvaging something out of my two years' personal effort and on behalf of the 15-odd biologists who will now have the opportunity to put their plans into effect. I hope that we can make a good showing." But in the middle of September 1969 Nierenberg had no choice but to order the return of the *Alpha Helix* to San Diego, leaving Buck in Madang to curtail or abandon parts of the research program, adjust operations on shore, and look out for local alternatives for seagoing operations. Fortunately, the fisheries trawler *Tagula* of the Papua New Guinea Government was put at the disposal of the bioluminescence expedition for a month. With operational facilities readied at Maiwara, the research work started in earnest late in September.

Bassot and John R. Paxton, an ichthyologist trained at the University of Southern California and affiliated with the Australian Museum in Sydney since 1968, took charge of the trawling operations on the *Tagula*. Paxton was interested in lanternfishes and later produced a monograph in which he discussed at length the evolutionary implications of the arrangements of photophores on the body of lanternfishes (Paxton, 1972). He suggested that the original function of ventro-lateral photophores was camouflage by counterillumination, and that their role in species recognition evolved later. Bassot was more interested in stomiid fishes, and he had brought with him an ultra-violet lamp to shine on the body of these fishes. In the science reports of the John B. Buck Papers, Bassot wrote that upon exposure to UV light the photophores of the majority of stomiatoid species display a blue fluorescence, close to the colour of their bioluminescence. In contrast, a few species (*Stomias*, *Chauliodus*) produce a red fluorescence, far from the colour of their bioluminescence. This was the first time such a puzzling phenomenon had been reported. I have seen spectacular colour photographs of the red fluorescence in *Chauliodus*, sent to me by Fernand Baguet, of the Université Catholique de Louvain, in 1976 (see Figure 5.6b), and the photophores look like shining cups of cherry sorbet.

The pair also caught luminous crustaceans. R.H. Kay (1965) at Oxford University had shown that a flash of artificial light elicited a long-lasting bioluminescent glow from euphausiids and that flashing the artificial light a second time while the krill is responding inhibits the ongoing glow. In addition, Kay found that the neurotransmitter serotonin not only elicited a glow but also hypersensitized the glow induced by artificial light. Bassot made similar experiments on a variety of shrimps collected by the *Tagula* and found that not all shrimp species respond to serotonin, thus suggesting that different control mechanisms may exist in crustacean luminescence. Bassot also chemically fixed a wide variety of luminous organisms aboard the *Tagula* for light and electron microscopy: ostracods, shrimps, squid, ponyfishes, and deep-sea fishes. However, due to lack of financial and logistical support back home in France, no publication from these efforts ever transpired.

Elsewhere other investigations were proceeding. On 6 October Buck reported: "Dr. James Case and [his student] Mr. Eldon Ball have been studying inhibition of flashing of fireflies by electrical stimulation of the brain and localized illumination of the eyes. They find that two local species of *Luciola* behave in opposite fashion, one being inhibited in predictable manner and the other either showing no inhibition or actual stimulation, depending on the general level of excitation of the nervous system." Their results, showing the key role of the eye-brain connection in flash inhibition, contradicted those of Franco Magni's (1930–2005) team in Pisa, Italy, who emphasized a hormonal involvement (Brunelli et al., 1968a,b). "Drs. Jack Shepherd and John Walsh," Buck continued, "have been making extensive collections of luminous fungi. They find many varieties, but not large numbers of any one. However, they are hopeful of getting a sufficient quantity to enable biochemical analyses." There is no evidence the two Australians ever completed these analyses.

Buck's interim report also mentioned that "Dr. Thomas Hopkins [of McElroys's lab at Johns Hopkins] has measured the spectral emissions of several species of fireflies, euphausid shrimps and fungi and the fluorescence spectrum of the luminous organ of the pony fish." No specific publication came out of this project either. Next, "Dr. Beatrice Sweeney has succeeded in culturing several luminous microorganisms collected from local waters, including a green Noctiluca with intracellular flagellated symbionts, and a Peridinium-like organism." The green *Noctiluca* was later the object of a publication (Sweeney, 1971). Frank Hanson, and John and Elizabeth Buck conducted their study of firefly synchrony mentioned in the previous chapter, which was published in 1971. And "Dr. James Lloyd has made an intensive study of the natural history, behavior, mating signals and periodicity of a local species of Luciola firefly." Lloyd delivered a paper on his extensive work in Papua New Guinea and other locations four years later (Lloyd, 1973b).

It is of the utmost historical importance here to open a large parenthesis about Jim Lloyd, his professional relationship with John Buck, and his scientific legacy. He was born in 1933 in New York State. Although Lloyd had an early aversion for school, it was a class project that stirred up his lifelong fascination with fireflies (Lewis, 2016). As a young adult he acted much like a drifter taking odd jobs here and there (Lewis, 2016). He eventually reconciled himself to higher education and in 1965 he earned his PhD at Cornell University with a thesis on field studies of the firefly *Photinus*. From the very start Lloyd's interest rested on flash communication between mates in their natural environment. He found that the different species of *Photinus* examined display species-specific flash signals shaped in the course of evolution by isolation mechanisms that kept populations geographically apart (1966). Lloyd took an academic position at the University of Florida, where his productive career studying fireflies was spent until his retirement.

When Buck invited Lloyd to join the expedition, he meant to include him in his overall firefly project. In a letter to Theodore Bullock dated 23 April 1968, Buck wrote: "Drs. Lloyd and Buck will record the flashing codes, test the responses of both sexes to programmed flashes of artificial light, test the tolerance of the code to changes in major period, flash intensity and duration, and 'phrasing.'" But no cooperation took place, and in his reports of activities Buck waxed lyrical about the firefly work of his team (Case, Hanson, Hopkins), yet seldom mentioned Lloyd. As Lewis (2016) emphasized, Lloyd was a loner, a self-proclaimed asocial. It is likely that Lloyd embarked in his studies of New Guinea fireflies with his solitary outlook, and never bothered to make overtures to Buck or the other members of the expedition. This was probably the beginning of the rift that developed over decades between Buck and Lloyd.

When I was a PhD student in Jim Case's lab in the mid-1970s, Case would make disparaging remarks about Lloyd and his approach to investigating firefly behaviour. This aptly underscores Sara Lewis's (2016) assessment of the "bitter feud" between Buck and his students versus Lloyd and his students. It rose to the point where manuscripts submitted to scientific periodicals by one lab were reviewed disparagingly by members of the other lab acting as peer referees. She explained the root of discord thus: "These two firefly researchers locked horns over an even more fundamental difference in their scientific perspectives [than lab versus field biology]. One man spent his career intently focused on asking 'how' questions, while the other focused on asking 'why' questions." The heated arguments between them over firefly synchrony must have originated in New Guinea, with Lloyd insisting on the "why" by approaching firefly signalling through evolutionary theory and Buck asking "how" questions through the lens of his physiological background.

In the end, the evolutionary ecology approach of Lloyd earned him disciples who have made their mark in firefly biology to this day. Outstanding among them are Sara M. Lewis and Marc A. Branham. Based at Tufts University, Lewis and her students gained "new insight into how firefly flash signals have been shaped by the dual evolutionary processes of sexual selection (mate choice) and natural selection (predation) (Lewis and Cratsley, 2008)." Her work has uncovered several exotic adaptations. For example, her team found that the male spermatophores penetrating deep in the female genitalia during copulation are nuptial gifts that provide extra nutritive supply to the developing eggs (Van Der Reijden et al., 1997). This phenomenon was shown to increase firefly female fecundity (Rooney and Lewis, 2002). Another adaptation discovered by Lewis's team is a new way for *Photuris* firefly *femmes fatales* to attack *Photinus* males by stealing them from spider orbs and feast on them (Faust et al., 2012). The benefit of predation on *Photinus*, apparently, is the acquisition of the latter's chemical defence substances (lucibufagins), which will help *Photuris* females to deter their own predators.

Branham, then early in his career, at the University of Kansas, made an important discovery (Branham and Greenfield, 1996). He noted that, contrary to received wisdom, males of *Photinus consimilis* show great variety in their flash rates. He asked what in the male bioluminescent signalling attracts the female for mating. He and his co-author, Michael Greenfield, found that females prefer the higher flash rates of the male's range. Females showed no particular interest in high-power light output by the male lantern. Later, while working at Ohio State University, Branham took up the challenge of tracing the evolution of bioluminescence in cantharoid beetles by phylogenetic (cladistics) analysis (Branham and Wenzel, 2001). The study suggested that bioluminescence arose once in an ancestor of the phengodid (railroad and other worms) line, which led eventually to the lampyrids (conventional fireflies). After losses of bioluminescence by groups along the way, bioluminescence was regained by the modern phengodid beetles. Two years later Branham and Wenzel (2003) reported another phylogenetic analysis proposing that bioluminescent signalling started in firefly larvae as a warning to potential predators that they taste bad. Only later in evolution did adult fireflies use bioluminescence as a sexual signal.

~~~

We now close the parenthesis and return to the *Tagula* expedition and to the animals other than fireflies that caught the field workers' fancy. Buck reported early in the expedition that Bassot "succeeded in isolating and culturing the luminous bacteria from the photic organ of the pony fish." Woody Hastings joined the expedition shortly after Buck filed that report, and proceeded, with his student George Mitchell, without consulting or coordinating with Bassot, to conduct intensive research both on the luminous bacteria and the behavioural physiology of ponyfish bioluminescence. Friction incited by this lack of consultation became the focus of a tense atmosphere at the work base, as Bassot did not hide his anger at what he regarded as ungentlemanly conduct. Buck, as chief scientist, tried to mediate the dispute, and he was sympathetic to Bassot who, after all, already had a headstart on Hastings. In the end, as surmised from a letter of Hastings to Buck, dated 5 November 1969, Hastings and Mitchell carried on with their work and Hastings's idea of making it up to Bassot was to give him "plates with all our cell cultures on them which he sent back to his lab [in France]."

Bassot's perception of being unfairly treated is not new to the history of bioluminescence research. His fellow Frenchman Raphäel Dubois had experienced something similar in his dealings with Harvey, as chronicled in chapters 8 and 10. That Hastings "suffered" a lapse of professionalism, possibly aided by unbridled ambition, in an otherwise brilliant scientific career, is unquestionable. Ironically, only a few months earlier, his former postdoctoral supervisor at Johns Hopkins University, William McElroy, was embroiled in a very public and embarassing incident in which his own professionalism was questioned (Boffey, 1969). McElroy's outstanding work on firefly bioluminescence had earned him accolades from his peers to the point that he was elected member of the Science Advisory Committee at the White House. Now, just as he was taking office as director of the National Science Foundation, he was accused of plagiarism. In a long review chapter on insect
~~~

bioluminescence (McElroy, 1964), McElroy had quoted verbatim more than 25 percent of the thirty-five-page text of an article by David S. Smith (1963) without proper credit. In fact the borrowed text "was not put in quotation marks and was not attributed clearly to Smith" (Boffey, 1969). Some colleagues of Smith at Cambridge University wondered "what would happen to a college student who tried to explain lack of attribution by saying he forgot to rework the material" (Boffey, 1969). But the high-profile McElroy got away with it and his tenure at NSF was not rescinded.

While Bassot had little to show for his efforts in terms of published results, the bioluminescence expedition was kinder to Hastings. Hastings and Mitchell (1971) showed that when the appropriate culture conditions exist in the ponyfish light organ, there occurs an "autoinduction" of bacterial luciferase synthesis and, consequently, luminescence competence. They remarked that when in doubt as to whether the luminescence of an organism has its source in luminous bacteria, one has only to assay for the presence of bacterial luciferase. This Haneda and Tsuji (1971) did in a contribution arising from the bioluminescence expedition. Although luminous bacteria were suspected to account for the luminescence of flashlight fishes, no previous investigator was able to culture such bacteria, and Haneda and Tsuji fared no better. However, they were able to demonstrate the presence of bacterial luciferase. They concluded that flashlight fish bacteria were of a primitive nature, which they called "bacteroid." Hastings had also conducted experiments with the ponyfish of Papua New Guinea which led to the discovery that ponyfishes may use their luminescence to camouflage their silhouette during daytime (Hastings, 1971).

The bioluminescence expedition ended 1 December 1969. The reviews were mixed. While it produced valuable scientific results, some of which given high visibility in the widely read journal *Science*, the goal of fertilizing the field by herding the specialists together was not achieved. A similar venture using the *Alpha Helix* and centred around the Central Indonesian Banda Islands was headed in 1975 by Jim Case and G. Adrian Horridge of the Australian National University. Not many new bioluminescent discoveries resulted from this second expedition either, but its success owed to substantial new knowledge on other aspects of the life of bioluminescent organisms.

Among the participants of the second expedition, Beatrice Sweeney, James Lloyd, John Paxton, and Eldon Ball were veterans of the first expedition. Sweeney published a paper on the ultrastructure of green symbionts of *Noctiluca* (Sweeney, 1978). New participants such as Bruce Robison, Peter Herring, and John Wampler had already made a name for themselves in bioluminescence or deep-sea research. Others were young and budding luminescence researchers like Jim Case's postdoctorate Anthony Barnes and James Michael Anderson, an associate of Milton Cormier and John Wampler at the University of Georgia. Anderson (1980) published a paper on the luminous centipede *Orphaneous brevilabiatus*, in which he established that a luciferin-luciferase reaction at an unusually low pH (4.6) was involved. And then there was the team that Horridge had assembled, which consisted entirely of young compound-eye researchers who had recently obtained their doctorates. One of them, the New Zealander Victor Benno Meyer-Rochow, was fortunate enough to observe and describe the spawning of the flashlight fish *Photoblepharon* in one of the aquariums he had set up on Banda Island's base camp (Meyer-Rochow 1976a), and to observe the dimming of *Anomalops*'s bacterial light organ when the fish were starved (Meyer-Rochow, 1976b).

These bioluminescence expeditions tested the limits of expectations in terms of logistics and returns on the scientific investment. They were never attempted again.

PART SIX

THE LEAP TO CURRENT UNDERSTANDING

16

Probing Oceanic Bioluminescence

We dive down slowly, in a darkness glittered with sparks.
–Théodore Monod (1991)

Modern oceanography has been instrumental in deciphering many of the remaining unknowns of the vast, unwieldy ecosystem that we call ocean. The development of the field has given impetus to the pace of learning about oceanic bioluminescence as a large-scale physical phenomenon and about the workings of specific bioluminescent organisms dwelling in the various depths of the ocean. Many scientists, academic institutions, and governmental organizations have participated in this eventful exploration over several decades, and this chapter will chronicle its unfolding.

"What we know about something," Bruce Robison (2004) aptly remarked, "is usually a function of how we learned it, and the evolution of our understanding of deep pelagic biology is strongly linked to the technologies that have been employed to investigate this difficult habitat." One innovative technology was the telerecording bathyphotometer, designed to record light events at various depths, particularly targeting the deep-scattering layer (DSL). During the Second World War navy sonars developed to detect enemy submarines were picking up false sea bottoms which kept rising at night and sinking at dawn. This not only puzzled US Navy officials but also raised concerns about the phenomenon's potential interference with military sea operations by "scrambling" intelligence signals. After the war the US Navy gave high priority to research into identifying the source of DSL.

Brian P. Boden and Elizabeth M. Kampa, a husband-and-wife team who both received PhDs in 1950 from the Scripps Institution of Oceanography, were contracted by the Office of Naval Research to conduct an investigation of DSL at the San Diego Trough (Kampa and Boden, 1953). The pair had noticed that euphausiids (krill) always showed up in large numbers in catches at shallow depths corresponding to the upper position of the DSL at night. Hilary B. Moore of the University of Miami had already reviewed indirect evidence that krill is associated with the DSL (Moore, 1950). As these euphausiids are luminescent (see Part Two of this book), Kampa and Boden followed the DSL with the lowered bathyphotometer, and recorded bioluminescence. Blue-green flashes were recorded at rates reaching a maximum (forty-two per minute) during the twilight (upward) migration of the DSL. The characteristics of this *in situ* bioluminescent activity were found in the laboratory to correspond to those of euphausiids, and possibly to pyrosomes known to occur in these zones.

The pioneering work of Kampa and Boden suggested also that living organisms associated with the DSL migrate in the water column in order to maintain an isolume; that is, to locate the depth at which ambient light has a fixed low intensity over a twenty-four-hour period. But their identification of the animals associated with the DSL in the San Diego Trough was circumstantial. The next step was to collect the fauna directly with a fishing gear. This was made possible by designing new, technically advanced trawling nets. One was the Isaacs-Kidd midwater trawl, introduced in 1951 by SIO scientists John D. Isaacs (1913–1980) and Lewis W. Kidd (1923–1982). Another was developed by their colleague Gordon H. Tucker (1909–1962); the Tucker opening-closing trawl was equipped with a time-release mechanism to close the net at specific depths. These trawls allowed discrete-depth sampling unparalleled in the first eighty years of oceanographic cruises.

The trawls were well suited for sampling organisms only within the DSL. In early studies using these gears, Tucker himself (1951) identified two different deep-scattering layers, one associated at a shallower depth with euphausiids and another, at a greater depth, with lanternfishes. In both cases the identified organisms are luminescent. Another significant discovery was made by a Southern California native, Eric G. Barham (1919–2002), an important figure in this story. Barham's PhD dissertation at Stanford University was concerned with surveying the fauna associated with the DSL in Monterey Bay. Barham (1957) showed that different organisms dominate the DSL according to season: a zone at 400 meters dominated by lanternfishes in the summer, a double sound-scattering layer in the early fall corresponding to

Tucker's just mentioned; and a variable single layer from mid-fall to late spring where shrimps abound. Years later, the Tucker trawl was also used in the Eastern Pacific (Dunlap, 1968), where strong evidence indicated that the DSL organisms, all bioluminescent, migrate along the DSL path for feeding purposes. Many of these organisms, especially lanternfishes and siphonophores, possess gas-filled chambers – swimbladders for fishes and pneumatophores for siphonophores – which were deemed responsible for bouncing the sonar signal (Marshall, 1960).

It was all very well to conduct a census of DSL organisms, but some oceanographers wanted also to see with their own eyes what went on with life in the DSL. Toward this goal an initial approach consisted in making "simultaneous observations with echo sounders and underwater television both in and out of deep scattering layers" (Backus and Barnes, 1957). Richard H. Backus (1922–2012), who piloted this study, was at the time a biological oceanographer at the Woods Hole Oceanographic Institution, where he spent his entire career. The television records confirmed that individual fishes account for the echo signals. However, at that time another technology took over which allowed direct sightings of the organisms.

Since the days of Beebe's bathysphere, chronicled in chapter 12, submersible technology had shown no tangible progress until the bathyscaphe was created in 1948 by the famous Swiss physicist and inventor Auguste Piccard (1884–1962). The prototype was designed to be autonomous – not requiring the cables that Beebe had to contend with – by an ingenious interplay of gazoline floater and ballast (Walsh, 2010). But it was not seaworthy at the sea surface, and an improved model was built in 1953 with the cooperation of the French Navy. Théodore Monod (1902–2000), a French naturalist, philosopher, and specialist of the African desert, dove with the bathyscaphe off the coast of Dakar and reported intermittent displays of bioluminescence in midwaters but very little at the bottom, 1,400 meters deep (Monod, 1954). That same year, 1954, a third version of the bathyscaphe, the *Trieste*, was launched by Auguste Piccard and his son Jacques. But its operating costs proved too high for the Piccards, and they sold it to the US Navy Office of Naval Research in 1958, and Jacques Piccard came on board as a consultant.

The director of the *Trieste* project in San Diego, Andreas B. Rechnitzer (1924–2005), wrote an account of its first three years of service, in which he compiled the sightings of bioluminescence (Rechnitzer, 1962). Using a more sophisticated bathyphotometer than Kampa and Boden had used, Clarke and Backus (1956) had found that bioluminescent flashing increases sharply at the level of the DSL, particularly during periods of rapid vertical migration. Visual observations by the *Trieste* occupants confirmed such findings. During the ascent of the submersible, "bioluminescent flashes are normal at great depths and do not necessarily require tactile stimulation by a source such as the moving bathyscaph." As to the intensity of luminescent activity, Rechnitzer noted that "the concentration of bioluminescence at any one time rarely exceeds the number of stars that can be seen in the heavens on a clear, dark night. Evidence of virtually incessant flashing in the viewing area was observed by [Jacques] Piccard (personal communication) in 1956." The sources of luminescence could not be made out, but unicellular organisms were suspected.

Eric Barham used the *Trieste* to observe the DSL of the San Diego Trough in 1962. In contrast to the trawling results, which emphasized euphausiids and fishes, Barham found that siphonophores are the primary cause of DSL, and he reasoned that siphonophores are underrepresented in trawl catches because their fragility ensures their destruction in net operations (Barham, 1963). The species he observed was *Nanomia bijuga*, a bioluminescent form. The gas bubbles of the siphonophore's flotation organ are ideal for bouncing the echo-sounding signal, as are the gas-filled swimbladders of fishes (Marshall, 1960). Subsequent dives by Barham in the *soucoupe plongeante* of Jacques-Yves Cousteau off Baja, California, forced a revision of his view (Barham, 1966). Cousteau's *soucoupe* was a more agile submersible than the bathyscaphe, which allowed the observer to follow the dynamics of migratory movements of the DSL organisms. He found that in the early morning the DSL splits into two distinct components as it undergoes its descent from the surface. The upper component is accounted for by lanternfishes, and the other component by siphonophores. Both types of organisms "are capable of rapid, extensive changes in depth." For example, Barham measured "a dramatic downward migration of approximately 300 m[eters] in 90 minutes."

Yet another submersible was developed, this time by the Woods Hole Oceanographic Institution, which featured further technical improvements, among which the ability to catch organisms *in situ* with a net rigged to the submersible. The *Alvin*, which came into public limelight for its underwater exploration of the *Titanic* wreckage in 1986, was first put to use in 1965 and recruited for a DSL study in 1967. Richard Backus and others conducted the study along the continental slope of the western North Atlantic. The DSL was found to be associated with densely populated schools of the lanternfish *Ceratoscopelus maderensis*, specimens of which were caught by the *Alvin* net (Backus et al., 1968).

Curiously, while Rechnitzer recorded many episodes of luminescence in descents of the bathyscaphe, neither Barham nor Backus mentioned any in their papers. In reply to my inquiry on the subject, Barham, in a letter dated 17 November 1969, stated that the "only mesopelagic fish I have ever observed bioluminescing from a deep submersible was a myctophid when it was hit by part of the craft's superstructure. This [lantern]fish produced a bluish flash strong enough to be seen against the submersible's light field. I had the impression the light was produced by the caudal gland." It is telling that this bioluminescence was a collision artifact. It underscores the fact that motorized submersibles added noise and vibration to the bulkiness of the vehicle and the searchlight, sufficient to scare away any luminescent creature. Ironically, Beebe thirty years before had it better: his bathysphere had no engine and when on the lookout for luminescence, he switched off the light.

But Barham's cumulative observations led him to an insight about the behaviour of deep-sea fishes that has important implications for bioluminescence (Barham, 1970). He observed that hatchetfishes are at times motionless but always oriented horizontally, dorsal side up, in the water column, whereas other fishes, and particularly bristlemouths and lanternfishes, are oriented vertically. Most lanternfishes lie vertically with head upward in late afternoon and downward in the morning. He distinguished two assemblages of lanternfish: active species with functional swimbladders, large eyes, firm silvery bodies and well-developed photophores which migrate to the surface at night, and lethargic species without functional swimbladder, with small eyes, soft dark bodies and less developed photophores which migrate up only to a point. It implies that active species make more efficient use of their luminescence than lethargic ones.

Now it is time to pause and examine the critical role of the US Navy in this unfolding story. At first, the Navy became concerned about the biological underpinnings of DSL, but the phenomenon of oceanic bioluminescent flashing recorded by bathyphotometers soon added a new layer of potential threat to sophisticated submarine operations. Bioluminescence could be an ally of the Navy. After all, ships and torpedo boats were detected and attacked during the Second World War because of the bioluminescent wakes resulting from mechanical stimulation of dinoflagellates (Galler, 1963). But it could also create confusion in visually guided naval operations. How did the Navy cope with these biological interlopers? They addressed the problem by conducting research in the Navy's own labs, but also by contracting university-based biologists. One venue for the latter approach was the Office of Naval Research.

The Office of Naval Research (ONR) was created in 1946 by Act of Congress originally to oversee and fund Navy-related research (Sapolsky, 1990). But as no alternative federal agency was created immediately after the war to finance basic research just when demobilized academic scientists crowded the university labs and American science blossomed, the ONR filled the void and supported basic research of all stripes. It took major infighting between Congress and President Truman's Office before NSF was established and mandated to support basic research in 1951. As Sapolsky showed, as far as the ONR was concerned, many Navy officials were uncomfortable with the "proclaimed intent to preserve the freedom of scientific inquiry while financing scientific investigations that would serve the operational needs of the Navy." On the other side of the fence, many academic scientists saw the dangers of striking a Faustian bargain with the Navy. Even some university administrators, Sapolsky noted, "feared that federal support for science would lead to centralized control of science." When the National Science Foundation took over basic research support in the early 1950s, the ONR did not totally evacuate the field, feeling that there still existed "problems in science of special interest to the Navy, requiring it to remain in contact with outstanding scientists."

One such problem in science, of course, was bioluminescence, and particularly marine bioluminescence. The Navy needed to assess the geographic distribution, type, and seasonal occurrence of bioluminescence for the planning of strategic and tactical naval operations (Galler, 1963). It also needed to understand the biochemical and physiological mechanisms of bioluminescence in the delusive hope of controlling the phenomenon. With a large Naval Base in San Diego, it is not surprising that the ONR funded the basic scientists in the Navy's own backyard, at Scripps in La Jolla. This is how Kampa and Boden's work became supported from the early 1950s to the late 1960s. Also in the 1950s the other marine biologists who measured bioluminescence in the sea with bathyphotometers, such as George L. Clarke and Richard H. Backus at Woods Hole, benefitted from the ONR's largesse. And of course Navy labs were also involved.

That the Navy took bioluminescence dead seriously is illustrated by the fact that the biological section of the US Naval Oceanographic Office had over time assembled approximately three thousand individual reports of luminous displays throughout most of the oceans of the world (Staples, 1966). These surveys were but one of the many top secret oceanographic surveys conducted by ships of all descriptions for the US Navy during the Cold War. The Medea Report, published in 1995, analysed these data, which had recently been declassified, and found that bioluminescence and the transmissibility of light in the ocean were among the sought-after data. These data were especially valuable for strategic purposes of submarine protection or antisubmarine tactics in war games with the Soviet Union.

The Naval Postgraduate School, based in Monterey, California, since 1951, conducted oceanographic research and granted master's degrees to Navy officers, a few of whom dealt with bioluminescence and the biology of luminous organisms. Lieutenant Calvin R. Dunlap and Lieutenant Commander Andrew J. Compton produced theses on luminous fishes and euphausiids of the DSL (Dunlap, 1968; Compton, 1974), none of which were published in scientific journals. Eric Barham, some of whose major contributions to oceanography were highlighted earlier in this chapter, although not a naval officer, worked for the Navy's Undersea Research and Development Center in San Diego. Similarly, David Lapota (1949–2014), who conducted several field studies of large-scale oceanic displays of dinoflagellate bioluminescence (Lapota et al., 1986; Lieberman et al., 1987; Lapota, 2012), was on the staff of the Space and Naval Warfare Systems Center in San Diego.

One of the major academic centres of bioluminescence research that received continuous support from the ONR was Jim Case's laboratory in Santa Barbara. Initially, Case's first marine interest at UCSB was the chemoreception of crabs, but he soon branched out to the bioluminescence of the midshipman fish, *Porichthys*, handing a postdoctoral fellow from Belgium, Fernand Baguet (1939–2017), the task of figuring out how photophores are controlled. Baguet designed an experimental chamber to accommodate individual excised photophores and for luminescence recording. He obtained results similar to those of Nicol (1957), but he added that excitation-luminescence coupling occurred even in very low ambient oxygen (Baguet and Case, 1971). The results, showing that maximal responses are obtained in response to two electrical pulses per second, supported a nerve-mediated response and were consistent with electron microscopic observations of nerve endings in the vicinity of *Porichthys* photocytes (Strum, 1969b).

The Baguet paper was the opening salvo of what Case called in his memoir "my long love affair with the Office of Naval Research over marine bioluminescence." It was followed by a project of greater import for the Navy brass. Lanternfishes are overwhelmingly abundant in the mesopelagic environment, and any detailed information on their luminescent behaviour was crucial to them. Case enlisted an outstanding postdoc to see the project through: Anthony T. Barnes, a local resident who had done his PhD at UCSB. Jim Case's tinkering tendencies, alluded to in chapter 14, flourished to a full embrace of the latest technologies that could propel the field of marine bioluminescence forward by giant steps. Case's contacts with the Navy gave him access to the US Army Night Vision Laboratory, which had developed the Starlight Scope, an image intensifier – analogous to a television cathode ray tube – which was mounted on rifles and used in the Vietnam War to detect enemies in almost complete darkness without the aid of infrared light. By combining optical and videotaping equipment with the Starlight Scope, Barnes was able to obtain records of the spatial and temporal dynamics of the luminescing organisms while also recording temporal traces of light emission with the conventional light detector, the photomultiplier tube.

Barnes had already tested the usefulness of the technology in a study of the fast discharge of luminescent secretions by the copepod *Gaussia princeps* (Barnes and Case, 1972). But lanternfishes were another challenge altogether. Their survival time, whether sampled at the surface at night or from several hundred meters deep, is such that experiments can only be performed immediately after capture in the ship's wet lab, in this instance the *R/V Velero*

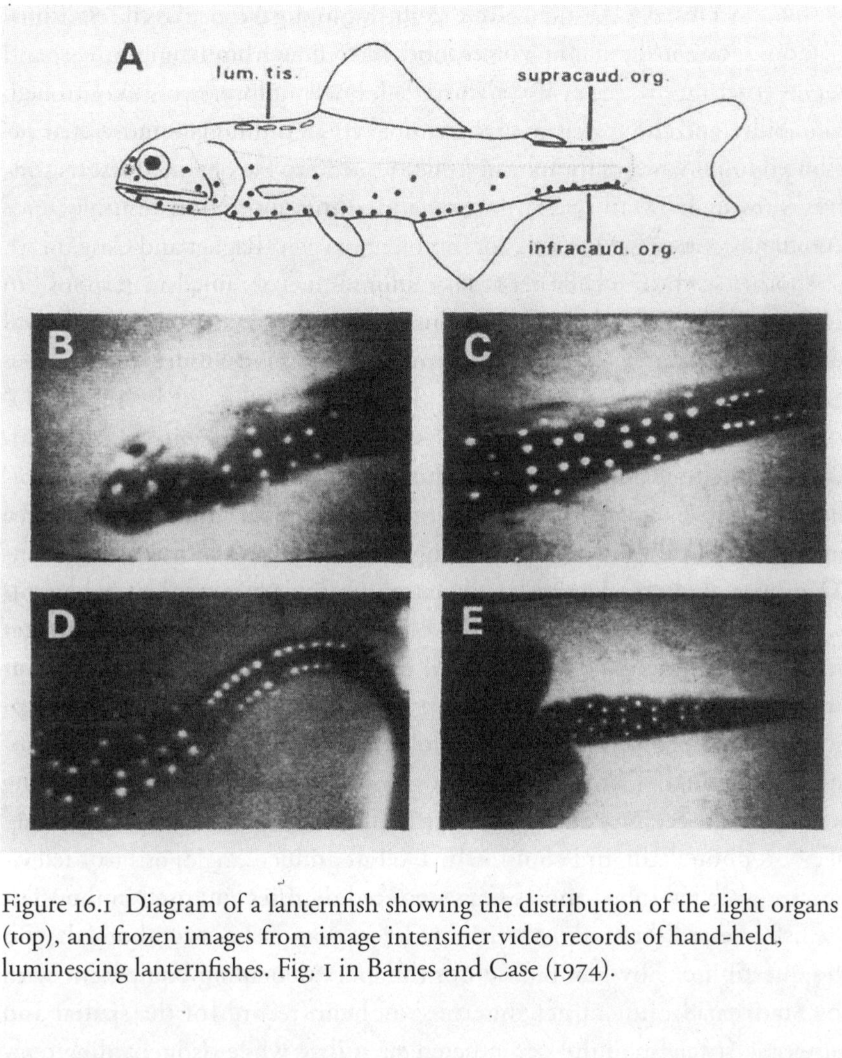

Figure 16.1 Diagram of a lanternfish showing the distribution of the light organs (top), and frozen images from image intensifier video records of hand-held, luminescing lanternfishes. Fig. 1 in Barnes and Case (1974).

IV of the University of Southern California. The reported results of the investigation were astounding (Barnes and Case, 1974). First, the lanternfish's body photophores and luminous patches (such as the caudal luminous organs) behaved differently. The photophores produced glows and the patches fast, facilitating flashes, all of a blue colour. Second, the photophores could light up simultaneously, and dim or glow in synchrony. The caudal luminous organs were triggered to flash when the fish's tail made thrashing swimming movements. These coordinations suggested a complex nervous control by spinal nerves. And third, its caudal luminous organs displayed functional subunits along its long axis, corresponding to major branches of the spinal nerves entering the organ along this axis (Anctil and Case, 1977). The high excitability and the response speed of the caudal luminous organs were attributed to the low-resistance gap junctions interconnecting the photocytes, thus allowing fast transmission of excitation throughout the photocyte mass (Anctil and Case, 1977).

Another scientific challenge that mobilized the techno-craving lab of Jim Case was testing the hypothesis that lanternfishes use their ventrolateral photophores to camouflage their silhouette by counterillumination. Barnes and Case (1974) had already noted that the light emitted by these photophores is preferentially directed downward, and two years earlier Eric J. Denton and his colleagues in Plymouth, UK, had calculated that the angular distribution of light emitted by the photophores of the hatchetfish and viperfish is ideally suited for a role of the bioluminescence in camouflage (Denton et al., 1972). Others had deduced that lanternfishes (Lawry Jr, 1974) and squids (Young, 1973) use their bioluminescence for counterillumination based on morphological data or crude visual observations, but Case and his team (1977) were the first to actually provide critical experimental data supporting the capability of lanternfish to deliver the performance. The experiments, conducted also on the *Velero IV*, involved exposures to changing artificial light regimes that revealed that lanternfishes adjust the intensity of photophore light emission to the level of ambient light, as recorded by photomultiplier. Image intensifier video records also demonstrated how fast the fish effected these adjustments, with less than half a second of delay. To the question of how the fish senses the variations of ambient light so as to make the adjustment of light emission, James Lawry Jr (1974) offered an answer: a special photophore situated such that it beams its light directly into the eye serves as comparator with the level of ambient light detected by the same eye.

That midwater shrimps are also equipped to counterilluminate was later demontrated by Case's team (Warner et al., 1979). Whereas lanternfishes cannot tilt their body in the water column without losing the efficiency of counterillumination, shrimps can, because their eyes, encased at the tip of eyestalks, and the posterior light organ make compensatory angular movements to ensure "that counter-illumination by *S*[*ergestes*] *similis* remains effective in inclined animals" (Latz and Case, 1982). The first author of this paper, Michael I. Latz, was one of Jim Case's students who, with Edith A. Widder and Steven H.D. Haddock, went on to enjoy highly successful careers studying marine bioluminescence. All three were schooled in the Santa Barbara ambience, which emphasized the design of innovative technologies to address specific scientific questions. They carried over these technical skills throughout careers that revealed bioluminescence as a key player in the ocean's ecosystem.

One technology in whose development both Widder and Latz participated in Case's lab was the computer-assisted Optical Multichannel Detection System (OMDS) to obtain sophisticated measurements of the spectral distribution (colour) of marine light emissions (Widder et al., 1983). Widder had just completed a thesis under Case's supervision on flash activity and the microsources of bioluminescence in a dinoflagellate, using a chamber of her own design to allow high-resolution observations of individual, restrained cells (Widder and Case, 1981, 1982). The OMDS allowed measurements of the most fleeting episodes of luminescence, which previous technologies could not achieve. The instrument was used to obtain emission spectra of seventy marine forms. The most exotic result obtained with the OMDS concerned the near infrared luminescence of the suborbital organs of two dragonfishes (Widder et al., 1984). The colour is astonishing, as the other light organs of these fishes emit blue flashes and it is difficult to imagine how their eyes can see such an extreme red light. But they do, thanks to a visual pigment sensitive to red light in their retina (Partridge and Douglas, 1995, Douglas and collaborators, 1999, 2000).

Another breakthrough technology was the use of submersibles equipped for stimulating nearby organisms and for videorecording luminescent displays. Widder and colleagues (1989) used such a submersible (Deep Rover program) in the Monterey Canyon and found that most displays originated in gelatinous organisms and many involved luminescent secretions. After her PhD Widder remained for a few years in Santa Barbara as a postdoctoral research biologist in what had become Case's "ONR Bioluminescence Program." She was a senior scientist and headed the Bioluminescence Department at the Harbor Branch Oceanographic Institution in Florida between 1989 and 2005. Since 2005 she has presided over the Ocean Research & Conservation Association, which she co-founded. Widder has been at the forefront of deep-sea investigations from submersibles of all types, and over the years she has garnered stunning images of organisms luminescing in their environment and sharp insights about life in the swarming deep-sea. More than anyone else in the field, she has brought bioluminescence stories into the limelight, especially through media interviews and television documentaries. She summed up her views on the amazing biochemical and ecological diversity of oceanic bioluminescence in a *Science* magazine article (Widder, 2010).

Lately Widder's e-videos (*Glowing Life in an Underwater World* and *The Weird and Wonderful Worlds of Bioluminescence*) have revealed to the world how rich and vibrant bioluminescent displays are in the sea. She was able to achieve these successes thanks to deep-sea probes which she designed to avoid the pitfalls that plagued William Beebe in the 1930s and the submersibles of the 1960s: bulky devices, bright lights, noise, and vibration. This kind of lying-in-wait seemed designed to scare away the resident marine animals. Widder came up with the Eye-in-the-Sea, a sensitive "camera fitted with a red light, thought to be invisible to the eyes of most deep-sea creatures" (Schrope, 2007). In addition, Schrope continued, it "is equipped with an LED lure, a cluster of tiny lights designed to flash and flicker in very specific patterns, all in the hope that something might respond." The LED lure originally used by Widder mimicked the luminescent pattern of the jellyfish *Atolla*.

Michael Latz, a contemporary of Edith Widder, did both his master's and PhD in Santa Barbara on the shrimp's use of bioluminescence for camouflage. Like Widder, he spent a few postdoctoral years in Santa Barbara, working on the kinetic and spectral characteristics of copepod luminescent secretions (Latz et al., 1987a) and on luminescence measurements of radiolarians and other organisms in the Sargasso Sea, using OMDS instrumentation (Latz et al. 1987b, 1988). He continued publishing on these animals and in 1991 he joined the SIO, where his lab became a centre of ocean bioluminescence, just as Boden and Kampa's had been decades earlier. His lab developed innovative instruments for the analysis of physical forces affecting bioluminescent output in the open ocean, especially for plankton organisms such as dinoflagellates (Latz et al., 1994, 2004). Using such techniques, his lab even analysed the flow of bioluminescence around a swimming dolphin (Rohr et al., 1998). An offshoot of his studies of shear stresses affecting dinoflagellate bioluminescence was the discovery of a stretch-activated ion channel in the plasma membrane of a dinoflagellate, which probably mediates the mechanical stimulation of bioluminescence (Jin et al., 2013).

Steve Haddock, the last of Jim Case's students considered here, completed his doctoral thesis at UCSB in 1997. He started out with a bang by adding a new phylum to the list of phyla harbouring luminescent species. It was "the first addition to the roster of luminescent phyla in more than 50 years (Haddock and Case, 1994)." The species in question was a bathypelagic arrow-worm (chaetognath) which, observed from a submersible, left "a plume of luminescence" as it darted away, suggesting a diversion role for escape behaviour. Another chaetognath species was eventually trawled, and its study led to the discovery that the luciferin in this group is coelenterazine, also present in several other phyla. The photocytes, located at the edge of the anterior fins, contain large paracrystalline bodies that appear to be the source of the emitted light.

Other exotic discoveries by Haddock and his colleagues include: (1) the "squid from hell," *Vampyroteuthis infernalis*, which luminesce at the tip of their arms (Robison et al., 2003); (2) a siphonophore living at great depths, which "possesses red-emitting bioluminescent appendages that may act as lures for fish prey" (Haddock et al., 2005); and (3) strange eyeless polychaete worms from even greater depths, whose light organs "suddenly burst into light" when shed by the animal, and are therefore called luminescent "bombs" (Osborn et al., 2009). All these discoveries were possible thanks to a new type of submersible, the remotely operated underwater vehicle (ROV), unmanned and controlled from aboard ship or from shore, and able to capture glimpsed organisms by special mechanical "grabbers" or by suction.

Haddock participated in Case's drive to measure the emission spectra of bioluminescence as a way to better assess the vertical distribution of oceanic gelatinous plankton (Haddock et al., 1995). As a result, he was able to show that the luminescence of comb-jellies has longer wavelengths than that of jellyfish, and that siphonophores have two different colours, depending on species (Haddock and Case, 1999). After his doctorate, Haddock took a position at the Monterey Bay Aquarium Research Institute in Moss Landing, California, where he is still active. Like the foundations in Florida where Edith Widder has worked, the Monterey Bay venue, under Bruce Robison's leadership, proved to be a hotbed of technological innovations for ocean research. There Haddock participated in the development of a multiplatform bathyphotometer for the study of coastal bioluminescence (Herren et al., 2005). With Jim Case and colleague Mark Moline he produced a fine review of his field (Haddock et al., 2010).

A phenomenon that attracted Haddock's interest was the "milky seas." These are eerie glows over large patches of sea surface which have been witsessed by mariners over many centuries and even by Darwin. Cases of "extreme bioluminescence" were even mentioned in popular novels such as Herman Melville's *Moby Dick* and Jules Verne's *Twenty Thousand Leagues Under the Sea*. The bioluminescent source of the milky sea seemed elusive, if only because the chance of a team of scientists sailing the open seas and happening on such fleeting displays were very slim. But finally such an encounter took place in the Arabian Sea and it was discovered that the most likely source was dense colonies of the luminescent bacterium *Vibrio harveyi* on blooms of microalgae (Lapota et al., 1988). Haddock participated with a Naval Research Laboratory scientist, Steven D. Miller, in the first observations ever, by satellite remote-sensing, of a glow "roughly the size of the state of Connecticut" (Miller et al., 2005, 2006). The display, on an unparalleled scale, lasted for three consecutive nights and was corroborated by shipboard observations.

One scientist who had tracked the history of "milky sea" spotting was Peter J. Herring (Herring and Watson, 1993). Herring's work emerged in the early 1970s and he soon became an undisputed leader in marine bioluminescence research. He created a strong focus for the field outside the United States, first at the National Institute of Oceanography in Wormley, Godalming, UK, and later at the National Oceanography Centre in Southampton. He was a hands-on, shipboard investigator, who signed on to many cruises of the RRS *Discovery*, commissioned in 1962 by the National Oceanographic Council of the UK (Herring and Partridge, 1992). Starting with a paper on the luminescence of the pearleye *Benthalbella* (Merrett et al., 1971), already the offspring of a *Discovery* cruise, Herring went on to examine the luminescent secretion of tubeshoulders (searsid fishes), which he concluded is a photoprotein similar to that of the tube-worm *Chaetopterus* (Herring,

1972). He diversified his interests to include the ocular light organ of a squid (Dilly and Herring, 1974) and the luminescence and photocytes of echinoderms (Herring, 1974). In the latter work Herring discovered that echinoderms other than brittle stars are also luminous: starfish, sea cucumbers, and sea lilies.

Early on, Herring made with the Dane Kjeld Hansen the important discovery that some anglerfishes possess two different light organs: the already known bulbous tip of the "fishing rod" above their head, which harbours bacteria and is of ectodermal origin, and the barbel light organ on the chin which is not of bacterial origin and appears derived from mesoderm (Hansen and Herring, 1977). Also in 1977 he described the digestive system–derived light organs of a new mesopelagic luminescent fish (Herring, 1977). Herring's emergence as an authority on bioluminescence gave him the confidence to edit a new book on the topic, which became an authoritative text for students interested in the field (Herring, 1978). He assembled experts who covered the different subfields, among whom Frank McCapra on basic chemistry, Milton J. Cormier on comparative biochemistry, Woody Hastings on bacterial and dinoflagellate luminescent systems, Herring himself on invertebrate systems other than insects, James Lloyd on insects, Herring and Jim Morin on fishes, Jim Case and his student Linda G. Strause on nervous control, J.A. Colin Nicol on the relations between vision and bioluminescence, and John Buck on the roles and evolution of the phenomenon. It was the most comprehensive book since Harvey's *Bioluminescence* (1952).

As Herring's career progressed, he paid more and more attention to aspects of marine luminescence that carry special ecological significance. One of these aspects is the role of colour in luminescent displays (Herring, 1983). Preceding Case's work on the topic by fifteen years, spectral measurements of light emissions by specific organisms were conducted over many cruises aboard the *Discovery*. "Species found in the pelagic environment," Herring wrote, "are mostly blue-emitting but there is some indication of a relative increase in green-emitting species in the benthic environment. Terrestrial organisms are predominantly yellow-green luminescent." There are adaptive values to the colour of bioluminescence when it matches the colour of the residual ambient light in the water column or the colour sensitivity of the eyes. Later, for example, John Turner et al. (2009) found that lanternfish eyes see their own light better than ambient light. Herring deflated the hopes of the Navy by noting that the wide range of luminescence spectral characteristics in the ocean precludes the possibility of identifying the source of a luminescent signal by its emission spectrum alone.

Herring also took stock of the distribution of luminous organisms among living organisms (1987). In creating a census of such organisms, he cautioned against including species whose ultraweak emissions merely reflect "the presence of active oxygen species in the cellular environment." The biological criterion for truly bioluminescent organisms is one where "the higher intensity light emission is itself observed (or more usually assumed) to have some adaptive value in the ecology of the organisms." Herring's census of luminous species within their phyletic groupings, imperfect as it may appear, provided an anchor for biologists eager to generate hypotheses on the evolution of bioluminescent capability. But he added that "the evolution of associated behaviour patterns and light organ morphologies, and the complementary evolution of the animals' visual systems" are also sources of fascination.

A long-standing interest of Herring's was the use of reflective materials as optical enhancers of the light emissions of such marine forms as crustaceans, cephalopods, and fishes. Before him, Eric J. Denton, one of his compatriots, had published on the importance of reflectors in marine animals (Denton, 1970). These can act as mirrors merely that reflect outward, with varying degrees of directionality, the light generated by the photocytes, or they can channel light through "light-pipes" (Herring, 2000). Guanine crystals are a recurrent material of these reflectors, although other materials such as collagen fibres are involved in some cases. Herring remarked that "the easy availability of biological materials which can be arranged into very efficient multiple interference reflecting systems has led to a plethora of reflector design for different purposes. This has probably had a profound influence on the evolution of bioluminescence, particularly in the deep ocean."

Herring's eclectism carried over to sexual dimorphism in marine bioluminescent species and the importance, to quote his pun, of "sex with the lights on" (Herring, 2007). The role that sexual differences of colour display in mate recognition and selection in terrestrial animals, Herring argued, may in luminous animals be played by sexual differences in photophore implements or patterns of light emission. It works for fireflies, so what about "sex with the lights on" for marine forms? There are many cases of sexual dimorphism in number, size, and body distribution of photophores, and in some cases it comes down to the absolute absence of light organs in one of the sexes. One can mention leiognathid fishes (McFall-Ngai and Dunlap, 1984), ostracods (Morin and Cohen, 2010) and syllid worms (Galloway and Welsh, 1911) as clear cases of sexual dimorphism. The question Herring asked is, of course, whether the dimorphism carries a sexual signal. Much of the evidence is ambiguous or inconclusive, but in a few cases there is little room for doubt. Among the invertebrates, the female, but not the male, of a few squids develops unique photophores at maturity which become resorbed after spawning. In contrast, in lanternfishes the males tend to possess caudal luminous organs or head patches, or if the females possess them also, the male's organs tend to be larger. Herring speculated that by virtue of the fast flash kinetics of these patches, which could have coding value, lanternfishes might use them in sex recognition and selection in a manner analogous to fireflies. A question not asked by Herring is what sexual dimorphism and bioluminescent courtship does for the evolutionary success of the group. This question was answered by Emily Ellis and Todd Oakley (2016), who found that "lineages with bioluminescent courtship, almost certainly a sexually selected trait, have more species and faster rates of species accumulation than their non-luminous relatives."

In addition to the numerous reviews on different aspects of marine bioluminescence, Herring and colleagues published many original papers on the structure of photophores of various crustaceans, cephalopods, and fishes. Herring also published a book on deep-sea biology in which a section covers bioluminescence (Herring, 2002).

~

Across the channel in Belgium, another European lab made significant contributions to marine bioluminescence. Fernand Baguet, after his stint in Jim Case's lab, returned to his country and took a position as *maître de recherche* of the Fonds National de la Recherche Scientifique (FNRS) at the Université Catholique de Louvain-la-Neuve. Baguet was trained as a muscle physiologist, but after his exposure to bioluminescence in California the latter topic gradually became a mainstay of his research activity. After pioneering the physiological study of isolated photophores in the midshipman fish, a coastal species, Baguet attempted to apply the approach to the more challenging deep-sea fishes. In order to do so, and not having a research vessel at his command, Baguet opted for the next best thing: the Straits of Messina between the coasts of continental Italy and Sicily. This is the site where the mythical rock shoal and whirlpool of Homer's *The Odyssey* played havoc with sailors and defined being squeezed between a rock and a hard place.

But the whirlpool is not quite a myth. Off the coast of Messina, tidal forces combine with unique rock topography to create the upwelling of deeper waters into the surface layer. As a result, deep-sea fishes are brought up near the sea surface and can be easily fished in the Straits; or they wash up on the beach, where they can be collected still alive. These are the exceptional circumstances that Baguet took advantage of for his bioluminescence research. With his Louvain colleague Georges Maréchal, a physiology professor, he travelled to Messina, where the laboratory facilities of a colleague at the local university, Sebastiano Genovese (1926–1983), were put at their disposal. In their first report, Baguet and Maréchal (1974) used the small experimental chamber that Baguet had designed in Jim Case's lab, equipped with resident stimulating microelectrodes and an RCA IP21 photomultiplier tube to detect light emissions and record the traces on a storage oscilloscope. They collected numerous hatchetfishes, excised the photophores, and placed them in the chamber. Their records showed extremely fast flashes in response to small electrical pulses; it took only three milliseconds after the stimulus for the flash to appear, and within the next seven milliseconds it was all over. They could make the organs produce twelve such flashes per second, but the photophores were soon fatigued.

In 1979 Baguet teamed up with Jacques Piccard for a series of submersible dives in the Straits of Messina aboard the *Forel*, a "mesoscaphe" designed by Piccard himself. The submersible was equipped with a photomultiplier light detector oriented upward, which allowed measurements of downwelling ambient light and the ventral bioluminescence of fishes (Baguet and Picard, 1981). They found that "*Argyropelecus* and Myctophids [lanternfishes] observed through the plexiglass dome of the mesoscaphe, show a sustained luminescence originating from the ventral photophores of fish and forming a luminous patch. On the other hand, [the bristlemouth] *Cyclothone braueri* were non-luminous." They also suggested that the counterillumination hypothesis, which had been validated four years earlier in laboratory conditions by Case and collaborators, was not supported by their field observations at depth. The light emitted from the ventral photophores was not deemed bright enough to match the intensity of the ambient light in the specific depth ranges where these fishes live, so the camouflage of their ventral silhouette would be inadequate.

With one of his students Baguet conducted experiments on isolated photophores and caudal luminous organs of lanternfishes (Christophe and Baguet, 1982). They showed that the glows of photophores are a property of the photocytes and not of their innervation. The caudal lumnous organs, in contrast, show fast flashes at very low thresholds of stimulation, thus suggesting that a different triggering mechanism is involved.

~~~~~

A student of Baguet gradually took over the leadership of the Louvain bioluminescence laboratory as Baguet's career waned. Jérôme Mallefet, an avid scuba diver and skilled underwater photographer, conducted his doctoral research under Baguet, concentrating on the relation between oxygen metabolism and bioluminescence in isolated fish photophores (Mallefet and Baguet, 1984, 1985). In the 1990s he switched to the little-known brittle stars as his main experimental animal (Mallefet et al., 1992). As Mallefet (2009) explained, many luminescent echinoderms are not readily accessible in the field, and it took a dedicated investment in field trips and a deep-sea cruise for Mallefet to add new brittle-star species to the list. But Mallefet's claim to prominence in the field of oceanic bioluminescence materialized in the new millenium when he and his students undertook an ambitious research program on lantern sharks (summarized in Claes and Mallefet, 2009a). Their work gave them visibility in the media as surely as Steve Haddock's did.

Luminous sharks had suffered the most glaring neglect from students of deep-sea bioluminescence. Nothing significant had emerged since the early studies of Leopold Johann (1899) and others in the early twentieth century (see chapter 9). In the chapter on fish bioluminescence in the book *Bioluminescence in Action* (1978), Peter Herring and Jim Morin are totally silent on the topic of sharks. All this changed when Mallefet's lab took on the challenge. It started when Mallefet and his student Julien Claes enlisted the help of Norwegians to collect the velvet belly shark, *Etmopterus spinax*, in the deep
~~~~~

waters of a fjord near Bergen. Their purpose was to follow the development of the photophores before and after birth (Claes and Mallefet, 2008). The fact that "ready-to-hatch embryos are already able to emit a downward blue light" was construed by the Belgian investigators as strong suggestive evidence that these sharks use their luminescence for counterillumination.

Their next discovery served as an illustration of nature's recycling ability. Shallow-water sharks camouglage their ventral silhouette by playing with their skin pigmentation screen to effect colour change. They found that deep-sea sharks recycle the skin pigmentation to adjust luminescence intensity for optimal counterillumination (Claes and Mallefet, 2009b, 2010). A comparison of the photophore patterns of twenty-one shark species worldwide, Claes and colleagues (2014) argued, "emphasizes the importance of bioluminescence as a speciation driver." They found that counterillumination takes second place to social luminescent signalling in species that run little predation risk from below, which means that their photophore patterns grew in complexity. In this regard, the presence of species-specific photophores on the body flanks of these sharks "may provide a way for reproductive isolation and hence may have facilitated speciation in the deep-sea" (Claes et al., 2015). A case of coexistence of two sets of photophores with different functional agenda was discovered in the velvet belly lantern shark (Claes et al., 2013). Besides the myriad ventral photophores serving in counterillumination, they found spine-shaped light organs on the dorsal fin, the luminescence of which seems to serve as a warning signal for predators that attacking the shark can be harmful.

~~~

If the investigations of Steve Haddock's and Jérôme Mallefet's teams revealed to science and the lay public wondrous adaptations involving bioluminescence in the ocean environment, the work of James G. Morin accomplished something similar, if on a smaller scale, for coastal bioluminescence. After graduating from UCSB, Morin pursued a master's degree and a PhD at Harvard under Woody Hastings and Ian Cooke, then an assistant professor of invertebrate physiology. With Hastings he conducted biochemical investigations on the luminescent system of numerous cteno-
~~~

phores and cnidarian species and concluded that in all cases a calcium-activated photoprotein was involved (Morin and Hastings, 1971a). An exception was the sea pansy *Renilla* in which Milton J. Cormier and his team (1970) had clearly demonstrated that a photoprotein is not involved, a finding that still holds true today.

Morin was the first to record "luminescence potentials," bioelectrical events in the animal which trigger a light emission. Usually such potentials originate in nerve fibres, but in the case of hydroids such as *Obelia geniculata* studied by Morin and Cooke (1971a,b) they originate in the electrically coupled epithelium harbouring the photocytes. The latter were localized by fluorescence microscopy in the endoderm epithelium (Morin and Reynolds, 1974), a location that proved common in all bioluminescent cnidarians.

After his doctorate Morin was appointed assistant professor at UCLA, and it was at about this time (1973) that he, his then wife Anne Harrington, Woody Hastings, and associates travelled to the Gulf of Eilat in Israel to study the Red Sea flashlight fish (Morin et al., 1975). Combining direct underwater observations and laboratory studies, Morin and his colleagues found an astonishing number of uses for the luminescence of the large subocular organ of these fishes; it is, in their own words, a "light for all reasons." In the field their task was made easier because the flashlight "fish, observed from the shore, by snorkeling, or by using scuba, were easily seen underwater from distances up to 20 m[eters]." The basic behaviour of the organ is blinking with the help of a muscular black lid that can extend over the light organ or retract from it, in which case the packed luminous bacteria of the organ are exposed. Infrequent blinking, Morin and his colleagues found, was associated with using the light to attract and see prey. The blink-and-run behaviour serves to deter predators. Finally, complex high-frequency blinking was associated with communication between mates.

In an incursion into luminescent symbioses, Morin and Kenneth Nealson's student Edward Ruby wanted to know how specific the symbiotic relationship between luminous bacteria and fish is. They discovered a temperature-related specificity (Ruby and Morin, 1978). Deep-sea macrourids harbour bacteria (*Photobacterium phosphoreum*) tolerant of the low-temperature waters of the deep, whereas shallow water forms such as ponyfishes and knightfishes host bacterial species sensitive to the cold (*P. leiognathi* and *P. fischeri*).

An evolutionary insight garnered from their study was that these bacterial differences, along with the unrelatedness between these fish families and differences in the structure and location of their luminescent organs, point to "an independent and multiple origin of each of these bacteria-fish associations."

Morin summed up what was currently known of coastal bioluminescence in a review (Morin, 1983). His objective was to put the topic in "a framework that has ecological perspective, so that we can better focus on the questions of the functions of luminescence in all light emitting organisms." He noted that, unlike oceanic organisms, very few coastal marine species can emit light (up to 2 percent), but "there are often enormous numbers of individual organisms within a small area that emit light in a variety of patterns." These include dinoflagellates, cnidarians, ctenophores, annelid worms, crustaceans, brittle stars, larvaceans, and fishes. He concluded that, while other functions or multiple functions are prevailed upon, "the majority of luminescence seen in coastal waters is aimed at deterring potential predators."

One luminescent form discussed in the review, the ostracods, became the focus of Morin's research. Up to that point whatever was known of these small crustaceans came mostly from *Vargula hilgendorfi* and its luciferin-luciferase system. Morin became engrossed in the luminescence of the group from the angle of their ecology and evolution. Between 1980 and the present, Morin has made field observations of many cypridinid species at different locations in the Caribbean (Morin, 1986; Morin and Cohen, 1991, 2010). Morin stressed that the luminescent secretions of these species have a dual role: bright, long-lasting clouds are associated with evasion from predators, and species-specific, complex series of pulses broadcast by males serve to court receptive females and to compete with other males for female attention. In one species, Morin observed, many males synchronized their cloud bursts, thus producing "a spectacular, reef-wide, sweeping luminescence that is reminiscent of a slow motion version of the flashing of synchronous fireflies from Southeast Asia."

Morin continued to enrich the understanding of these displays in papers with his students (Cohen and Morin, 2003, 2010; Rivers and Morin, 2008). They have described several new genera and many new species from the Caribbean. Morin also examined the sexual luminescent displays in an evolutionary framework, noting that sexual selection for cloud burst signaling went hand in hand with selection for other characteristics, such as male copulation organs (Morin and Cohen, 2010). He and his former student Trevor Rivers also made an analysis of the energy expenditure of luminescent activities, finding that luminescent displays aimed at self-defence released fifty times more luminescent secretion than displays aimed at mating (Rivers and Morin, 2012). It is a persuasive illustration of Morin's and Cohen's quip that "it's all about sex."

Recently, luminescent ostracods were highlighted in a BBC documentary on bioluminescence (Attenborough's *Life That Glows*, 2016), in which Trevor Rivers and Gretchen Gerrish, two of Morin's former students, are seen recording ostracod activities in their natural habitat, which they could previously monitor only in the laboratory. Their recordings were made possible thanks to a camera designed by wildlife filmmaker Martin Dohrn, which split the low-light image captured by a lens into two channels, one giving an infrared view of the background and the other projecting an image-intensified view of bioluminescent events. The two images are then fused into one, so that bioluminescent events are viewed in their natural context. Thanks to this technology, the pair could see a constellation of a thousand ostracods in which males compete among each other for a female's favour by disrupting ongoing courtships – the equivalent of a bar brawl over a woman! The show is amazing. Technological innovations have come a long way toward peeling off the historical layers of mystery about bioluminescence in the sea.

17
Understanding How Light Sources Are Controlled

> From a philosophical point of view, it is clear that future developments in bioluminescence will involve … the manner in which the simple components are manipulated by the cell or the organism, to achieve highly sophisticated terminal effects.
> –Jean-Pierre Henry and A. Michael Michelson (1978)

While marine bioluminescence was enjoying a revival thanks to technological breakthroughs and media attention, progress in the study of physiological control of living light sources was slower and hidden from the public gaze. The regulation of light emission, a precondition for shaping the behavioural and ecological significance of luminescent displays, can be looked at from two levels of biological organization: the cell and the organism. Cells of physiological relevance here can be free-ranging, such as luminescent bacteria and dinoflagellates, or part of a tissue/epithelium or an organ in multicellular organisms (photocytes). At the organismic level, several factors can exercise control over the light output, depending on the complexity of the light organ or the organism. Let us examine first bacteria and dinoflagellates.

The first breakthrough in understanding how bacterial luminescence is controlled originated in the lab of Woody Hastings at Harvard. One of his students, Kenneth Nealson, discovered that a synthesis of bacterial luciferase occurs only when the population of bacteria in a culture medium has reached a critical mass (Nealson et al., 1970). A conditioning factor is then released in the medium, which turns on the DNA coding for luciferase synthesis. This control process was called "autoinduction," and it was studied in detail in the 1970s (Nealson, 1977; Nealson and Hastings, 1979). The substance released by the bacteria which is responsible for the autoinduction, called autoinducer or pheromone, was identified in 1981 by Anatol Eberhard and his team at Cornell University; it is a small organic compound that is species-specific, being active only in the *Photobacterium* species which produces it, but not in the others.

The concept of collective behaviour exemplified by bacteria in the case of autoinduction found application in other spheres of bacterial life and is now called "quorum sensing" (Williams et al., 2007). It defines "bacterial cell-to-cell communication mechanisms which co-ordinate gene expression usually, but not always, when the population has reached a high cell density." Nealson and Hastings (2006) themselves saw in the milky seas, discussed in the preceding chapter, a clear example of quorum sensing on a grand scale. But examples on a more modest scale appear in the bacterial colonies nested symbiotically in light organs of squids and fishes. Bacterial growth in these organs is promoted by a rich blood vessel supply which facilitates the procurement of nutrients and oxygen.

Luminous bacteria are present in sea water but, except in milky seas, in insufficient density to enable luminescence. But some squids and fishes have exploited the autoinduction property of luminous bacteria to fashion light organs based on a symbiotic relationship. Nealson, then at Scripps Institution of Oceanography in La Jolla, and his student Edward Ruby were the first to study the physiology of the relationship (Ruby and Nealson, 1976). They found that the bacteria (*Photobacterium fischeri*) respond to an inducer of luciferase which accumulates in the host light organ of the knightfish *Monocentris*. Light production is optimal under low, growth-limiting concentrations of oxygen. And when grown on glucose, these bacteria excrete pyruvic acid into the medium, which plays an important role in regulating energy metabolism between bacteria and host. All in all, the two organisms are well adapted to each other. If one of the two partners is not playing the game anymore, as when a flashlight fish is starved for three weeks (Meyer-Rochow, 1976b), the bacteria eventually cease to emit light for lack of nutrients.

A major contributor to our understanding of the developmental and evolutionary aspects of luminescent bacterial symbiosis is Margaret McFall-Ngai. She obtained her PhD at UCLA under the supervision of Jim Morin, studying the light diffusion mechanism (McFall-Ngai, 1983) and the sexual dimorphism (McFall-Ngai and Dunlap, 1984) of the light organs of the ponyfish *Leiognathus*. Only a few years later did she demonstrate interest in the luminescent bacterial symbiosis in the light organ of ponyfishes with her colleague

Paul Dunlap. In their paper (Dunlap and McFall-Ngai, 1987), they stated: "We have begun to address questions on the specificity and development of the association and on the regulation of bacterial activities in the symbiosis." Rarely has such a statement of intention been followed to the letter for a lifetime as by McFall-Ngai. After moving to the University of Southern California she began her highly productive research program on the Hawaiian bobtail squid *Euprymna scolopes* (McFall-Ngai and Ruby, 1991). In this paper the importance of the symbiotic relationship was emphasized by the conclusion that "the initiation of symbiosis influences, and is perhaps a prerequisite for, the normal developmental program of the juvenile host."

Juvenile bobtail squids acquire luminous bacteria after the light organ has started its development (Ruby, 1996). Mary K. Montgomery and McFall-Ngai (1993) have characterized the genesis of the symbiosis: "(i) embryonic development, during which the light organ rudiment first forms and the stage is set for initiation of the symbiosis; (ii) early post-hatch development, during which the light organ undergoes morphological and biochemical changes that accompany infection with and subsequent growth of the bacterial symbionts; and (iii) late post-hatch development, which results (within two weeks) in a fully differentiated light organ that primarily functions to maintain the symbiosis and control light emission." It was subsequently found that bacterial symbionts, once they have infected the light organ rudiment, produce signal molecules that profoundly transform the structure of the light organ until it reaches its final form in the adult (Doino and McFall-Ngai, 1995). The morphogenetic signal, it appears, is controlled by the *lux* genes involved in the synthesis of enzymes participating in the bacterial luminescent reaction itself (Visick et al., 2000).

Whatever tortuous pathway is taken to achieve the symbiosis, the end-result is that bacterial light emission tends to be continuous and it is up to the host to provide control mechanisms for shaping an intermittent light ouput. Shutters are used by both the cardinalfish *Siphamia* (Dunlap and Nakamura, 2011) and flashlight fishes (Johnson and Rosenblatt, 1988) to intermittently hide the bacterial light. In flashlight fishes it involves jaw muscles, ligaments, and cartilages that operate from outside the light organ, the nervous control of which has not been investigated so far.

Dinoflagellates, the other form of luminescent unicellulars, differ from bacteria in that they possess intrinsic control mechanisms. Central to the excitation-luminescence control pathway of these protists is mechanical stimulation as the entry point. We have mentioned earlier how Michael Latz's team has highlighted the role of shear stresses – breaking waves, animals swimming about – in stimulating dinoflagellate flash production. Experimental evidence suggested that dinoflagellate luminescence serves to startle and repel grazers such as copepods, who themselves mechanically cause the luminescence by swimming in the midst of dense dinoflagellate populations (Esaias and Curl, 1972; White, 1979). But an alternative role, originally suggested out of the blue by Martin D. Burkenroad (1910–1986), later gained currency.

Burkenroad was born to a family of Louisiana eccentrics from whom he inherited a headstrong and mercurial character (Schram, 1986). This led to the many falling-outs with colleagues and employers that punctuated his career. The main thread of his career in marine biology was the shrimp-fishing industry. His only incursion into the field of bioluminescence happened when he worked at Yale University during the Second World War. Perhaps he had the leisure to let his creativity loose in the office, what with the war effort stealing away the focus of scientific projects. Be that as it may, he then articulated the "burglar alarm theory" (Burkenroad, 1943), according to which dinoflagellates respond to copepod predation by producing luminescence, which alerts the copepod's predators – fishes – to the presence of their prey. To take an anthropocentric shortcut, dinoflagellates use their flashes to enlist a third party to get their predators off their back.

It took half a century for Burkenroad's hypothesis to be tested experimentally. Canadian biologists Mark V. Abrahams and Linda D. Townsend (1993) showed that dinoflagellate luminescence increases copepod mortality in the grip of fishes as third party. And Kellie J. Fleischer in Jim Case's Santa Barbara lab demonstrated "the ability of squid to use dinoflagellate bioluminescence to locate and capture nonluminous prey" (Fleischer and Case, 1995). Either way, the burglar alarm theory appeared validated. However, although the case looked strong, methodological flaws in these studies may have weakened the validation (Valiadi and Iglesias-Rodriguez, 2013). A variant of the theory has recently surfaced, according to which bacterial bioluminescence can be used as bait (Zarubin et al., 2012). In this case, luminescent bacteria are ingested by zooplankton transparent enough for the colonized bacterial luminescence to shine through and for the zooplankton to become visible to their predators, nocturnal fishes.

Now what happens downstream in the cell's control cascade? Shear stresses affect the dinoflagellate outer membrane by activating a class of cell membrane receptors known as G protein-coupled receptors (Chen et al., 2007). This action results in elevated calcium levels inside the cell (Dassow and Latz, 2002), thus causing the vacuole membrane inside the cell to depolarize and generate an action potential. Because the vacuole is intimately associated with the luminescent source, the scintillon, the vacuolar action potential, activates voltage-gated proton channels in the scintillon membrane, thereby lowering the pH inside the scintillon. The lower pH is optimal for initiating the luciferin-luciferase reaction, which generates the flash (Valiadi and Iglesias-Rodriguez, 2013).

Among multicellular organisms, one distinguishes animals without light organs – possessing clusters of photocytes often inserted in a simple epithelium – from those with variously complex light organs. Since Nicol's contribution discussed in chapter 14, progress in understanding how luminescent epithelia or organs are controlled has been steady but slow, and curiously the topic has not been thoroughly reviewed since the 1980s (Case and Strause, 1978; Anctil, 1987). What had been hypothesized in these reviews was that nervous control overwhelmingly predominates as a mode of regulation over hormonal control. Also, it was apparent that few experimental models were investigated, leaving out a large number of less-accessible animal groups in the realm of the unknown. And finally, our understanding of what happens inside the photocytes in the way of control processes was sketchy at best.

~~~

The luminescent animals without organs include cnidarians (jellyfish, siphonophores, hydroids, and colonial anthozoans), ctenophores (comb-jellies), annelid worms (earthworms, polychaete worms), echinoderms (brittle stars), and protochordates (pyrosomes, tunicates, and acorn worms). Some of these groups were touched upon in previous chapters, so we will concentrate here on groups or aspects as yet unexamined.

One question related to control in these forms is whether luminescence is controlled by a nerve net or by a neuroid conduction pathway. By neuroid
~~~

or epithelial conduction is meant the propagation of impulses across low-resistance (gap) junctions between epithelial cells, as an alternative mechanism to nerve impulse propagation (Anderson, 1980). Epithelial conduction of luminescence signals and/or the presence of gap junctions was demonstrated in jellyfish (Mackie and Mackie, 1963), siphonophores (Bassot et al., 1978), hydroids (Dunlap et al., 1987), ctenophores (Anctil, 1985), and scale-worms (Bilbaut 1978, Herrera 1979). Nervous control co-exists with neuroid conduction in the luminescent system of ctenophores and scale-worms, but neural involvement in jellyfish and siphonophores has not yet surfaced. Clear cases of luminescence coordination by basiepithelial nerve nets are sea pens and other colonial anthozoans (Satterlie et al., 1980), tube-worms (Martin and Anctil, 1984), and acorn worms (Baxter and Pickens, 1964).

One approach to revealing the involvement of nervous control is to examine the effects of known neurotransmitter systems on luminescence. In the sea pansy *Renilla* a type of beta-adrenergic receptor was associated with excitation of bioluminescence (Awad and Anctil, 1993), and the molecular evidence suggested that a suspected adrenergic-like compound acts indirectly, through granular cells abutting the photocytes (Awad and Anctil, 1994). Evidence that luminescence is dependent on conduction between photocytes and neighbouring endodermal cells (Germain and Anctil, 1995), led to the view that adrenergic-like nerve net neurons activate granular cells which, in turn, transmit their signal to photocytes by epithelial conduction.

In the luminescent system of the tube-worm *Chaetopterus*, it appears that acetylcholine acts as an excitatory neurotransmitter and the amino acid GABA as an inhibitor of luminescence (Anctil, 1981). Likewise, the luminescence of the polychaete worm *Tomopteris* is under nervous control and acetylcholine appears to be the neurotransmitter involved (Gouveneaux and Mallefet, 2013). In contrast, the luminescent scales (elytra) of scale-worms contain or can take up monoamines (catecholamines and serotonin), which are localized in nerves and in the luminescent epithelium (Miron et al., 1987). In the echinoderm *Amphipholis*, acetylcholine triggers luminescence (De Bremaeker et al., 1996), whereas amino acid transmitters (GABA, glycine) and neuropeptides modulate the acetylcholine-induced luminescent response (De Bremaeker et al., 1999).

The phenomenon of luminescence inhibition by light has attracted the interest of investigators ever since Harvey drew attention to it, as recorded in chapter 10. This phenomenon can be viewed as a control mechanism to prevent waste of the luminescent material during daytime when bioluminescence is least likely to fulfill its biological role. Already in 1959 Sweeney and colleagues had analysed the phenomenon in the dinoflagellate *Gonyaulax* and had concluded that the pigment involved in dinoflagellate photosynthesis was a probable cause, along with an unidentified, more red-sensitive pigment. As explained in chapter 14, Nicol had shown that photoinhibition of *Renilla* luminescence occurs either at the nerve-photocyte "synapse" or in the control cascade inside the photocytes, but not in the luciferin-luciferase regulation process. In contrast, it is precisely the photoprotein of the ctenophore *Mnemiopsis* which photoinhibition targets.

The calcium-activated photoprotein called mnemiopsin (Ward and Seliger, 1974), is considered one of the most light-sensitive proteins in existence and "is a pre-charged enzyme already containing bound luciferin [coelenterazine] and oxygen" (Ward and Seliger, 1976). A puzzle arose when attempting to reactivate mnemiopsin in the dark after inactivation by light. Reactivation was evident in the intact animal, but all attempts to obtain it in the test tube had failed. In seeking the solution to the puzzle, Anctil and Shimomura (1984) discovered first that, contrary to a previous view (Ward and Seliger, 1976), both oxygen and coelenterazine become dissociated from the photoprotein upon light exposure – a result later vindicated through the use of more sophisticated techniques (Powers et al., 2013). To recharge the system, it was found that oxygen, and a pH environment (9.0) creating highly charged (ionized) molecules, is required for coelenterazine to bind and regenerate photoinactivated mnemiopsin (Anctil and Shimomura, 1984). This finding suggested to the authors that the photocytes of *Mnemiopsis*, when exposed to sunlight, keep the photoprotein inactivated "by limiting the supply of free coelenterazine, or possibly by slightly offsetting the pH inside the photocytes from the optimum pH of 9.0."

But is mnemiopsin directly destroyed by sunlight? Or does the ctenophore need to "sense" the light for the inactivation to occur? Ctenophores, unlike some jellyfishes, do not possess eyes, and until recently no evidence of light-sensing epithelial cells had surfaced. Christine Schnitzler and colleagues (2012), using tools of molecular genetics, made the remarkable discovery that opsins, the class of light-receptive proteins required in our retina for vision, co-exist with mnemiopsin in the photocytes of *Mnemiopsis*. The authors raise the tantalizing hypothesis "that light production and light reception may be functionally connected in ctenophore photocytes." It remains to be tested whether this connection means that sunlight must be sensed first by the photocyte through an opsin for the inactivation of mnemiopsin to unfold.

As for animals endowed with light organs, the question of nervous control was principally investigated in fireflies and fishes. The firefly story related to physiological control merits telling if only for its saga of twists and turns. Nervous control had never been in doubt since the investigations of John Buck and Jim Case in the early 1960s (see chapter 14). Soon afterward, a former student of Jim Case conducted a pharmacological analysis the results of which were "interpreted to mean that normal neural excitation of the light organ is mediated by an adrenergic transmitter located in the nerve endings" (Smalley, 1965). Another of Case's former students came to a similar conclusion with regard to control of bioluminescence in the larva of the firefly (Carlson, 1968). Even exotic luminescent beetles, the phengodids, were examined in Case's lab and, except for the railroad worm, their luminescence was found to be under nervous control and susceptible to the effect of the neurotransmitter noradrenaline (Halverson et al., 1973).

Case followed up on these findings by assigning a student, Donata Oertel, the task of further pursuing the problem of luminescence control in the larval firefly. She found that each of the two larval light organs is supplied by a nerve branch containing only two axons, which branch profusely into the organ (Oertel et al., 1975). Contrary to the adult light organ, in which nerve synapses are seen only on tracheal end-cells, nerve terminals make synapses directly on photocytes in the larva. But there lies a glaring contradiction: in spite of the direct innervation, the firefly larvae produce slow-starting, long-lasting glows in response to nerve stimulation (Oertel and Case, 1976), while the adult, whose innervation is indirect, produce fast flashes. Oertel also characterized adrenergic receptors in the larval organ, which are known to respond slowly to adrenergic transmitters, thus at least partly explaining why larvae produce glows instead of flashes.

Just as Oertel was preparing her 1976 paper for publication, she became aware of a paper fresh off the press in which the results directed the search for the firefly light organ neurotransmitter in quite another direction. Albert Carlson was a student of Case's who had moved on to a tenure-track position at the State University of New York in Stony Brooks. He too was engrossed by the problem of the nervous control of firefly luminescence, to the point that he did not mind entering into a friendly competition with his former mentor. Carlson had heard the buzz that adrenaline or noradrenaline may not exist as neurotransmitters in insects after all, but that another chemical, octopamine, present only in insignificant traces in the mammalian brain, is the insect substitute for adrenergic neurotransmission. The investigator who had championed this idea, Harold A. Robertson, was a Canadian then posted at the Psychiatric Research Unit of the University of Saskatchewan Hospital in Saskatoon. Carlson enlisted him to make the case for octopamine in the firefly light organ.

The result of the collaboration was a paper in which biochemical analyses showed that octopamine, but not adrenaline or noradrenaline, was detectable in the adult firefly light organ (Robertson and Carlson, 1976). Carlson did not let the matter stand. He soon enrolled a graduate student, Thomas A. Christensen, to dissect further the intricacies of nerve control. Using axon-tracking dyes, Christensen was able to identify four large neurons in the posterior ganglia of the abdomen, the Dorsal Unpaired Median (DUM) neurons, which send axons bifurcating to both right and left light organs (Christensen and Carlson, 1981). "The lantern nervous system," they concluded, "is organized in an arrangement capable of synchronizing the excitation of all the lantern photocytes. This neural organization could aid in the control of the complex flash pattern displayed by male *Photuris versicolor* fireflies." Christensen also discovered in the firefly larva that stimulating any of these DUM neurons produced simultaneous responses from the light organs on both sides of the body, and demonstrated "that the four DUM neurons represent the sole photomotor output from the central nervous system to the larval lanterns" (Christensen and Carlson, 1982).

DUM neurons were already known to contain octopamine in other insects, so Christensen and colleagues (1983) had no difficulty demonstrating that the DUM neurons of the larval firefly light organs also contain octopamine. To finally put the cherry on the cake, Carlson obtained experimental evidence that octopamine was released from the nerve terminals in the light organs when stimulated in such a way as to elicit luminescence

(Carlson and Jalenak, 1986). The circuitry for luminescence control in firefly larvae seemed nicely elucidated, but the story was far from over regarding adult fireflies.

I have alluded above to the incongruity of having nerve terminals abutting tracheolar end-cells instead of the photocytes in adults, but the quandary was more wide-ranging. The consensus was that flash control is regulated by the oxygen supply to the luminescent sources (peroxisomes) within the photocytes. Mitochondria are localized near the plasma membrane of the photocytes, and for this reason had been considered to act as gatekeepers for oxygen access to the more centrally located peroxisomes. Because neurons make synapses on tracheolar end-cells situated at the end of the tracheal air supply, photocyte activation must involve a signal passing from the tracheolar end-cells to the peroxisomes, a distance too long (about 17 micrometers) to reconcile with the speed of the firefly flash. At the turn of the new millennium, Barry A. Trimmer, a Cambridge University–trained British physiologist, and Sara Lewis at Tufts University near Boston, thought of an agent that could speedily cross this barrier – the gas nitric oxide (NO). NO had already been implicated as a gaseous signal transmitter in the nervous system of insects (Bicker, 1998), so Trimmer and his colleagues (2001) just went a small step further and examined whether it played a role in firefly flash control.

They found that the enzyme that generates NO, called NO synthase, is present in the tracheolar end-cells. Exposing the firefly light organs to NO induced glows, and an agent known to remove NO in tissues prevented the octopamine-induced luminescence from appearing. Thus, Trimmer and his colleagues felt justified in proposing that "the role of NO is to transiently inhibit mitochondrial respiration in photocytes [the gatekeeper] and thereby increase O_2 levels in the peroxisomes [microsources]." Later they provided experimental evidence of the capability of NO to inhibit mitochondrial respiration (Aprille et al., 2004). Although the involvement of NO is undeniable, Helen Ghiradella and John T. Schmidt (2004) argued that a rise in hydrogen peroxide levels in the light sources, the peroxisomes, is the critical factor at the end of the signalling cascade in triggering the flash.

Another kind of luminous insect shares with fireflies a role for octopamine in luminescence control. The New Zealand glow-worm *Arachnocampa luminosa* possesses a light organ in the strangest of locations, as a swollen tip of the Malpighian tubules, which are the insect equivalent of a kidney. The nerve supply of this light organ originates from a posterior ganglion of the ventral nerve cord, and "fine nerve terminals, axons, and glial cells can be seen in close proximity to the basal surface of the cells of the light organ" (Green, 1979). The New Zealand glow-worm differs from the firefly in that the removal of brain centres either does not affect or enhances luminescence instead of eliminating it (Meyer-Rochow, 2007), a characteristic that points to an inhibitory role for innervation. In the Australian next-of-kin (*Arachnocampa flava*), a pharmacological analysis led to the conclusion that "[octopamine] is involved in the regulation of repression, either by acting directly on the cells of the light organ or through regulation of oxygen access to the cells (Rigby and Merritt, 2011). The implication is that octopamine is the neurotransmitter mediating inhibitory innervation, acting in an opposite way to its action in the firefly light organ.

~~~

Adrenergic-like neurotransmission was also the focus of investigations of nervous control of fish photophores. The midshipman fish, *Porichthys*, emerged as the favourite experimental model thanks to its ease of capture and maintenance in laboratory tanks. The Belgian team of Fernand Baguet and Jérôme Mallefet in Louvain-la-Neuve was instrumental in propelling this research program forward, not only for *Porichthys*, but also for deep-sea fishes. With regard to *Porichthys*, Bernard Christophe and Baguet (1983) gained insights by examining adrenergic control at the level of isolated luminous cells (photocytes) rather than intact photophores. They found that noradrenaline triggered fast luminescent responses and adrenaline a slow response superimposed on an initial fast response. This suggested that the two response types were mediated by different adrenergic receptors on the photocyte membrane, and that noradrenaline is the likely neurotransmitter, whereas adrenaline might act by hormonal channels. This hypothesis was corroborated by immunohistochemical evidence showing that noradrenaline is concentrated in nerve endings of the photophores, whereas adrenaline is more diffusely distributed (Mallefet and Anctil, 1992).

Likewise, evidence of adrenergic control was obtained for the hatchetfish *Argyropelecus* (Baguet and Maréchal, 1978), and both adrenaline and nora-
~~~

drenaline were detected in the photophores of this deep-sea fish (Salpietro et al., 1998). A surprising discovery by a team of Swedes and Belgians was that the gas NO, echoing the firefly control mechanism, modulates the adrenergic luminescent response of the hatchetfish, probably by affecting mitochondrial respiration in the photocytes (Krönström et al., 2005). NO synthase activity was found in the nerve endings of the photocytes, and the researchers speculated that NO modulation may serve to adjust luminescent output for counterillumination.

So little was understood of shark luminescence for such a long time that the outburst of discoveries by Jérôme Mallefet's team caused surprise and attracted the attention of the media. Mallefet and his student Julien M. Claes conducted a study showing that the luminescence of the lantern shark (*Etmopterus spinax*) is entirely under hormonal control (Claes and Mallefet, 2009). Contrary to bony fishes, this shark harbours thousands of minute photophores which carry no innervation. Pharmacological screening failed to reveal the involvement of neurotransmitters. Instead, it turned out that the hormones that control shark skin colour – melatonin and prolactin – also control luminescence through different pathways in the photocytes. A third hormone, α-melanocyte stimulating hormone, modulates the luminescence triggered by the two other hormones.

Among the crustaceans, only for euphausiids (krill) do we have a fair grasp of their luminescence control. What fascinates here is the compelling role of external light in triggering their luminescence. But the role of light is more complex than that. Their photophores are mounted on short muscular stalks as are the eyes. Eye rotations as the eyes follow a light source are accompanied by photophore rotations in the same direction (Grinnell et al., 1988). This eye-photophore motor coordination is controlled by the central nervous system. Excitation of giant axons – known for their fast speed of conduction – in the ventral nerve cord elicits one-second flashes after a delay of only one sixth of a second (Fregin and Wiese, 2002). The photophores are innervated, but the small nerves contain only three nerve fibres which end on the sphincter muscle of blood vessels, suggesting "that light generation is probably controlled by alteration of the blood flow through the lantern, and that this in turn is under nervous control" (Herring and Locket, 1978). The neurotransmitter serotonin is known to stimulate luminescence (Doyle and Kay, 1967), and it was later found that sphincter muscle innervation contains serotonin and that serotonin relaxes the sphincter, thus allowing oxygen access to the photocytes, thereby triggering flash activity (Krönström et al., 2009). By analogy with the hatchetfish, NO modulates the flash response of euphausiids by inhibiting the response to serotonin (Krönström et al., 2007).

One important group where the study of physiological control has made no significant inroads is the luminous squids. Because squids are capable of producing fast skin colour changes, an activity under strict nervous control, future investigators should perhaps emulate the work of Jérôme Mallefet's team on sharks and examine if the nerve supply and its neurotransmitters of the colour effector system are co-opted by the photophores. The only recent development worth mentioning is an oddity reminiscent of the photocytes of the ctenophore *Mnemiopsis*. The symbiotic light organ of the squid *Euprymna scolopes* contains cells that have the molecular machinery for light reception, including an opsin (Tong et al., 2009). These cells, localized in the epithelium of the crypt housing the bacterial colony, respond to light by generating membrane depolarization signals. These cells, Tong and colleagues explained, "serve as extraocular photoreceptors, with the potential to perceive directly the bioluminescence produced by their bacterial partners." They added that this process provides a mechanism whereby the squid can compare the ambient light perceived by the eyes with the light generated by the bacterial symbionts for camouflage purposes (counterillumination).

Ultimately, luminescence control inside the photocytes is predicated on the compartmentalization of the luminescent material, be it luciferin, luciferase, or a photoprotein. Just as in human society containment is the prevalent principle for crowd management, so is compartmentalization for the final act in the luminescence control cascade. The notion of dispersing the luminescent material everywhere in the luminous cell was counterintuitive to the investigators, who struggled to understand how the chemical participants get together to light up in the cell. Their tools of the trade for addressing this problem are fluorescence and electron microscopic observations of resting or luminescing photocytes.

When Beatrice Sweeney reviewed the topic of intracellular sources of bioluminescence in 1980, few bioluminescent organelles had been discovered in photocytes, and the field was wide open. The paracrystalline bodies of scale-worms were known, and their packed arrays of membranes suggested that the luminescent material, a pre-charged photoprotein as it turned out, was membrane-bound. The firefly peroxisomes, discussed earlier, contain luciferase, luciferin, and ATP – all ingredients for chemiluminescence – and they only need inflow of oxygen to light up. In the Californian midshipman fish photophores, the photocytes are filled with vesicles containing a dense flocculent material (Anctil and Case, 1976). But in the same fish in Puget Sound, Washington State, which does not luminesce and lacks luciferin, the vesicles are largely devoid of flocculent material (Strum, 1969a; Case and Strause, 1978). "These observations," Case and Strause summed up, "support the idea that the vesicles are storage sites of luciferin and that luminescence involves mobilization from such sites."

More recently, Mallefet's team has followed the intracellular dynamics of bioluminescent organelles in the course of luminescent episodes. In the photocytes of the brittle star *Amphipholis*, Dimitri Deheyn and his colleagues (2000) identified four types of vesicles, two of which represent storage sites of the bioluminescent reactants, the other two, representing transient phases associated with ongoing luminescence, which lead to the accumulation of paracrystalline bodies in these vesicles. In the photocytes of the lantern shark there are two zones of inclusions (organelles): a zone of vesicles and another of membrane-free dense granules (Renwart et al., 2014). Only the granular zone undergoes changes in size and composition during luminescent episodes (Renwart et al., 2015).

Clearly, new molecular imaging tools will be necessary to refine the information on these intracellular processes, so that a clearer picture can emerge of how compartmentalized luminescent reactants are released to interact with each other.

18
Unravelling Molecular Mechanisms

The chemistry of all known bioluminescence reactions remains punctuated with question marks.
–Thérèse Wilson and J. Woodland Hastings (1998)

After the early successes of Harvey's disciples, re-enactment in the test tube of the molecular interactions of luminescent reactants was to occupy the minds and lab benches of investigators for years to come. The field was ripe for breakthroughs in the fine points of the molecular underpinnings of bioluminescence, but few structures and properties of luciferins, luciferases, or photoptoteins were identified to everyone's satisfaction. A formidable inventory task mobilized biochemists to gain as complete a picture of the extent of diversity of these bioluminescent reactants in the natural world. More than just an inventory was at stake, as efforts to understand how bioluminescence as a chemical phenomenon emerged and evolved in the course of evolution depended on an abundance of biochemical data.

Central to the drive to characterize new luciferin-luciferase complexes and photoproteins was the figure of Osamu Shimomura. By the time we left Shimomura in chapter 11, we had reached the mid-1960s and, with the exception of the firefly and *Cypridina* luminescent systems, only partial purification of the luminescent reactants had been achieved and the structural identity of luciferins remained unknown. The luminescent protein of jellyfish was characterized, and it was in that time period, in fact, that the neologism "photoprotein" was coined, when Shimomura and Frank Johnson partially purified the peroxide-activated luminescent protein of the tubeworm *Chaetopterus* (Shimomura and Johnson, 1966). More biochemists also contributed significantly to the chemical identification of other bioluminescent systems.

Shimomura, in his highly informative book on bioluminescent systems (Shimomura, 2006), provides a timeline of discoveries. In 1968 the crystallization of the *Chaetopterus* photoprotein was achieved and the requirement of two new organic co-factors for its bioluminescence was discovered (Shimomura and Johnson, 1968a). In the same year the chemical structure of the luciferin of the New Zealand freshwater limpet, *Latia*, was established (Shimomura and Johnson, 1968b). It is a form of aliphatic aldehyde reminiscent of the one involved in bacterial bioluminescence. The quantum yield (or light emission efficiency) of *Latia* luciferin is very low compared to the luciferin-luciferase reactions of other luminescent organisms (Ohmiya et al., 2005). In 1974 several long-chain aldehydes were identified in luminous bacteria (Shimomura et al., 1974). In 1975 the luciferase of the boring mollusc *Pholas dactylus* was characterized (Henry et al., 1975), but only twelve years later was the photoprotein pholasin purified by Anthony Campbell's team at the University of Wales (Roberts et al., 1987). In 1976 the luciferin of the earthworm *Diplocardia* joined the club of aldehydes (Ohtsuka et al., 1976). Recently, however, it was found that the luciferin of another earthworm, *Fridericia heliota*, is dramatically different from that of *Diplocardia* or of any other known luciferin (Petushkov et al., 2014); it is a highly unusual short peptide.

Between 1984 and 1986 the new technique of molecular cloning was incorporated into the methodological arsenal of bioluminescence studies. The firefly luciferase was cloned by the team of Marlene DeLuca (1936–1987). DeLuca received her PhD in 1958 from the University of Minnesota, where she gained expertise in bioenergetics (Stanley, 1989). She joined Bill McElroy's laboratory at Johns Hopkins in 1962 for postdoctoral studies on properties of firefly luciferase. She eventually married McElroy, but she was no mere appendix to her husband, creating for herself an independent, tenure-track professorial career. She had moved with McElroy to the University of California in San Diego, where she did the cloning. Philip Stanley (1989) recalled how the project went and its outcome:

> Marlene started thinking about cloning firefly luciferase in 1976 but at that time the rules and regulations concerning genetic engineering were very stringent and she was not able to go ahead with her famous work for about seven or eight years. Together with Keith Wood, Jeff de Wet

and Don Helinski, Marlene successfully cloned cDNA encoding firefly luciferase in *Eschericia coli* and showed that it could be used as reporter gene in studies with animal and plant cells and we will all remember her slides of glowing tomato plants! Marlene's laboratory has also cloned cDNA's encoding luciferases from click beetles (*Pyrophorus*) into *E. coli* which produce colonies with at least four colours of light.

Her paper reporting the firefly cloning and the successful bioluminescent expression of the firefly luciferase cDNA in colonies of the bacterium *E. coli* (de Wet et al., 1985) opened up a new field whereby reporter genes could be used to follow biological processes by non-invasive bioluminescent imaging. The glowing tomato plants DeLuca produced demonstrate how quickly she saw the possibilities of her firefly genes and acted on them. It also underscores the fact that academics moved ahead of private industry in applying the technology that allowed living lights to monitor gene expression (Root, 1988).

Unfortunately, DeLuca died prematurely two years later in the prime of her career. Soon afterward, the cloning of the luxA gene coding for the alpha subunit of bacterial luciferase (Cohn et al., 1985) and the luxB gene coding for the beta subunit (Johnston et al., 1986) expanded the field further. Later, Anthony Campbell's lab cloned other firefly luciferases and found that luciferases of the European glow-worm *Lampyris* share a high degree of sequence homology with the North American *Photinus*, but not with the luciferases of the click beetle *Pyrophorus* (Sala-Newby et al., 1996).

Another early practitioner of bioluminescence gene reporting was also an academic scientist. Thomas Baldwin, a former postdoctoral student of Woody Hastings and then professor at Texas A&M University, cloned the bacterial luciferase genes from *Vibrio harveyi* and expressed them in *Escherichia coli* (Baldwin et al., 1984). He obtained a patent for these clones. Within two years Baldwin's team elegantly demonstrated the usefulness of bacterial luciferase in monitoring gene expression in soybeans (Legocki et al., 1986). Between DeLuca's tobacco plants and Baldwin's soybeans, the benefits of these bioluminescent probes for agriculture seemed very promising. As Michael Root (1988) reported:

> Baldwin, Szalay, and co-workers have already placed the bacterial luciferase genes into soybean (*Bradyrhizobium japonicum*) chromosomes. The genes are inserted next to promoters for the production of nitrogenase, an enzyme that catalyzes the reactions involved in nitrogen fixation within legume root nodules. The nitrogenase genes are activated only when the plant is deficient in nitrogen. If nitrogen fertilizer is required by the crops, luciferase is synthesized along with nitrogenase. The amount of nitrogenase produced (an indication of how much fertilizer is needed) is indirectly determined by measuring the amount of luciferase made. "The assay is exquisitely sensitive," Baldwin says.

Concurrently the gene for the photoprotein aequorin was also cloned. The field of molecular cloning had become so hot that two laboratories reported the aequorin cloning in the same year: Milton Cormier's lab at the University of Georgia in Athens (Prasher et al., 1985) and a Japanese team coordinated by Frederick Tsuji, then at the Marine Biology Research Division of the Scripps Institution of Oceanography in La Jolla. The Athens team expressed the gene in *E. coli*, but only the La Jolla team analysed the sequence of the protein in which they found three binding sites for calcium activation. Cloning other photoproteins, such as pholasin from the bivalve mollusc *Pholas*, showed that there is very little homology between them, a factor that strongly supports the idea that they had independent origins in evolution (Dunstan et al., 2000).

In other developments over the years, the krill luciferin structure was identified by a team led by Shimomura (Nakamura et al., 1988). The next year Woody Hastings joined that team to elucidate the structure of dinoflagellate (*Gonyaulax*) luciferin (Nakamura et al., 1989). It turned out that krill and dinoflagellate luciferins are almost identical, both being bile pigments derived from chlorophyll. Earlier, before the chemical structure was completely known, the labs of both Hastings and Shimomura had suggested in a joint paper that krill obtain their luciferin by ingesting dinoflagellates (Dunlap et al., 1980).

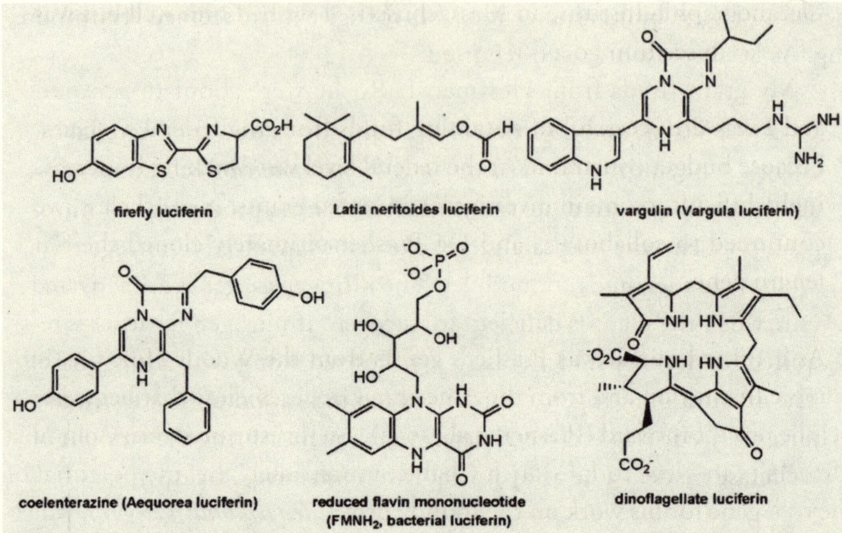

Figure 18.1 Chemical structure of the best-known luciferins.

Green fluorescent protein (GFP), mentioned in chapter 11, took centre stage again. Wherever GFP is present in bioluminescent systems, it usually serves as an interloper by converting the original blue light source into a final greenish luminescent broadcast. The story of GFP deserves special attention for its human pathos as well as for its science. The GFP cloning work was done by Douglas Prasher, and Milton Cormier recalled the circumstances of the project (Cormier, 2007):

> After Dr. Prasher had successfully cloned the gene coding for aequorin, I suggested to him that he attempt to clone the gene coding for the green fluorescent protein (GFP) from the jellyfish Aequorea. I already had the contacts he needed to make arrangements to collect these animals in large numbers at Friday Harbor, Washington.
>
> Within about two years, Dr. Prasher succeeded in isolating a clone of the GFP gene. After determining its structure, we realized it was a partial clone; that is, we had only about 70% of the gene, not the entire structure. At this point, though, Dr. Prasher left UGA [University of Georgia in Athens] to take a promising position at the Woods Hole

Oceanographic Institute in Massachusetts. I wished him well but was very sad to see him go.

My grant funds from Hoffman-LaRoche were about to run out, and I was unsuccessful in obtaining funds from the federal agencies, because budget constraints at the federal level were making it increasingly difficult to obtain investigator-initiated grants. Nevertheless, we continued to collaborate, and Dr. Prasher ultimately cloned the full length gene.

As it turned out, it was Prasher's grants from the Woods Hole Oceanographic Institution and from the American Cancer Society that helped propel the project forward (Prasher et al., 1992). An investigator from Columbia University in New York, Martin Chalfie, immediately saw the potential of the GFP gene for his work on the roundworm *Caenorhabditis elegans*, which had become an important genetic model for all kinds of studies because of its disarmingly simple structure, being composed of only about one thousand cells. And another plus was that the worm is transparent, so any fluorescence would show through. In an interview Chalfie explained what was at stake:

"It didn't take much to realize that if I put that fluorescent protein inside this transparent animal, I would be able to see the cells that were making it," he said. "And that's what we set out to do."

He thought that the fluorescent protein could be made to serve as a biological marker by splicing the gene that makes the protein into an organism's DNA next to a gene switch or another gene.

"That serves as a lantern," Dr. Chalfie said, and biologists would be able to see when specific genes turn on or off and where different proteins are produced.

He was not able to pursue the idea until Douglas C. Prasher, a scientist then at the Woods Hole Oceanographic Institution in Massachusetts, found the G.F.P. gene and shared it with Dr. Chalfie in 1992. Dr. Chalfie said that within a month his group was able to insert the gene into E. coli bacteria.

In 1994, Dr. Chalfie and his collaborators reported that they had inserted the protein into six cells of the C. elegans worm. When placed

> under ultraviolet light, those cells shined green, revealing their location. (Chang, 2008)

The paper reporting these results (Chalfie et al., 1994) explained why GFP surpasses other methods of monitoring gene activity and protein expression inside cells: it requires no complicated procedure to insert it into cells; its strong fluorescence is easily detected by near UV or blue light and is not easily quenched even by laser light; it does not interfere with the cell's growth and other activities; it is a small enough protein to facilitate its diffusion even in fine processes like neuronal axons; and its fluorescence is preserved when tissues are chemically fixed for histological purposes. The paper's advocacy of GFP was so persuasive that it soon became an unavoidable tool of molecular biology at large.

Chalfie acknowledged Prasher's help by adding his name to the list of authors. But by the time the paper came out, Prasher had already run out of research funds two years earlier and, when his contract with the Woods Hole Oceangraphic Institution was not renewed, he had to eke out a living not in his own projects but in government projects such as those of the US Department of Agriculture and the NASA Science Program. When that ended, he was unemployed for a year before taking a job as a shuttle driver for a car dealer (Sherwell, 2008). Prasher was at this low-paying job and in debt in 2008 when he heard that the Nobel Prize for Chemistry was awarded to Osamu Shimomura, Martin Chalfie, and Roger Tsien. The latter (1952–2016), a professor at the UCSD School of Medicine in San Diego, had imitated Chalfie in asking Prasher for his GFP gene in 1992. Tsien's contribution to the GFP story consisted in tinkering with the protein's fluorescence intensity and colour to diversify the uses it could be put through as a molecular marker (Ormö et al., 1996).

Had the research grant agencies recognized the potential of Prasher's work, the latter could have been in a position to do the work Chalfie and Tsien did, instead of ending up giving them his hard won GFP gene, and he could have been a contender for the Nobel. But not for the first time, and certainly not for the last, a research grant panel showed a dismal lack of foresight. Prasher expressed "hopes that the burst of publicity about green fluorescent protein could give him a final chance at a job back in the scientific field" (Sherwell 2008). His wish was granted thanks to Roger Tsien, who took him as research associate in 2012. However, Prasher apparently left Tsien's lab a year before Tsien's untimely death in 2016. Prasher's future is again uncertain.

It was all very nice what GFP could accomplish, but what about the photoproteins – aequorin, mnemiopsin, obelin, and the like? Could they compete with GFP, which uses a vast array of colours in gene monitoring? As Laura Rowe, Emre Dikici, and Sylvia Daunert (2009) of the University of Kentucky explained:

> Since GFP was first discovered, researchers have successfully mutated it and the Anthozoa-derived FPs to create a vast array of colors, pH stabilities, half-lives, aggregation tendencies, reduced cytotoxicities, and other desired properties, greatly expanding the FP's range of analytical applications. Researchers now can "pick and choose" the FPs that are most suited for a given study. But bioluminescent proteins have not undergone the same intensive mutagenesis, and as a result, there are few color varieties available. However, this limitation is being gradually overcome by recent mutagenesis studies on both aequorin and luciferase.

Thus mutations in the genes for both aequorin and obelin brought in a diversity of colours of luminescence that increased the usefulness of these proteins in gene monitoring. But there were other ways in which photoproteins coud be useful – as molecular switches. Rowe et al. (2009) explained:

> For example, the DNA sequence coding for the glucose binding protein (GBP) was inserted into the aequorin sequence; upon binding of glucose, the GBP portion underwent a significant conformational change, which brought the two halves of the aequorin molecule close enough together to form a bioluminescent product. Thus, this molecular switch functioned as an "on-off" switch for detecting glucose between 1.0 × 10-7 and 1.0 × 10-2 M, which is well within the physiologically relevant range of 2-20 mM. The switch was selective for glucose over several significant mono- and disaccharide molecules.

Another fascinating story worth telling concerns the ubiquitous luciferin, coelenterazine. The story starts in 1975 when Shimomura and Frank Johnson in Princeton took a comparative look at various organisms to try and make sense of their light-emitting molecules. The coelenterates – including jellyfish, sea pens and comb-jellies – were the focus of their search for "the structures of light-producing compounds, such as luciferin, and the light-producing chromophore of a native photoprotein, as well as the structures of light-emitters derived therefrom" (Shimomura and Johnson, 1975). They found that these coelenterates all share in one form or another an imidazolopyrazine as the luciferin structure, which they called coelenterazine. It shares with *Cypridina* luciferin an indole ring, which is found in many molecules, including the neurotransmitter serotonin. An important realization was that coelenterazine is also the chromophore of the photoprotein aequorin; calcium binding to the photoprotein triggers the oxidation of coelenterazine, resulting in light emission and the formation of coelenteramide.

But there was more to the story than it seemed. Shimomura and Johnson noted in their paper that coelenterazine was discovered in the firefly squid by Shoji Inoue and colleagues at the University of Nagoya (Inoue et al., 1975). The following year, Inoue and Hisae Kakoi (1976) established that the luciferin of the shrimps *Oplophorus* and *Heterocarpus* is also coelenterazine. Further exploration found coelenterazine in lanternfish and hatchetfish (Shimomura et al., 1980), copepods (Campbell and Herring, 1990), arrow-worms (Haddock and Case, 1994), and more shrimps (Thomson et al., 1995). The variety of unrelated luminous organisms carrying coelenterazine was bewildering. Could it be that coelenterazine was shared among deep-sea animals through the diet just as *Cypridina* luciferin was shared among shallow-water forms? The buzz among investigators favoured a yes to the question, but the issue as a whole was picked up by Catherine M. Thomson, Peter Herring, and Anthony K. Campbell (1997).

They expanded the number of genera in which coelenterazine has been detected from fifty-two to about ninety. Shimomura (1987) had already found that many non-luminescent deep-sea forms contain coelenterazine, but not luciferase, and this finding clearly supported the thesis that they acquired coelenterazine by ingesting luminescent prey. But for luminescent animals, as Catherine Thomson and her colleagues pointed out, it was not so easy to determine whether they acquired their coelenterazine solely through diet or by biosynthesis, or a combination of both. The diet hypothesis found support in the observation that in squid and fish coelenterazine is stored in large amounts in the liver or digestive glands, sources from which photophores can draw their luciferin supply as needed through the bloodstream. In comparison, non-luminescent forms had much smaller stores of coelenterazine in their livers.

It was all very well to document the involvement of the ocean's food web in disseminating coelenterazine, but the latter had to be synthesized by a seed organism in the first place before it found its way in the food web. The way to go about it is not to rely on circumstantial evidence as had been done so far, but to seek experimental evidence. Such evidence already existed; in decapod shrimps Thomson and colleagues (1995) found that species expelling luminescent secretions have coelenterazine supplies in excess of what could be assimilated from the food chain. Later, evidence of synthesis of coelenterazine from amino acid precursors came from the copepod *Metridia* (Oba et al., 2009). Did it mean that some luminescent crustaceans synthesize coelenterazine, but not coelenterates after which the luciferin was named?

Steve Haddock and his colleagues (2001) answered this question with experiments showing that jellyfish "reared on a luciferin-free diet are unable to produce light unless provided with coelenterazine from an external source." They cited evidence that the luminescence of other hydrozoans dimmed considerably when they were kept in captivity. The authors of the paper concluded that "the name 'coelenterazine' itself may be a misnomer, as the only evidence for production of this molecule in the ocean comes from crustaceans." However, Haddock's team itself later found molecular evidence suggesting that coelenterazine is synthesized in comb-jellies (Francis et al., 2015), but decisive evidence is still lacking. From the forms where evidence was not lacking we have an indication of how few luciferins there are in nature. Besides the dinoflagellate/krill luciferin, the *Cypridina*/shallow water fish luciferin, and the coelenterazine shared by so many marine phyla, we are left with only the bacterial, firefly, limpet, and earthworm luciferins, and a variety of photoproteins.

In 1995 Woody Hastings provided a snapshot of progress in our understanding of chemical and molecular mechanisms of bioluminescence. In his view the chemical constituents share many features, among which the fact that "all known luciferases are oxygenases that utilize molecular oxygen to oxidize a luciferin," a process followed by further molecular steps leading to the chemiluminescence that are not completely understood. Hastings (1995) debated the prevailing view, expounded by Frederick Tsuji among others, that photoproteins behave differently than luciferases. Hastings argued instead that aequorin may "be viewed as a stable enzyme-bound peroxide intermediate whose further reaction is triggered by calcium. The quasistable bacterial luciferase peroxide intermediate is analogous and has all the properties of a 'photoprotein' … instead of Ca2+ it requires aldehyde for its further reaction."

Hastings cited the work of Keith V. Wood, which delved into the fascinating molecular basis for the colours of bioluminescence in beetles. While all luminescent beetles – lampyrid fireflies, phengodid beetles and elaterid (click) beetles – share the same luciferin and ATP-dependent light reaction, there are almost as many luciferases as there are beetle species that light up (Day, 2009). Wood, who earned his PhD on this topic in McElroy's lab at UCSD, recalled his involvement in the doctoral project (Liebert, 2007):

> I approached the leading laboratory in bioluminescence for firefly chemistry – which was using a traditional enzymology approach at that time – and I proposed applying what were then new molecular biology tools to understand the mechanism of this very curious chemistry. Although somewhat reluctant at first, they agreed to allow me to work on this project, and, as a result, we cloned the luciferase gene and went on to develop many applications that generated a lot of interest.

What Wood accomplished was the cloning of four isoforms of click beetle luciferase which, when expressed in *E. coli*, emit upon luciferin oxidation green, yellow-green, yellow, or orange lights (Wood et al., 1989; Wood, 1995). Small substitutions of a few amino acids in the protein of the beetle luciferases are sufficient to shift the colour of luminescence. "Because of the different colors," Wood and his co-authors (1989) concluded, "these clones may be useful in experiments in which multiple reporter genes are needed."

Wood soon moved to a private company, Promega Corporation, to implement applications for his findings, especially the development of reporter gene technology for the pharmacological industry. Companies like Promega were mushrooming and attracting ambitious young investigators away from academia as biotechnology took off in different directions: gene expression analysis, protein dynamics monitoring, mapping of signal transduction pathways, and drug discovery. Wood summarized the technological timeline (Glaser, 2007):

> After we cloned the firefly luciferase gene in 1984, it quickly became apparent that there were some very useful applications for this chemistry, most notably as a reporter gene. Its popularity began to take off at about the same time that HTS [high throughput screening] was beginning to take root in the pharmaceutical industry. To put this in perspective, in the mid-1980s, high throughput was considered 10,000 samples a week, the 96-well plate was a novelty, and robotics was just starting to be applied … Luciferase was, at that time, becoming known as a very sensitive, easy-to-use reporter gene technology and became the obvious way to tap into this rich genetic resource for measuring cell physiology. Thus, this concept of using firefly luciferase in HTS began. It took off in the early 1990s and became one of the major methods for measuring, in particular, G-protein coupled receptors in high throughput screens.

By 2007 Wood was looking forward to building "luciferases that directly respond to intracellular events in a real-time fashion that would be of interest to cell physiology research and drug development. We would be able to see an instantaneous change in the luminescent signal, reporting on the exact state of the luciferase-sensing molecule within living cells." Other researchers used luciferases to illuminate gene expression and physiological processes, and this was chronicled as early as 1988 by Michael Root. The firefly luciferase gene was spliced into the DNA of plants and so were bacterial luciferase genes, thus giving a boost to agricultural applications. At the other end of the spectrum, Root foresaw that bioluminescence was poised to assist cancer research by providing sensitive assays to determine under what conditions oncogenes are activated. Lighting up the disease process was already looking up as a major tool of biomedicine. This approach was implemented a decade later with imaging techniques (Benaron et al., 1997). Likewise, imaging brain activity during natural behaviour became another area where bioluminescence found application (Naumann et al., 2010). In this way, the activity pattern of groups of neurons genetically modified to generate GFP and aequorin could be monitored by their luminescence in real time over several days.

If Raphäel Dubois had been able to see these developments – he who inaugurated the chemical exploration of bioluminescence and showed a deep interest in bioluminescent applications – he would have been enthralled.

All this unfurling of molecular mechanisms associated with bioluminescence was bound to attract interest in the favourite speculative playground of biologists: the evolutionary origin of the phenomenon. Indeed, how did bioluminescence come about eons ago, and how did its molecular machinery evolve? In broad brushstrokes E. Newton Harvey and John Buck had already alluded to the independent origin of many bioluminescent systems. Woody Hastings revisited this speculative space with an emphasis on the chemical protagonists of luminescent reactions (Hastings, 1983). On the basis of the great diversity of luciferases and, to a lesser extent, luciferins, Hastings "estimated that bioluminescence may have arisen independently as many as 30 times in the course of evolution."

But if bioluminescent systems evolved so many separate times, the problem is compounded by the evolutionary history of each of the organisms that display luminescent capability. A case in point is luminescent fishes, which constitute a larger assemblage of species than any of the other luminescent forms, so one should expect to gain interesting insights by examining their evolution. Recently, such insights were gained by Matthew Davis (St. Cloud State University in Minnesota), John Sparks (American Museum of Natural History in New York), and W. Leo Smith (University of Kansas in Lawrence). Exploiting the tools of molecular phylogeny, they found a striking panorama of their evolution. According to Davis and his colleagues (2016), bioluminescence arose independently twenty-seven times in the fourteen major fish groups represented, all events occurring between the Early

Cretaceous, about 150 million years ago, and the Cenozoic, about sixty million years ago. Over half of the 1,510 known bioluminescent fish species have intrinsic (not bacterial) luminescence, but fish with symbiotic luminous bacteria have evolved independently seventeen times. The deep-sea dragon- and viperfishes are the oldest to have evolved and they, along with the midshipman fishes, the flashlight fishes, the pinecone fishes, and the lantern bellies, display the richest diversity of luminous species. "Our findings and these additional studies investigating the evolution and function of bioluminescence and biofluorescence in marine systems," Davis and his colleagues concluded, "highlight how much remains to be discovered regarding the potential impacts of bioluminescence, and luminescent signaling in general, on the evolutionary history and ecology of marine fishes."

Beyond the multiplicity of evolutionary events, what could have caused the emergence of these bioluminescent systems? As early as 1962, McElroy and Howard Seliger had proposed that bioluminescence was originally a detoxifying mechanism that removed the nascent oxygen harmful to the early anaerobic forms of life:

> This struggle for anaerobic conditions led to the selection of organisms having specific oxidases (luciferase) which catalyzed the rapid removal of oxygen; all primitive forms were therefore potentially luminescent. The gradual selection and evolution of electron transport processes in which oxygen was reduced stepwise to form water gave rise finally to the aerobic forms. With the appearance of the latter the luminescent, oxidative reaction was no longer of selective advantage. Thus we argue that bioluminescence is a vestigial system of organic evolution, but that through secondary evolutionary processes the luminescent system has been preserved in various and unrelated organisms by virtue of the fact that it has been adapted for other useful purposes. (McElroy and Seliger, 1962)

Curiously, when Seliger revisited the subject thirteen years later, he had profoundly modified his and McElroy's views. His alternative theory no longer required "the presumption of a detoxification mechanism for molecular oxygen" (Seliger, 1975). "Primitive luciferases," he now argued, "evolved in order to utilize oxygen directly as an electron acceptor at the low oxygen concentrations of the primitive atmosphere. The selective advantage was therefore in increased metabolic capability. Light emission was then, and in present day hydroxylase systems are now, an adventitious, non-essential, event. Only much later, subsequent to the development of vision, might there have been a selective advantage for efficient bioluminescence and for the secondary adaptation to light organs and nervous regulation of a common oxygenase mechanism." So, to Seliger the advent of light-detecting cells and organs was the paramount selective pressure to develop physiologically significant bioluminescent systems. An add-on to his theory was that all luciferases evolved from oxygenases associated with the metabolism of toxic or pigment molecules (Seliger, 1993).

At this juncture, a team of Belgian and French biologists led by Jean-François Rees – a former student of Fernand Baguet – produced a thesis that turned the consensual concept on its head. Luciferins, they contended, "are the evolutionary core of most systems, while luciferases, the enzymes catalysing the photogenic oxidation of the luciferin, serve to optimise the expression of the endogenous chemiluminescent properties of the luciferin" (Rees et al., 1998). The basis for this contention is the ability of luciferins, and particularly coelenterazine – the most widespread luciferin in marine environments – to react with molecular oxygen and its derivatives, superoxide radicals and peroxides. In fact, coelenterazine appears as efficient an antioxidant as ascorbic acid. "The postulated reaction mechanism of imidazolopyrazinones with ROS [reactive oxygen species]," Rees and his colleagues pointed out, "is similar to that of the bioluminescent reaction and includes the formation of a dioxetanone intermediate, which breaks down to excited oxyluciferin and CO_2."

The theory of the Franco-Belgian team would explain the wide distribution of coelenterazine in luminescent and non-luminescent marine species, and within the bodies of these organisms. The predilection of the luciferin for digestive glands and livers would signal not only its having a storage space, but its use as an antioxidant in organs where oxidative metabolism is intense and likely to generate potentially harmful reactive oxygen species. From all these considerations Rees and his colleagues formulated the following evolutionary scenario:

> We suggest that the primary function of coelenterazine was originally the detoxification of the deleterious oxygen derivatives. The functional shift from its antioxidative to its light-emitting function might have occurred when the strength of selection for antioxidative defence mechanisms decreased. This might have been made possible when marine organisms began colonising deeper layers of the oceans, where exposure to oxidative stress is considerably reduced because of reduced light irradiance and lower oxygen levels. A reduction in metabolic activity with increasing depth would also have decreased the endogenous production of reactive oxygen species. Therefore, in these organisms, mechanisms for harnessing the chemiluminescence of coelenterazine in specialised organs could have developed, while the beneficial antioxidative properties were maintained in other tissues. (Rees et al., 1998)

The origin and evolution of bioluminescent systems in specific groups of organisms were also objects of speculation. Dennis O'Kane and Douglas Prasher (1992) proposed that the ancestral prototype of bacterial luciferase was the product of a single lux gene and that a two-gene system evolved later to form the current luciferases. Anderson G. Oliveira and colleagues (2012) noted that the seventy-one known species of luminous mushrooms are distributed among four distantly related lineages. They asked "whether the mechanism of bioluminescence is the same in all four evolutionary lineages suggesting a single origin of luminescence in the Fungi, or whether each lineage has a unique mechanism for light emission implying independent origins." Their data supported a supposition of a single evolutionaty origin for the luciferin-luciferase system of these mushrooms, a claim corroborated by the team of Russian biochemist Ilia V. Yampolsky (Purtov and colleagues, 2015).

In another recent study, Vadim R. Viviani (2002) sought to make evolutionary sense of the diversity of bioluminescent systems in beetles, including fireflies. He hypothesized that the luminescent reaction is a two-step enzymatic process in which a non-luciferase enzyme (ligase) acts first, followed by the luciferase, which carries a penultimate step before light emission. Viviani constructed a scenario in which beetle luciferases are derived from ancestral ligases that are able to produce a weak luminescence when exposed to luciferin and ATP. A mutation in this ligase would conceivably have improved luciferin interaction and enhanced the light emission yield.

In this chapter we have seen unfold series of molecular data from which hypotheses about the origin and evolution of bioluminescence were generated – about how the science of today can help us fathom the "deepest of times." It is not always clear how testable some of these hypotheses are, but human curiosity will not rest. Indeed, question marks continue to beckon us forward – and further and further back – on the winding path to understanding.

As we take stock of the journey in time that brought us from the naïve wonderment of the early observers of luminous creatures to the current use of these creatures' luminescence for the betterment of mankind, it is gratifying to note that the scientists who made the field of bioluminescence what it is today never lost that sense of wonderment, however informed and stripped of naïveté it has become. In this scientists join the non-scientist public in perceiving the aesthetic richness of bioluminescent events. It is this aesthetic sense, this appreciation of beauty in the performance of luminous organisms, that stirred the curiosity of natural historians and modern scientists alike. Undoubtedly, this kind of wonderment and curiosity will continue to inspire researchers of luminous creatures in the future.

Epilogue

"The history of science, as the history of our culture in general," Joseph Agassi (2008) argued, "is the history of noble and heroic efforts to push away a little the darkness in the middle of which we are doomed to live. In brief, the history of science, as the history of our culture in general, is the history of noble and wise errors." The story that emerges in the course of this book can also be construed as a drive through a murky terrain of "noble and wise errors," to finally reach a fair understanding of the how and why of living lights. This story follows that of all scientific endeavours, in that its developments are pretty much in step with the identified historical landmarks of science: the "pre-history" of Antiquity, the middle Ages, the Scientific Revolution and the Age of Enlightenment, the Modern era, and the technological revolutions of the contemporary age.

It is striking that for most of the timeline of this story, the study of living lights progressed slowly by dint of painstaking, sporadic observations on the part of naturalists who on the whole paid but flickering attention to the phenomenon. These investigators were in awe of the light displays, but so were they for plenty of other natural phenomena, and their attention was divided. Taking into account the inaccuracies, confusions, and preconceptions that corrupted many of their observations, it is hardly surprising that it took countless observations over centuries to reach a modest grasp of the meaning of it all. The turning point in the pace of progress, beginning in the 1870s, coincided with the rise of two sets of scholars who laboured side by side.

Those associated with the great pioneering oceanographic expeditions revealed to the world the vast oceanic expanses where numerous, unsuspected luminous creatures dwell. Those associated with a deeper examination of coastal or terrestrial species dedicated their careers to the study of living lights and produced large, cumulative bodies of work from their academic bases and research laboratories. Both currents converged to elevate the study of bioluminescence to a scientific discipline in itself.

The oceanographic expeditions did more than merely stage on paper the jaw-dropping spectacles of luminous oceanic monsters. The scientists who spearheaded these expeditions wrote popular accounts, with stunning iconographies to boot, as a way of repaying their patrons and sponsors. But the science behind these accounts morphed into massive monographs in which deep-sea animals and the distribution of their light organs were minutely described. There followed other monographs detailing the anatomy and histology of the individual light organs. Armed with this abundance of data, some, like the *Valdivia* scientists Carl Chun and August Brauer, were able to gain penetrating insights into the roles played by bioluminescence in the life of individual animals and in the ecological scheme of things.

Meanwhile, on land, the strong personalities of Paolo Panceri, Raphäel Dubois, and E. Newton Harvey were instrumental in building upon the century-old knowledge base of littoral and terrestrial luminous species to develop new approaches to studying their luminescence. Fireflies, pholad clams, ostracods, and polychaete worms became prime experimental models for locating and characterizing light emissions, and for unlocking the secret of the chemistry behind the luminescent sources. They earned accolades from biologists around the world precisely for their contributions to the advancement of bioluminescence research, through which they enhanced the visibility of the field in the scientific world. Harvey's Princeton School turned out brilliant disciples who not only reached a deep understanding of the biochemical and molecular intricacies of bioluminescence but in the process also introduced new principles to the field of biochemistry.

The impact of bioluminescence as a field of study on other fields cannot be overestimated. The story of how the parallel progress of the fields of molecular biology and bioluminescence led to their intercourse through the development of monitoring tools to follow gene and protein activities is fascinating in itself. Even though molecular biology emerged a long time ago – in the 1940s – as a result of the marriage of biochemistry with genetics (Morange, 1998), the field went through long gestation and maturation periods before achieving a fair understanding of how genes are activated to encode proteins and regulate cell activities. It was not until late in the twentieth century – in the 1980s – that luciferases, photoproteins, and green fluorescent proteins could be used to track gene activity by means of their strong light. Today, it is a key biotechnology driving progress in medicine, pharmacology, and even agriculture. There is no reason to doubt that it will continue to be so.

Similarly, the entry of the US Navy's oceanographic program into the realm of bioluminescence led to technological developments in support of military strategy as well as the management of the ocean's resources. Basically, American bioluminescence investigators going as far back as Harvey received significant financial support from the US Navy through the Office of Naval Research. Although such Navy sponsorship raised a few academic eyebrows on ethical grounds with regard to the potential military applications of investigators' findings, the open nature of the relationship – with no strings attached for the publication of the results – gave moral sanction to the transaction. The "contract" between the two parties was a two-way street. Researchers increased the resources of their laboratories and received gifts or loans of equipment – image intensifiers, for example – developed by the Navy or Army to assist military operations. In return, the Navy gained access to a font of knowledge about marine bioluminescence which complemented their own secret intelligence and helped them assess better the threat or benefit that bioluminescence might represent for naval operations. The symbiosis with the Navy, and later with other organizations, led to the development of the sophisticated technology used by current investigators.

How does the field of bioluminescence fare today? Paradoxically, the field has reached a point where it has become a victim of its own success. For indeed, the applied science of bioluminescence now overshadows fundamental research. How did this happen? After all, who promoted and successfully implemented the use of bioluminescent proteins for gene monitoring but the likes of Marlene DeLuca, Thomas Baldwin, Martin Chalfie, and Roger Tsien, all active in basic research at the time, in the 1980s and 1990s? Perhaps by allowing the technological offsprings of bioluminescent systems to fall ripe into the lap of private industry, these basic researchers unwittingly positioned fundamental bioluminescence research for a decline. Young investigators, when confronted with a choice of career after their PhD or postdoctoral studies, may have opted for the easy money and perks of industrial ventures at a time when institutional funding of basic research for such an exotic topic as bioluminescence was dwindling. One can only conjecture. Valuable and creative basic research is still being conducted, but fewer laboratories are active than in previous decades, and ecological or evolutionary aspects of the field have taken precedence over the biochemical and physiological aspects.

In contrast, the buzz has shifted to the ever-increasing technological applications of the fruits of basic bioluminescence research, as Jérôme Mallefet recalled in a recent communication to me:

> [In] the early 1980s, when I started to study the phenomenon of living light in fishes, it was the golden age of physiology, biochemistry and genetics; in short, fundamental research was popular even though potential applications of bioluminescence were creeping in. The following decade witnessed an explosion of research using luminous reactions as substitutes for improving the sensitivity or the localization of compounds inside cells or tissues of interest … The reign of genes and molecules was heralded; one watched as research teams rivalled in ingeniousness to modify gene sequences, manipulate luminescent molecules inside cells. It was the advent of the "luminoscope" … Bioluminescence was billed to serve humans with powerful diagnostic tools and genetic manipulations, and these objectives are still pursued today. But we forgot that basic research formed the basis of all these appplications and I fear that the inspired spark in the eyes of curiosity-driven researchers becomes extinguished by the irresistible attraction of applications … I am reminded of Henri Poincaré's remark, that the scientist does not study nature because of its usefulness, but rather because he finds pleasure in it, and he finds this pleasure because nature is beautiful. And as my compatriot Christian de Duve pointed out, "as the results of basic research are unpredictable, let the researchers do what they want, because that's what they will do best."

There is always the risk that technological innovation will dry up unless a renewal of the resources for basic research comes to the rescue in the next generation of dedicated scientists. Let us hope that what is happening is part of a continuous cycle alternating periods of basic research and its technological applications. But if a new crest of basic research exploration is to emerge in the future, where will it lead us? There are still unresolved mysteries to be tackled. The molecular mechanisms of neglected luminous groups are ripe for investigation. Research on ecological and evolutionary aspects of bioluminescence will expand in years to come, not only for their own sake but also as a way of gaining insights into ecological and evolutionary principles at large. And a marriage of convenience between molecular biology and ecology will blossom, whereby bioluminescent or fluorescent molecules will throw light on the ecological usefulness of the living lights; a recent example of this is the discovery of a role for the green fluorescent protein as a prey attractant by Steven Haddock and Casey Dunn (2015). The conditions under which future research will be conducted are also changing. As Hastings and Wilson (2013) noted: "The field of bioluminescence is now a collection of fruitful collaborations between scientists of very different backgrounds." Large-scale projects assembling multiple collaborators, instead of individually funded researchers, seem the way of the future.

"It is easy to be on the side of science because it brings about worldly success," remarked Joseph Agassi (2008). "It is somewhat harder, it seems, to advocate it as an adventure, much less as the spiritual adventure that it is." Considering all the events told in this story, I project that the study of living lights, in the mindscapes of the protagonists, will show the trappings at once of a scientific pursuit and, if not always a spiritual, certainly an intellectual adventure.

Bibliography

Abrahams, Mark V. and Linda D. Townsend (1993). "Bioluminescence in dinoflagellates: a test of the burglar alarm hypothesis." *Ecology* 74: 258–60.

Adamowsky, Natascha (2015). *The Mysterious Science of the Sea, 1775–1943*. London and New York: Routledge (Taylor & Francis Group).

Agassi, Joseph (2008). *Science and Its History: A Reassessment of the Historiography of Science*. Dordrecht, Netherlands: Springer Science+Business Media BV.

Agassiz, Alexander (1892). "General sketch of the expedition of the 'Albatross' from February to May 1891." *Bulletin of the Museum of Comparative Zoology* 23: 1–89.

– (1906). "Reports on the scientific results of the expedition to the eastern tropical Pacific, in charge of Alexander Agassiz, by the U.S. Fish Commission Steamer 'Albatross', from October, 1904, to March, 1905, Lieut. Commander L.M. Garrett, U.S.N., commanding. V. General report of the expedition." *Bulletin of the Museum of Comparative Zoology* 33: 1–75.

Agassiz, George R. (1913). *Letters and Recollections of Alexander Agassiz, With a Sketch of His Life and Work*. Boston and New York: Houghton and Mifflin Company.

– (1917). "Sir John Murray (1841–1914)." *Proceedings of the American Academy of Arts and Sciences* 52(13): 853–9.

Agassiz, Louis (1850). "On the structure of the halcyonoid polyps." *Proceedings of the American Association for the Advancement of Science* 3: 207–13.

Albert de Monaco, Prince (1889). "Sur un appareil nouveau pour la recherche des organismes pélagiques à des profondeurs déterminées." *Comptes rendus hebdomadaires des séances et mémoires de la Société de biologie*, Neuvième série, 1: 459–64.

– (1921). "Discours sur l'océan." *Bulletin de l'Institut Océanographique de Monaco* 392: 1–16.

Aldrovandi, Ulisse (1602). *De animalibus insectis. Libri Septem*. Bologna: Joannes Baptista Bellagamba.

Allen, Louis (2003). "Japan as occupying power: the revision of history." In Adriana Boscaro, Franco Gatti, and Massimo Raveri, eds., *Rethinking Japan, Volume II: Social Sciences, Ideology and Thought*. London: Japan Library (Curzon Press). 60–70.

Allman, George J. (1851). "On the emission of light by *Anurophonus fimetareous* Nicholi." *Proceedings of the Royal Irish Academy* 5: 125–6.
– (1862). "Note on the phosphorescence of Beröe." *Proceedings of the Royal Society of Edinburgh* 4: 518–19.
Altmann, Richard (1890). *Die Elementarorganismen und ihre Beziehungen zu den Zellen*. Leipzig: Veit & Co.
Alvarez, Luis W. (1980). "Alfred Lee Loomis 1887–1975." *Biographical Memoirs of the National Academy of Sciences* 51: 309–41.
Anctil, Michel (1979). "The epithelial luminescent system of *Chaetopterus variopedatus*." *Canadian Journal of Zoology* 57: 1290–1310.
– (1981). "Luminescence control in isolated notopods of the tube-worm *Chaetopterus variopedatus*: effects of cholinergic and GABAergic drugs." *Comparative Biochemistry and Physiology* 68C: 187–94.
– (1985). "Ultrastructure of the luminescent system of the ctenophore *Mnemiopsis leidyi*." *Cell and Tissue Research* 242: 333–40.
– (1987). "Neural control mechanisms in bioluminescence." In M. Ather Ali, ed., *Nervous Systems in Invertebrates*. New York: Plenum Press. 573–602.
– (2015). *Dawn of the Neuron: The Early Struggles to Trace the Origin of Nervous Systems*. Montreal: McGill-Queen's University Press.
Anctil, Michel and James F. Case (1976). "Pharmacomorphological study of denervation induced by 6-hydroxydopamine in *Porichthys* photophores." *Cell and Tissue Research* 166: 365–88.
– (1977). "The caudal luminous organs of lanternfishes: general innervation and ultrastructure." *American Journal of Anatomy* 149: 1–22.
Anctil, Michel and Osamu Shimomura (1984). "Mechanism of photoinactivation and re-activation in the bioluminescence system of the ctenophore *Mnemiopsis*." *Biochemical Journal* 221: 269–72.
Anderson, James M. (1980). "Biochemistry of centipede luminescence." *Photochemistry and Photobiology* 31: 179–81.
Anderson, Peter A.V. (1980). "Epithelial conduction: its properties and functions." *Progress in Neurobiology* 15: 161–203.
Anderson, Peter A.V. and James F. Case (1975). "Electrical activity associated with luminescence and other colonial behavior in the pennatulid *Renilla köllikeri*." *Biological Bulletin* 149: 80–95.
Anderson, Rupert S. (1933). "The chemistry of bioluminescence. I. Quantitative determination of luciferin." *Journal of Cellular and Comparative Physiology* 3: 45–59.
– (1935). "Studies on bioluminescence. II. The partial purification of *Cypridina* luciferin." *Journal of General Physiology* 19: 301–5.
Anderson, Thomas R. and Tony Rice (2006). "Deserts on the sea floor: Edward Forbes and his azoic hypothesis for a lifeless deep ocean." *Endeavour* 30, 4: 131–7.
Aprille, June R., Christopher J. Lagace, Josephine Modica-Napolitano, and Barry A. Trimmer (2004). "Role of nitric oxide and mitochondria in control of firefly flash." *Integrative and Comparative Biology* 44: 213–19.
Arditti, Joseph (1989). "Kwan Koriba: Botanist and Soldier." *The Garden's Bulletin, Singapore* 42: 1–17.

Audouin, Victor (1840). "Remarques sur la phosphorescence de quelques animaux articulés, à l'occasion d'une lettre de M. Forester sur la phosphorescence des Lombrics terrestres." *Comptes rendus hebdomadaires des séances de l'Académie des sciences* 11: 747–9.

Awad, Edward W. and Michel Anctil (1993). "Identification of β-like adrenoceptors associated with bioluminescence in the sea pansy *Renilla koellikeri*." *Journal of Experimental Biology* 177: 181–200.

– (1994). "Distribution of β2-like adrenergic receptors in the cnidarian *Renilla koellikeri* as revealed by autoradiography and in situ hybridization." *Cell and Tissue Research* 278: 207–15.

Azara, Felix de (1809). *Voyages dans l'Amérique Méridionale*. Vol. 1. Paris: Dentu, Imprimeur-Libraire.

Baba, Kikutorô (1974). "The late Prof. emer. Dr. Hiroshi Ohshima (Kyushu University); his life and his works on zoology." *Proceedings of the Japanese Society of Systematic Zoology* 47: 4–10 (in Japanese).

Backus, Richard H. and Harold Barnes (1957). "Television-echo sounder observations of midwater sound scatterers." *Deep Sea Research* 4: 116–18.

Backus, Richard H., J.E. Craddock, R.L. Haedrich, D.L. Shores, J.M. Teal, and A.S. Wing (1968). "*Ceratoscopelus maderensis*: peculiar sound-scattering layer identified with this myctophid fish." *Science* 160: 991–3.

Baguet, Fernand and James Case (1971). "Luminescence control in *Porichthys* (Teleostei): excitation of isolated photophores." *Biological Bulletin* 140: 15–27.

Baguet, Fernand and Georges Maréchal (1974). "Bioluminescence of a bathypelagic fish, *Argyropelecus hemigymnus* Cocco." *Archives Internationales de Physiologie et de Biochimie* 82: 754–6.

– (1976). "Bioluminescence of bathypelagic fish from the Strait of Messina." *Comparative Biochemistry and Physiology* 53C: 75–82.

– (1978). "The stimulation of isolated photophores (*Argyropelecus*) by epinephrine and norepinephrine." *Comparative Biochemistry and Physiology* 60C: 137–43.

Baguet, Fernand and Jacques Piccard (1981). "The counterlighting hypothesis: in situ observations on *Argyropelecus hemigymnus*." In Marlene DeLuca and William D. McElroy, eds., *Bioluminescence and Chemiluminescence*. New York: Academic Press. 517–23.

Baird, William (1843). "Note on the luminous appearance of the sea, with descriptions of some of the entomostracous insects by which it is occasioned." *The Zoologist* 1: 55–61.

Baker, Henry (1743). *The Microscope Made Easy*. London: R. and J. Dodsley.

– (1753). *Employment for the Microscope: In Two Parts*. London: R. Dodsley.

Baldwin, Thomas O. and collaborators (1984). "Cloning of the luciferase structural genes from *Vibrio harveyi* and expression of bioluminescence in *Escherichia coli*." *Biochemistry* 23: 3663–7.

Bange, Christian and Renée Bange (1994). "Les recherches physiologiques à la Station maritime de Biologie de Tamaris (Var) de 1920 à 1950." In C. Debru, J. Gayon, and J.F. Picard, eds., *Les sciences biologiques et médicales en France 1920–1950*. Paris: CNRS Editions. 55–69.

– (2010). "A l'ombre ou en marge de Claude Bernard: la physiologie à Paris à la fin du XIXème siècle (1878–1905)." *Bulletin d'Histoire et d'Épistémiologie des Sciences de la Vie* 17: 51–68.

Banks, Sir Joseph (1768–71). *The Endeavour Journal of Sir Joseph Banks*. Project Gutenberg of Australia.

Barham, Eric G. (1957). *The Ecology of Sonic Scattering Layers in the Monterey Bay Area, California*. PhD dissertation, Stanford University.

– (1963). "Siphonophores and the deep scattering layer." *Science* 140: 826–8.

– (1966). "Deep scattering layer migration and composition: observations from a diving saucer." *Science* 151: 1399–1403.

– (1970). "Deep-sea fishes: lethargy and vertical orientation." In G. Brooke Farquhar, ed., *Proceedings of an International Symposium on Biological Sound Scattering in the Ocean*. Washington, DC: Maury Center for Ocean Science. 100–16.

Barnes, Anthony T. and James F. Case (1972). "Bioluminescence in the mesopelagic copepod, *Gaussia princeps* (T. Scott)." *Journal of Experimental Marine Biology and Ecology* 8: 53–71.

– (1974). "The luminescence of lanternfish (Myctophidae): spontaneous activity and responses to mechanical, electrical, and chemical stimulation." *Journal of Experimental Marine Biology and Ecology* 15: 203–21.

Bartholomew, James R. (1978). "Japanese modernization and the Imperial universities, 1876–1920." *Journal of Asian Studies* 37, 2: 251–71.

Bassot, Jean-Marie (1958). "Caractères histochimiques des cellules glandulaires dans les organes lumineux du téléostéen *Maurolicus pennanti* Walbaum." *Comptes rendus des séances de l'Académie des sciences* 247: 1909–11.

– (1959a). "Les structures annexes des organes photogènes de *Maurolicus pennanti* Walbaum." *Comptes rendus des séances de l'Académie des sciences* 248: 297–9.

– (1959b). "Caractères histochimiques des organes photogènes du siphon chez *Pholas dactylus* (L.)." *Comptes rendus des séances de l'Académie des sciences* 249: 1267–9.

– (1960a). "Caractères cytologiques des cellules lumineuses chez quelques téléostéens." *Comptes rendus des séances de l'Académie des sciences* 250: 3878–80.

– (1960b). "Données cytologiques et histochimiques sur les photophores d'Euphausiacés. I. Les cellules lumineuses et les structures en bâtonnets." *Comptes rendus des séances de l'Académie des Sciences* 251: 444–6.

– (1960c). "Données cytologiques et histochimiques sur les photophores d'Euphausiacés. II. Les structures annexes du photophore." *Comptes rendus des séances de l'Académie des sciences* 251: 599–601.

– (1963). "Aspects du cycle sécrétoire des photocytes chez le Téléostéen *Maurolicus mülleri* (Gmelin)." *Comptes rendus des séances de l'Académie des sciences* 256: 4732–5.

– (1964). "Présence, dans les photocytes des Annélides Polynoinae, d'une forme paracristalline de réticulum endoplasmique." *Comptes rendus des séances de l'Académie des sciences* 259: 1549–52.

– (1966a). "Histology and ultrastructure of the luminous organs of the siphon of *Pholas*." *Zeitschrift für Zellforschung und mikroskopische Anatomie* 74: 474–504.

– (1966b). "Données histochimiques et histologiques sur l'organe lumineux des élytres d'Annélides Polynoinae." *Cahiers de Biologie Marine* 7: 39–52.

– (1966c). "Une forme microtubulaire et paracristalline de réticulum endoplasmique dans les photocytes des Annélides Polynoinae." *Journal of Cell Biology* 31: 135–8.
– (1966d). "On the comparative morphology of some luminous organs." In Frank. H. Johnson and Yata Haneda, eds., *Bioluminescence in Progress*. 557–610.
– (1978). "Les corps noirs, cellules géantes du Diptère mycétophilide lumineux *Platyura fultoni* et leur sécrétion mitochondriale." *Comptes rendus des séances de l'Académie des sciences* 286: 623–6.
– (1987). "A transient intracellular coupling explains the facilitation of responses in the bioluminescent system of scale worms." *Journal of Cell Biology* 105: 2235–43.
Bassot, Jean–Marie and André Bilbaut (1977a). "Bioluminescence des éıytres d'*Acholoe*. III. Déplacement des sites d'origine au cours des émissions." *Biologie Cellulaire* 28: 155–62.
– (1977b). "Bioluminescence des élytres d'*Acholoe*. IV. Luminescence et fluorescence des photosomes." *Biologie Cellulaire* 28: 163–8.
Bassot, Jean-Marie and Micheline Martoja (1968). "Présence d'un organe lumineux transitoire chez le Gastéropode Pulmoné, *Hemiplecta weinkauffiana* (Crosse et Fischer)." *Comptes rendus des séances de l'Académie des sciences* D266: 1045–7.
Bassot, Jean-Marie and Gisèle Nicolas (1987). "An optional dyadic junctional complex revealed by fast-freeze fixation in the bioluminescent system of the scale worm." *Journal of Cell Biology* 105: 2245–56.
Bassot, Jean-Marie and Ivan Polunin (1967). "Synchronously flashing fireflies in the Malay Peninsula." *Science Repports of the Yokosuka City Museum* 13: 18–20.
Bassot, Jean-Marie, André Bilbaut, George O. Mackie, Leonard M. Passano, and Max Pavans de Ceccaty (1978). "Bioluminescence and other responses spread by epithelial conduction in the siphonophore *Hippopodius*." *Biological Bulletin* 155: 473–9.
Baxter, Charles H. and Peter E. Pickens (1964). "Control of luminescence in hemichrodates and some properties of a nerve net system." *Journal of Experimental Biology* 41: 1–14.
Beddall, Barbara G. (1975). "Un naturaliste original: Don Felix de Azara, 1746–1821." *Journal of the History of Biology* 8: 15–66.
Beebe, William (1924). *Galapagos, World's End*. New York: G.P. Putnam's Sons.
– (1926). *The Arcturus Adventure*. New York: G.P. Putnam's Sons.
– (1932). "Nineteen new species and four post-larval deep-sea fishes." *Zoologica, NY*, 13: 47–107.
– (1933). "New data on the deep-sea fish *Stylophthalmus* and *Idiacanthus*." *Science* 78: 390.
– (1934). *Half Mile Down*. Chicago: Cadmus Books, E.M. Hale and Company.
– (1937). "Preliminary list of Bermuda deep-sea fishes." *Zoologica, NY*, 22: 197–208.
Beebe, William and M. Vander Pyl (1944). "Eastern Pacific expeditions of the New York Zoological Society. XXXIII. Pacific Myctophidae (fishes)." *Zoologica, NY*, 29: 59–95.
Benaron, David A., Pamela R. Contag, and Christopher H. Contag (1997). "Imaging brain structure and function, infection and gene expression in the body using light." *Philosophical Transactions of the Royal Society of London* B352: 755–61.

Benham, W. Blaxland (1899). "Phosphorescent earthworms." *Nature* 60: 591.

Bennett, Frederick Debell (1840). *Narrative of a Voyage Round the Globe from the Year 1833 to 1836*. London: Richard Bentley.

Beolens, Bo, Michael Watkins, and Michael Grayson (2011). *The Eponym Dictionary of Reptiles*. Baltimore: Johns Hopkins University Press.

Bernoulli, Christoph (1803). *Ueber das Leuchten des Meeres, mit besondrer Hinsicht auf das Leuchten*. Göttingen: Heinrich Dieterich.

Berry, S. Stillman (1920a). "Light production in cephalopods. I. Introductory survey." *Biological Bulletin* 38: 141–69.

Berry, S. Stillman (1920b). "Light production in cephalopods. II. Introductory survey." *Biological Bulletin* 38: 171–95.

Bertelsen, Erik and Julius Grøntved (1949). "The light organs of a bathypelagic fish *Argyropelecus olfersi* (Cuvier) photographed by its own light." *Videnskabelige Meddelelser fra Dansk Naturhistorisk Forening* 111: 163–7.

Bicker, Gerd (1998). "NO news from insect brains." *Trends in Neuroscience* 21: 349–55.

Bilbaut, André (1978). "Excitabilité et conductibilité des cellules épithéliales photogènes et non photogènes de l'élytre chez *Acholoe astericola* (Annélide, Polynoinae)." *Comptes rendus des séances de l'Académie des sciences* D286: 985–8.

– (1980). "Excitable epithelial cells in the bioluminescent scales of a polynoid worm; effects of various ions on the action potentials and on the excitation-luminescence coupling." *Journal of Experimental Biology* 88: 219–38.

Boffey, Philip M. (1969). "W.D. McElroy: an old incident embarrasses new NSF Director." *Science* 165: 379–80.

Boisvert, Henri, R. Chatelain, and Jean-Marie Bassot (1967). "Étude d'un *Photobacterium* isolé de l'organe lumineux de poissons *Leiognathidae*." *Annales de l'Institut Pasteur* 112: 520–4.

Bongardt, Johannes (1903). "Beiträge zur Kenntnis der Leuchtorgane einheimischer Lampyriden." *Zeitschrift für wissenschaftliche Zoologie* 75: 1–45.

Bonhomme, Charles (1940). "Nouvelles recherches biologiques et cytologiques sur la luminescence de quelques Polynoïnos." *Bulletin de l'Académie des sciences et lettres de Montpellier* 70: 78–84.

Bosc, Louis Augustin Guillaume (1801). *Histoire naturelle des vers. Tome second*. Paris: Deterville.

Bouchier, Ian A. (1983). "Whales and whaling: contributions by the medical profession." *Medical History* 27(2): 182–4.

Boyle, Robert (1666). "New experiments concerning the relation between light and air (in shining wood and fish)." *Philosophical Transactions of the Royal Society of London* 2: 581–600.

Boyle, Robert (1672). "Some observations about shining flesh." *Philosophical Transactions of the Royal Society of London* 7: 5108–16.

Brandes, Georg (1898). "Die Leuchtorgane der Tiefseefische *Argyropelecus* und *Chauliodus*." *Zeitschrift für Naturwissenschaften* 71: 447–52.

Branham, Marc A. and Michael D. Greenfield (1996). "Flashing males win mate success." *Nature* 381: 745–6.

Branham, Marc A. and John W. Wenzel (2001). "The evolution of bioluminescence in cantharoids (Coleoptera: Elateroidea)." *Florida Entomologist* 84: 565–86.

– (2003). "The origin of photic behavior and the evolution of sexual communication in fireflies (Coleoptera: Lampyridae)." *Cladistics* 19: 1–22.

Branner, John C. (1885). "The reputation of the lantern fly." *American Naturalist* 19, 9: 835–8.

Brauer, August (1906). *Wissenschaftliche Ergebnisse der Deutschen Tiefsee–Expedition auf dem Dampfer "Valdivia" 1898–1899. Fünfzehnter Band: Die Tiefsee-Fische. 1: Systematischer Teil.* Jena: Verlag von Gustav Fischer.

– (1908). *Wissenschaftliche Ergebnisse der Deutschen Tiefsee-Expedition auf dem Dampfer "Valdivia" 1898–1899. Fünfzehnter Band: Die Tiefsee-Fische. 2: Anatomischer Teil.* Jena: Verlag von Gustav Fischer.

Brehm, Paul and James G. Morin (1977). "Localization and characterization of luminescent cells in *Ophiopsila californica* and *Amphipholis squamata* (Echinodermata: Ophiuroidea)." *Biological Bulletin* 152: 12–25.

Brehm, Paul, James Leichleiter, Stephen Smith, and Kathleen Dunlap (1989). "Intercellular signalling as visualized by endogenous calcium-dependent bioluminescence." *Neuron* 3: 191–8.

Bresadola, Marco (2011). "Carlo Matteucci and the legacy of Luigi Galvani." *Archives italiennes de biologie* 149 (Suppl.): 3–9.

Breslau, Lloyd R. (1959). *Triggered Underwater Cameras for Marine Biology.* M.Sc. Thesis, Massachusetts Institute of Technology, Cambridge, MA.

Browne, Ethel N. (1909). "The production of new hydranths in Hydra by the insertion of small grafts." *Journal of Experimental Zoology* 8, 1: 1–23.

Bruguière, Jean Guillaume (1792). "Sur la qualité phosphorique du ver de terre, dans certaines circonstances, et sur des ossemens humains trouvés fossiles." *Journal d'histoire naturelle* 2: 267–8.

Brunelli, M., M. Buonamici, and F. Magni (1968a). "Mechanisms for photic inhibition of flashing in fireflies." *Archives italiennes de biologie* 106: 85–99.

– (1968b). "Effects of castration on the inhibition of flashing fireflies." *Archives italiennes de biologie* 106: 100–12.

Buchner, Paul (1915). "Sind die Leuchtorgane Pilzorgane?" *Zoologischer Anzeiger* 45: 17–21.

Buck, John B. (1935). "Synchronous flashing of fireflies experimentally induced." *Science* 81: 339–40.

– (1937a). "Studies on the firefly. I. The effects of light and other agents on flashing in Photinus pyralis, with special reference to periodicity and diurnal rhythm." *Physiological Zoology* 10: 45–58.

– (1937a). "Studies on the firefly. II. The signal system and color vision in Photinus pyralis." *Physiological Zoology* 10: 412–19.

– (1938). "Synchronous rhythmic flashing of fireflies. *Quarterly Review of Biology* 13: 301–14.

– (1948). "The anatomy and physiology of the light organ in fireflies." *Annals of the New York Academy of Science* 49: 397–482.

– (1955). "Some reflections on the control of bioluminescence." In Frank H. Johnson, ed., *The Luminescence of Biological Systems*. Washington, DC: American Association for the Advancement of Science. 323–32.

– (1958). "Light on light." *Quarterly Review of Biology* 33: 59–61.

– (1961). "Speculations on the interrelations and evolution of photic organs." In William D. McElroy and Bentley Glass, eds., *Light and Life*. Baltimore, Maryland: Johns Hopkins Press. 754–70.

– (1966). "Unit activity in the firefly lantern." In Frank. H. Johnson and Yata Haneda, eds., *Bioluminescence in Progress*. 459–74

– (1978). "Functions and evolutions of bioluminescence." In Peter J. Herring, ed., *Bioluminescence in Action*. 419–60.

Buck, John and Elizabeth Buck (1966). "Biology of synchronous flashing of fireflies." *Nature* 211: 562–4.

– (1968). "Mechanism of rhythmic synchronous flashing of fireflies." *Science* 159: 1319–27.

– (1978). "Toward a functional interpretation of synchronous flashing by fireflies." *The American Naturalist* 112: 471–92.

Buck, John and James F. Case (1961). "Control of flashing in fireflies. I. The lantern as a neuroeffector organ." *Biological Bulletin* 121: 234–56.

Buck, John, James F. Case, and Frank E. Hanson (1963). "Control of flashing in fireflies. III. Peripheral excitation." *Biological Bulletin* 125: 251–69.

Buck, John B., J. Woodland Hastings, William D. McElroy, Osamu Shimomura, and Frederick I. Tsuji (1995). "In memoriam Yata Haneda." *Journal of Bioluminescence and Chemiluminescence* 10: 323.

Buhl, Peter Neerup (1995). *H.J. Hansen, Dr. phil: zoolog og polemiker*. Copenhagen: Danskerens forlag, Den Danske Forening.

Burckhardt, Rudolf (1900). "On the luminous organs of selachian fishes." *Annals and Magazine of Natural History*, 7th series, 6: 558–68.

Burkenroad, Martin D. (1943). "A possible function of bioluminescence." *Journal of Marine Research* 5: 161–4.

Burstyn, Harold L. (1972). "Pioneering in large-scale scientific organisation: the *Challenger* Expedition and its Report. I. Launching the Expedition." *Proceedings of the Royal Society of Edinburgh. Section B. Biology*, 72: 47–61.

Calder, Dale R. (2004). "From birds to hydroids: Charles Cleveland Nutting (1858–1927) of the University of Iowa, USA." *Hydrobiologia* 530: 13–25.

– (2015). "George James Allman (1812–1898): pioneer in research on Cnidaria and freshwater Bryozoa." *Zootaxa* 4020(2): 201–43.

Calman, William T. (1947). "Stanley Wells Kemp. 1882–1945." *Obituary Notices of Fellows of the Royal Society* 5(15): 467–76.

Campbell, Anthony K. (2008). "Jean-Marie Bassot (1933–2007): a life of unquenched curiosity." *Luminescence* 23: 187–90.

– (2012). "Darwin shines light on the evolution of bioluminescence." *Luminescence* 27: 447–9.

Campbell, Anthony K. and Peter J. Herring (1990). "Imidazolopyrazine bioluminescence in copepods and other marine organisms." *Marine Biology* 104: 219–25.

Capanna, Ernesto (2008). "Battista Grassi entomologist and the Roman School of Malariology." *Parassitologia* 50: 201–11.

Carillo, Jesús (2002). "The Historia General y Natural de las Indias by Gonzalo Fernández de Oviedo." *Huntington Library Quarterly* 65(3/4): 321–44.
Carlson, Albert D. (1968). "Effects of adrenergic drugs on the lantern of the larval *Photuris* firefly." *Journal of Experimental Biology* 48: 381–7.
Carlson, Albert D. and Mary Jalenak (1986). "Release of octopamine from the photomotor neurones of the larval firefly lanterns." *Journal of Experimental Biology* 122: 453–7.
Carpenter, Edward, J. Gwyn Jeffreys, and Charles Wyville Thomson (1869). "Preliminary report of the scientific exploration of the deep Ssa in H.M. Surveying-Vessel 'Porcupine,' during the summer of 1869." *Proceedings of the Royal Society of London* 18: 397–492.
Carpine-Lancre, Jacqueline (1998). *Albert Ier, prince de Monaco: des oeuvres de science, de lumiere et de paix.* Monaco: Palais de S.A.S. le Prince.
Case, James F. (1984). "Vision in mating behaviour of fireflies." In Trevor Lewis, ed., *Insect Communication.* New York: Academic Press. 195–222.
Case, James F. and John Buck (1963). "Control of flshing in fireflies. II. Role of central nervous system." *Biological Bulletin* 125: 234–50.
Case, James F. and Frank E. Hanson (2004). "The luminous world of John and Elizabeth Buck." *Integrative and Comparative Biology* 44: 197–202.
Case, James F. and Linda Strause (1978). "Neurally controlled luminescent systems." In Peter J. Herring, ed., *Bioluminescence in Action.* 331–66.
Case, James F., Jon Warner, Anthony T. Barnes, and Mark Lowenstine (1977). "Bioluminescence of lantern fish (Myctophidae) in response to changes in light intensity." *Nature* 265: 179–81.
Chalfie, Martin, Yuan Tu, Ghia Euskirchen, William W. Ward, and Douglas C. Prasher (1994). "Green fluorescent protein as a marker for gene expression." *Science* 263: 802–5.
Chang, Kenneth (2008) "Man who set stage for a Nobel now lives a life outside science." *The New York Times,* 16 October 2008.
Chase, Aurin M. (1948). "The chemistry of *Cypridina* luciferin." *Annals of the New York Academy of Sciences* 49, 3: 353–75.
Chase, Aurin M. and James H. Gregg (1949). "Analysis of Cypridina luciferin for nitrogen." *Journal of Cellular and Comparative Physiology* 33: 67–71.
Chase, Laurence B. (1967). "Bioluminescence: A Princeton Discipline." *Princeton Alumni Weekly* 67, 22: 10–14, 22.
Chen, Antony K., Michael I. Latz, Peter Sobolewski, and John A. Frangos (2007). "Evidence for the role of G-proteins in flow stimulation of dinoflagellate bioluminescence." *American Journal of Physiology* 292: R2020-7.
Christensen, Thomas A. and Albert D. Carlson (1981). "Symmetrically organized dorsal unpaired median (DUM) neurones and flash control in the male firefly, *Photuris versicolor.*" *Journal of Experimental Biology* 93: 133–47.
– (1982). "The neurophysiology of larval firefly luminescence: direct activation through four bifurcating (DUM) neurons." *Journal of Comparative Physiology* 148: 503–14.
Christensen, Thomas A., T.G. Sherman, R.E. McCaman, and Albert D. Carlson (1983). "Presence of octopamine in firefly photomotor neurons." *Neuroscience* 9: 183–9.

Christophe, Bernard and Fernand Baguet (1982). "Luminescence of isolated photophores and supracaudal gland from *Myctophum punctatum*: electrical stimulation." *Comparative Biochemistry and Physiology* 71A: 131–6.

Christophe, Bernard and Fernand Baguet (1983). "Luminescence of isolated photocytes from *Porichthys* photophores: adrenergic stimulation." *Journal of Experimental Biology* 104: 183–92.

Chun, Carl (1888). "Die pelagischer Thierwelt in grösseren Meerestiefen und ihre Beziehungen zu der Oberflächenfauna." *Bibliotheca Zoologica* 1: 1–66.

Chun, Carl (1893). "Leuchtorgan und Facettenauge. Ein Beitrag zur Theorie des Sehens in grossen Meerestiefen." *Biologisches Zentralblatt* 13: 544–71.

– (1896). "Die Leuchtorgane der Euphausiden." *Zoologica* (Stuttgart) 7: 196–212.

– (1903). "Über Leuchtorgane und Augen von Tiefsee-Cephalopoden." *Verhandlungen der Deutschen Zoologischen Gesellschaft* 13: 67–91.

– (1905). *Aus den Tiefen des Weltmeeres.* Second Edition. Jena: Gustav Fischer.

– (1908). "Über Cephalopoden der Deutschen Tiefsee-Expedition." *Zoologischer Anzeiger* 33: 86–9.

– (1910). *Wissenschaftliche Ergebnisse der Deutschen Tiefsee-Expedition auf dem Dampfer "Valdivia" 1898–1899. Achtzehnter Band: Die Cephalopoden. 1: Oegopsida.* Jena: Verlag von Gustav Fischer. (Plates published in 1914.)

– (1915). *Wissenschaftliche Ergebnisse der Deutschen Tiefsee-Expedition auf dem Dampfer "Valdivia" 1898–1899. Achtzehnter Band: Die Cephalopoden. 2: Myopsida, Octopoda.* Jena: Verlag von Gustav Fischer.

Claes, Julien M. and Jérôme Mallefet (2008). "Early development of bioluminescence suggests camouflage by counter-illumination in the velvet belly lantern shark *Etmopterus spinax* (Squaloidea: Etmopteridae)." *Journal of Fish Biology* 73: 1337–50.

– (2009a). "Hormonal control of luminescence from lantern shark (*Etmopterus spinax*) photophores." *Journal of Experimental Biology* 212: 3684–92.

– (2009b). "Bioluminescence of sharks: first synthesis." In V. Benno Meyer-Rochow, ed., *Bioluminescence in Focus – A Collection of Illuminating Essays.* Trivandrum (Kerala): Research Signpost. 51–67.

– (2010). "The lantern shark's light switch: turning shallow water crypsis into midwater camouflage." *Biology Letters* 6: 685–7.

Claes, Julien M., Mason N. Dean, Dan-Eric Nilsson, Nathan S. Hart, and Jérôme Mallefet (2013). "A deepwater fish with 'lightsabers' – dorsal spine-associated luminescence in a counterilluminating lanternshark." *Scientific Reports* 3: 1308.

Claes, Julien M., Dan-Eric Nilsson, Nicolas Straube, Shaun P. Collin, and Jérôme Mallefet (2014). "Iso-luminance counterillumination drove bioluminescent shark radiation." *Scientific Reports* 4: 4328.

Claes, Julien M., Dan-Eric Nilsson, Jérôme Mallefet, and Nicolas Straube (2015). "The presence of lateral photophores correlates with increased speciation in deep-sea bioluminescent sharks." *Royal Society Open Science* 2: 150219.

Clarapède, René-Edouard (1864). *Glanures zootomiques parmi les annélides de Port-Vendre (Pyrénées orientales).* Genève, Bâle: H. Georg.

Clarke, George L. and Richard H. Backus (1957). "Measurements of light penetration in relation to vertical migration and records of luminescence of deep-sea animals." *Deep Sea Research* 4: 1–4.

Clement-Guyader, Nelly (1992). *La bibliothèque personnelle du naturaliste rouennais Henri Gadeau de Kerville (1858–1940)*. DEA Sciences de l'information, École Nationale Supérieure des Sciences de l'information et des Bibliothèques, 1–137.
Cocco, Anastasio (1838). "Su di alcuni Salmonidi del mare di Messina." *Nuovi annali delle scienze naturali di Bologna* 2: 161–94.
Cohen, Anne C. and James G. Morin (2003). "Sexual morphology, reproduction and the evolution of bioluminescence in Ostracoda." *Paleontological Society Papers* 9: 37–70.
Cohn D.H., A.J. Mileham, M.I. Simon, K.H. Nealson, S.K. Rausch, D. Bonam, and T.O. Baldwin (1985). "Nucleotide sequence of the luxA gene of *Vibrio harveyi* and the complete amino acid sequence of the alpha subunit of bacterial luciferase." *Journal of Biological Chemistry* 260: 6139–46.
Collett, Robert (1895). "Poissons nouveaux provenant des campagnes de l'*Hirondelle*." *Bulletin de la Société zoologique de France* 14: 291–3.
Compton, Andrew J. (1974). *A Study of the Bioluminescence of a Deep Scattering Layer Organism (Euphausia pacifica) in Monterey Bay, California*. Master of Science thesis, Naval Postgraduate School, Monterey, California.
Copeland, Jonathan and Maryellen M. Daston (1989). "Bioluminescence in the terrestrial snail *Dyakia* (*Quantula*) *striata*." *Malacologia* 30: 317–24.
– (1993). "Adult and juvenile flashes in the terrestrial snail *Dyakia striata*." *Malacologia* 35: 1–7.
Corfield, Richard (2004). *The Silent Landscape: In the Wake of* HMS *Challenger 1872–1876*. London: John Murray.
Cormier, Milton J. (2007). *My Journey. Into, Through and Beyond the Ivory Tower*. Chandler, Arizona: OPA Publishing and Distribution.
Cormier, Milton J. and Bernard H. Strehler (1953). "The identification of KFC: requirement of long-chain aldehydes for bacterial extract luminescence." *Journal of the American Chemical Society* 75: 4864–5.
Cormier, Milton J., Jules M. Crane, Jr., and Yo Nakano (1967). "Evidence for the identity of the luminescent systems of *Porichthys porosissimus* (fish) and *Cypridina hilgendorfii* (crustacean)." *Biochemical and Biophysical Research Communications* 29: 747–52.
Cormier, Milton J., Yashwant D. Karkhanis, and Kazuo Hori (1970). "Evidence that Renilla luciferase is not a calcium-triggered photoprotein." *Biochemical and Biophysical Research Communications* 38: 962–4.
Cornalia, Emilio (1877). *Commemorazione del Prof. Paolo Panceri*. Milan: Giuseppe Bernardori.
Corner, John E.J.H. (1946). "Japanese men of science in Malaya during Japanese occupation." *Nature* 158: 63.
– (1950). "Descriptions of two luminous tropical agarics (*Dictyopanus* and *Mycena*)." *Mycologia* 42: 423–31.
– (1981). *The Marquis – A Tale of Syoyan-to*. Singapore: Heinemann Asia.
– (1990). "His Majesty Emperor Hirohito of Japan, K.G. 29 April 1901–7 January 1989." *Biographical Memoirs of the Fellows of the Royal Society* 36: 242–72.
Corner, John K. (2013). *My Father in His Suitcase. In Search of E.J.H. Corner the Relentless Botanist*. Singapore: Landmark Books.

Crawshay, Lionel R. (1935). "Possible bearing of a luminous syllid on the question of the landfall of Christopher Columbus." *Nature* 136: 559–60.

Dahlgren, Ulric (1915a). "The production of light by animals. Introduction." *Journal of the Franklin Institute* 180, 5: 513–37.

– (1915b). "The production of light by animals. Unicellular organisms." *Journal of the Franklin Institute* 180, 6: 711–27.

– (1916a). "The production of light by animals. Luminous plants." *Journal of the Franklin Institute* 181, 1: 109–25.

– (1916b). "The production of light by animals. Porifera and coelenterates." *Journal of the Franklin Institute* 181, 2: 243–61.

– (1916c). "The production of light by animals. Luminosity in the echinoderms." *Journal of the Franklin Institute* 181, 3: 377–400.

– (1916d). "The production of light by animals. Light production in cephalopods." *Journal of the Franklin Institute* 181, 4: 525–56.

– (1916e). "The production of light by animals. Light production as seen in worms." *Journal of the Franklin Institute* 181, 5: 659–96.

– (1916f). "The production of light by animals. The luminous crustaceans." *Journal of the Franklin Institute* 181, 6: 805–43.

– (1917a). "The production of light by animals. The luminous insects." *Journal of the Franklin Institute* 183, 1: 79–94.

– (1917b). "The production of light by animals. Light production by the elaterid beetles." *Journal of the Franklin Institute* 183, 2: 211–20.

– (1917c). "The production of light by animals. The fire-flies of Lampyridae." *Journal of the Franklin Institute* 183, 3: 323–48.

– (1917d). "The production of light by animals. Histogenesis and physiology of the light-tissues in lampyrids." *Journal of the Franklin Institute* 183, 5: 593–624.

– (1917e). "The production of light by animals. Tunicates and luminosity in fishes." *Journal of the Franklin Institute* 183, 6: 735–54.

– (1928). "The bacterial light organ of Ceratias." *Science* 68: 65–6.

Damkaer, David M. (1995). "Hans Jacob Hansen (10 August 1855–26 June 1936)." *Journal of Crustacean Biology* 15, 4: 795–6.

Darwin, Charles (1839). *Narrative of the Surveying Voyages of His Majesty's Ships Adventure and Beagle Between the Years 1826 and 1836. Volume III: Journal and Remarks 1832–1836.* London: Henry Colburn.

– (1859). *On the Origin of Species by Means of Natural Selection.* London: J. Murray.

Dassow, Peter von and Michael I. Latz (2002). "The role of Ca2+ in stimulated bioluminescence of the dinoflagellate *Lingulodinium polyedrum*." *Journal of Experimental Biology* 205: 2971–86.

Davenport, Demorest and J.A. Colin Nicol (1955). "Luminescence in hydromedusae." *Proceedings of the Royal Society of London* B144: 399–411.

– (1956). "Observations on luminescence in sea pens (Pennatulacea)." *Proceedings of the Royal Society of London* B144: 480–96.

Davenport, Demorest, Donald M. Wootton, and John E. Cushing (1952). "The biology of the Sierra luminous millipede, *Luminodesmus sequoiae*, Loomis and Davenport." *Biological Bulletin* 102: 100–10.

Davis, Matthew P., John S. Sparks, and W. Leo Smith (2016). "Repeated and widespread evolution of bioluminescence in marine fishes." *PLOS One* 11, 6: 1–11.

Day, John (2009). "Beetle bioluminescence: a genetic and enzymatic research review." In Victor B. Meyer-Rochow, ed., *Bioluminescence in Focus: A Collection of Illuminating Essays*. Kerala, India: Research Signpost. 255–73.

De Bersaques, Jean (2011). "Wilhelm Gottlieb Tilesius – a forgotten dermatologist." *Journal der Deutschen Dermatologischen Gesellschaft* 9: 563–70.

De Bremaeker, Nancy, Jérôme Mallefet, and Fernand Baguet (1996). "Luminescence control in the brittlestar *Amphipholis squamata*: effect of cholinergic drugs." *Comparative Biochemistry and Physiology* 115C: 75–82.

De Bremaeker, Nancy, Fernand Baguet, Michael C. Thorndyke, and Jérôme Mallefet (1999). "Modulatory effects of some amino acids and neuropeptides on luminescence in the brittlestar *Amphipholis squamata*." *Journal of Experimental Biology* 202: 1785–91.

Dees, Bowen C. (1997). *The Allied Occupation and Japan's Economic Miracle*. UK: Richmond, Surrey, Japan Library (Curzon Press).

Deheyn, Dimitri, Jérôme Mallefet, and Michel Jangoux (2000). "Cytological changes during bioluminescence production in dissociated photocytes from the ophiuroid *Amphipholis squamata* (Echinodermata)." *Cell and Tissue Research* 299: 115–28.

Denton, Eric J. (1970). "On the organization of reflecting surfaces in some marine animals." *Philosophical Transactions of the Royal Society of London* B258: 285–313.

Denton, Eric J., John B. Gilpin-Brown, and Paul G. Wright (1970). "On the 'filters' in photophores of mesopelagic fish and on a fish emitting red light and especially sensitive to red light." *Journal of Physiology* 208: 73–4P.

– (1972). "The angular distribution of the light produced by some mesopelagic fish in relation to their camouflage." *Proceedings of the Royal Society of London*, B, 182: 145–58.

De Sa, Richard, J. Woodland Hastings, and A.E. Vatter (1963). "Luminescent 'crystalline' particles: an organized subcellular bioluminescent system." *Science* 141: 1269–70.

Devaraj, Balasigamani, Mashashi Usa, and Umio Inaba (1997). "Biophotons: ultraweak light emission from living systems." *Current Opinion in Solid State & Materials Science* 2: 188–93.

De Wet, Jeffrey R., Keith V. Wood, Donald R. Helinski, and Marlene DeLuca (1985). "Cloning of firefly luciferase CDNA and the expression of active luciferase in *Escherichia coli*." *Proceedings of the National Academy of Science* 82: 7870–3.

Dilly, Peter N. and Peter J. Herring (1974). "The ocular light organ of *Bathothauma lyromma* (Mollusca: Cephalopoda)." *Journal of Zoology* 172: 81–100.

Doino, Judith A. and Margaret McFall-Ngai (1995). "A transient exposure to symbiosis-competent bacteria induces light organ morphogenesis in the host squid." *Biological Bulletin* 189: 347–55.

Douglas, Ronald H., Conrad W. Mullineaux, and Julien C. Partridge (2000). "Long-wave sensitivity in deep-sea stomiid dragonfish with far-red bioluminescence: evidence for a dietary origin of the chlorophyll-derived retinal photosensitizer of *Malacosteus niger*." *Philosophical Transactions of the Royal Society of London* B355: 1269–72.

Douglas Ronald H., Julien C. Partridge, Kanwaljit S. Dulai, David M. Hunt, Conrad W. Mullineaux, and Paavo H. Hynninen (1999). "Enhanced retinal longwave sensitivity using a chlorophyll-derived photosensitiser in *Malacosteus niger*, a deep-sea dragon fish with far red bioluminescence." *Vision Research* 39: 2817–32.

Doyle, J.D. and R.H. Kay (1967). "Some studies on the bioluminescence of the euphausiids, *Meganyctiphanes norvegica* and *Thysanoessa raschii*." *Journal of the Marine Biological Association of the UK* 47: 555–63.

Dubois, Raphaël (1884). "Note sur la physiologie des Pyrophores." *Comptes rendus des séances de la Société de biologie*, Series 8, 1: 661–4.

– (1885). "Fonction photogénique des Pyrophores." *Comptes rendus des séances de la Société de biologie*, Series 8, 2: 559–62.

– (1886a). *Les élatérides lumineux.* Meulan: Imprimerie de la Société Zoologique de France.

– (1886b). "De la fonction photogénique chez les Myriapodes. " *Comptes rendus des séances de la Société de biologie*, Series 8, 3: 518–22.

– (1886c). "De la fonction photogénique chez les Podures." *Comptes rendus des séances de la Société de biologie*, Series 8, 3: 600–3.

– (1887a). "De la fonction photogénique chez le *Pholas dactylus*." *Comptes rendus des séances de l'Académie des sciences* 105: 690–2.

– (1887b). "Les vacuolides. " *Comptes rendus des séances de la Société de biologie*, Series 8, 39: 9–16.

– (1888). "Sur la production de la lumière chez le *Pholas dactylus*." *Comptes rendus des séances de la Société de biologie*, Series 8, 5: 451–3.

– (1889). "Sur le mécanisme des fonctions photodermatique et photogénique, dans le siphon du *Pholas dactylus*." *Comptes rendus des séances de la Société de biologie*, Series 9, 1: 233–5.

– (1892). *Anatomie et physiologie comparées de la pholade dactyle.* Paris: G. Masson.

– (1893). "Sur le mécanisme de la production de la lumière chez l'Orya barbarica d'Algérie." *Comptes rendus des séances de la Société de biologie*, Series 9, 5: 184–6.

– (1898). *Leçons de physiologie générale et comparée.* Paris: Georges Carré et C. Naud.

– (1911). "Nouvelles recherches sur la lumière physiologique chez *Pholas dactylus*." *Comptes rendus des séances de l'Académie des sciences* 153: 690–2.

– (1914). *La vie et la lumière.* Paris: Librairie Félix Alcan.

– (1917). "Étude critique de quelques travaux récents relatifs à la biophotogénèse." *Annales de la Société Linnéenne de Lyon* 64: 65–117.

– (1918). "La paix par la science et le protectorat rhénan." *Bulletin de l'Académie du Var* 86: 1–40.

– (1919). "Symbiotes, vacuolides, mitochondries et leucites." *Comptes rendus des séances de la Société de biologie* 82: 473–7.

Dudek, Peter (2013). *"Wir wollen Krieger sein im Heere des Lichts": reformpädagogische Landerziehungsheime im hessischen Hochwaldhausen 1912–1927.* Bad Heilbrunn, Germany: Julius Klinkhardt.

Dugès, Antoine Louis (1837). "Nouvelles observations sur la zoologie et l'anatomie des Annélides abranches sétigères." *Annales des sciences naturelles*, série 2, 8: 15–35.

Dunlap, Calvin R. III (1968). *An Ecological Reconnaissance of the Deep Scattering Layers in the Eastern Tropical Pacific*. Master of Science thesis, Naval Postgraduate School, Monterey, California.

Dunlap, Jay C., J. Woodland Hastings, and Osamu Shimomura (1980). "Crossreactivity between the light-emitting systems of distantly related organisms: novel type of light-emitting compound." *Proceedings of the National Academy of Sciences* 77: 1394–7.

Dunlap, Kathleen, K. Takeda, and Paul Brehm (1987). "Activation of a calcium-dependent photoprotein by chemical signalling through gap junctions." *Nature* 325: 60–2.

Dunlap, Paul V. and Margaret J. McFall-Ngai (1987). "Initiation and control of the bioluminescent symbiosis between *Photobacterium leiognathi* and leiognathid fish." *Annals of the New York Academy of Science* 503: 269–83.

Dunlap, Paul V. and Masaru Nakamura (2011). "Functional morphology of the luminescence system of *Siphamia versicolor* (Perciformes: Apogonidae), a bacterially luminous coral reef fish." *Journal of Morphology* 272: 897–909.

Dunstan, Sarah L., Graciela B. Sala-Newby, Alexandra Bermudez Fajardo, Kathryn M. Taylor, and Anthony K. Campbell (2000). "Cloning and expression of the bioluminescent photoprotein pholasin from the bivalve mollusc *Pholas dactylus*." *Journal of Biological Chemistry* 275: 9403–9.

Eberhard, A., A.L. Burlingame, C. Eberhard, G.L. Kenyon, K.H. Nealson, and N.J. Oppenheimer (1981). "Structural identification of autoinducer of luciferase." *Biochemistry* 20: 2444–9.

Eckert, Roger (1965). "Bioelectric control of bioluminescence in the dinoflagellate *Noctiluca*: I. Specific nature of triggering events. II. Asynchronous flash initiation by a propagated triggering potential. *Science* 147: 1140–5.

Eckert, Roger and George T. Reynolds (1967). "The subcellular origin of bioluminescence in *Noctiluca miliaris*." *Journal of General Physiology* 50: 1429–58.

Ehrenberg, Christian Gottfried (1835). *Das Leuchten des Meeres*. Berlin: Königlichen Akademie der Wissenschaften.

Ellis, Emily A. and Todd H. Oakley (2016). "High rates of species accumulation in animals with bioluminescent courtship displays." *Current Biology* 26: 1916–21.

Emery, Carlo (1884a). "Intorno alle macchie splendenti della pelle nei pesci del genere Scopelus." *Mittheilungen aus der Zoologischen Station zu Neapel* 5: 471–82.

– (1884b). "Untersuchungen über Luciola italica L." *Zeitschrift für wissenschaftliche Zoologie* 40: 338–55.

– (1886). "La lumière de la *Luciola italica* observée au microscope." *Archives italiennes de biologie* 7(2) : 274–8.

Enenkel, Karl A.E. (2014). "Die antike Vorgeschichte der Verankerung in Politik und Religion: Plinius' Zoologie und der römische Imperialismus." In Karl A.E. Enenkel and Paul J. Smith, eds., *Zoology in Early Modern Culture: Interactions of Science, Theology, Philology and Political and Religious Education*. Boston: Brill. 15–53.

Ernster, Lars and Gottfried Schatz (1981). "Mitochondria: a historical review." *Journal of Cell Biology* 91: 227–55s.

Esaias, Wayne E. and Herbert C. Curl, Jr. (1972). "Effect of dinoflagellate bioluminescence on copepod ingestion rates." *Limnology and Oceanography* 17: 901–6.

Evans, David (2015). *Marine Physiology Down East: The Story of the Mt. Desert Island Biological Laboratory*. New York, Heidelberg, Dordrecht, and London: Springer.
Fabre, Jean-Henri (1855). "Recherches sur la cause de la phosphorescence de l'agaric de l'olivier." *Annales des sciences naturelles*, Series 4, 4: 179–97.
Falcon, Andrea (2005). *Aristotle and the Science of Nature: Unity without Uniformity*. Cambridge and New York: Cambridge University Press.
Faust, Lynn, Raphael De Cock, and Sara Lewis (2012). "Thieves in the night: kleptoparatisitism by fireflies in the genus Photuris Dejean (Coleoptera: Lampyridae)." *Coleopterists Bulletin* 66: 1–6.
Findlen, Paula (1996). *Possessing Nature: Museums, Collecting, and Scientific Culture in Early Modern Italy*. Berkeley and Los Angeles: University of California Press.
– (2004). "The last man who knew everything … or did he? Athanasius Kircher, S.J. (1602–80) and his world." In Paula Findlen, ed., *Athanasius Kircher: The Last Man Who Knew Everything*. New York and London: Routledge. 1–48.
Flaugergues, Honoré (1771). "Lettre sur le phosphorisme des vers de terre." *Journal de physique, de chimie, d'histoire naturelle et des arts* 16: 311–13.
Fleisher, Kellie J. and James F. Case (1995). "Cephalopod predation facilitated by dinoflagellate luminescence." *Biological Bulletin* 189: 263–71.
Fontaine, Maurice (1980). "From the physiology of marine organisms to oceanographic physiology or physiological oceanography." In Mary Sears and Daniel Merriman, eds. *Oceanography: The Past*. Berlin-Heidelberg-New York: Springer-Verlag. 353–9.
Forbes, Edward (1848). *A Monograph of the British Naked-Eyed Medusae*. London: The Ray Society.
Förster, Johannes (1914). "Über die Leuchtorgane und das Nervensystem von Pholas dactylus." *Zeitschrift für wissenschaftliche Zoologie* 109: 349–92.
Francis, Warren R., Nathan C. Shaner, Lynne M. Christianson, Meghan L. Powers, and Steven H.D. Haddock (2015). "Occurrence of Isopenicillin-N-Synthase homologs in bioluminescent ctenophores and implications for coelenterazine biosynthesis." *PLOS One* 10, 6: 1–20.
Frank, Jr, Robert G. (1987). "American physiologists in German laboratories" In Gerald L. Geison, ed., *Physiology in the American Context, 1850–1940*. New York: Springer. 11–46.
Fraser, James (1962). *Nature Adrift. The Story of Marine Plankton*. London: G.T. Foulis & Co.
Fredj, Gaston and Michel Meinardi (2007). *L'ange et l'orchidée: Risso, Vérany & Barla, une lignée de savants de renommée mondiale à Nice au XIXe siècle*. Nice: Serre Éditeur.
Fregin, Torsten and Konrad Wiese (2002). "The photophores of *Meganyctiphanes norvegica* (M. Sars) (Euphausiacea): mode of operation." *Helgoland Marine Research* 56: 112–24.
French, Richard D. (1970). "Darwin and the physiologists, or the medusa and modern cardiology." *Journal of the History of Biology* 3: 253–74.
French, Roger K. (1994). *Ancient Natural History. Histories of Nature*. London and New York: Routledge.

Freycinet, Louis de (1827). *Voyage autour du monde. Tome premier*. Paris: Chez Pillet Aîné, Imprimeur-Libraire.

Fröhlich, Friedrich W. (1923). "Max Verworn." *Zeitschrift für allgemeinee Physiologie* 20: 185–92.

Gadeau de Kerville, Henri (1881). *Les insectes phosphorescents*. Rouen: Léon Deshays.

– (1890). *Les animaux et les végétaux lumineux*. Paris: J.B. Baillière et Fils.

Galler, Sydney R. (1963). "Office of Naval Research." *Bulletin of the American Institute of Biological Sciences* 13: 25–7.

Galloway, Thomas W. and Paul S. Welch (1911). "Studies on a phosphorescent annelid, Odontosyllis enopla Verrill." *Transactions of the American Microscopical Society* 30, 1: 13–39.

Galt, Charles P. (1978). "Bioluminescence: dual mechanism in a planktonic tunicate produces brilliant surface display." *Science* 200: 70–2.

Garman, Samuel (1899). "Reports on an exploration off the west coasts of Mexico, Central and South America, and off the Galapagos Islands, in charge of Alexander Agassiz, by the U.S. Commission steamer 'Albatross,' during 1891, Lieut. Commander Z.L. Tanner, U.S.N., commanding. The Fishes." *Memoirs of the Museum of Comparative Zoology* 24: 1–431.

Gatti, Michele (1899). "Ricerche sugli organi biofotogenetici dei pesci. Parte 2: Organi di tipo ellettico. Parte 3: Sviluppo degli organi dei due tipi." *Rendiconti Atti della Reale Academia dei Lincei*, Series 5, 8: 81–7.

– (1904). "Ricerche sugli organi luminosi dei pesci." *Annali di agricoltura, Lavori eseguiti nella Reale stazione di piscicoltura di Roma*, 233: 7–126.

Gegenbaur, Carl, Albrecht von Kölliker, and Heinrich Müller (1853). "Bericht über einige im Herbste 1852 in Messina angestellte vergleichend-anatomische Untersuchungen." *Zeitschrift für wissenschaftliche Zoologie* 4: 299–370.

Geipel, Erich (1915). "Beiträge zur Anatomie der Leuchtorgane tropischer Käfer." *Zeitschrift für wissenschaftliche Zoologie* 112: 239–90.

Geison, Gerald L. (1978). *Michael Foster and the Cambridge School of Physiology*. Princeton: Princeton University Press.

Gerbi, Antonello (2010). *Nature in the New World: From Christopher Columbus to Gonzalo Fernández de Oviedo*. Pittsburgh: University of Pittsburgh Press.

Germain, Guy and Michel Anctil (1995). "Evidence for intercellular coupling and connexin-like protein in the luminescent endoderm of *Renilla koellikeri* (Cnidaria, Anthozoa)." *Biological Bulletin* 191: 353–66.

Gesner, Conrad (1555). *De raris et admirandis herbis quae sive quod noctu luceant, sive alias ob causas, Lunariae nominantur et obiter de alias etiam rebus, quae in tenebris lucent, commentariolus*. Zurich: Tiguri.

Ghiradella, Helen and John T. Schmidt (2004). "Fireflies at one hundred plus: a new look at flash control." *Integrative and Comparative Biology* 44: 203–12.

Giesbrecht, Wilhelm (1895). "Über das Leuchten der pelagischen Copepoden und das thierische Leuchten im Allgemeinen." *Mittheilungen aus der Zoologischen Station zu Neapel* 11: 648–94.

– (1896). "Über den Sitz der Lichtentwicklung in den Photosphaerien der Euphausiden." *Zoologischer Anzeiger* 19: 486–90.

Giglioni, Enrico (1870). "La fosforescenza des mare." *Bollettino della Società geografica Italiana* 4: 105–28.

Gilchrist, John D.F. (1919). "Luminosity and its origin in a South African earthworm (*Chilota* sp.?)." *Transactions of the Royal Society of South Africa* 7: 203–12.

Girard, Maurice (1857). *F. Péron: Naturaliste, Voyageur aux Terres Australes*. Paris: J.B. Baillière et Fils.

Gissing, Judith (2003). *Rassenhygiene und Schule im Dritten Reich*. Dissertation/Habilitation, Westfälische Wilhelms-Universität, Münster. 155–7.

Glaser, Vicki (2007). "An interview with Keith V. Wood, Ph.D." *Assay and Drug Development Technologies* 5: 1–6.

Godeheu de Rouville (1760). "Mémoire sur la mer lumineuse." *Mémoires de l'Académie Royale des Sciences* 3: 269–76.

Gould, Carol Grant (2004). *The Remarkable Life of William Beebe: Explorer and Naturalist*. Washington, DC: Island Press.

Gouveneaux, Anaïd and Jérôme Mallefet (2013). "Physiological control of bioluminescence in a deep-sea planktonic worm, *Tomopteris helgolandica*." *Journal of Experimental Biology* 216: 4285–9.

Greeff, Richard (1882). "Über die rosettenförmigen Leuchtorgane der Tomopteriden und zwei neue Arten von Tomopteris." *Zoologischer Anzeiger* 5: 384–7.

– (1885). "Über die pelagische Fauna an den Küsten der Guinea-Inseln." *Zeitschrift für wissenschaftliche Zoologie* 42: 432–47.

Green, Arda A. and William D. McElroy (1956). "Crystalline firefly luciferase." *Biochimica Biophysica Acta* 20: 170–6.

Green, Louise F.B. (1979). "The fine structure of the light organ of the New Zealand glow-worm *Arachnocampa luminosa* (Diptera: Mycetophilidae)." *Tissue and Cell* 11: 457–65.

Greene, Charles W. (1899). "The phosphorescent organs in the toad-fish Porichthys notatus Girard." *Journal of Morphology* 15, 3: 667–96.

Greene, Charles W. and Harold H. Greene (1924). "Phosphorescence of Porichthys notatus, the California singing fish." *American Journal of Physiology* 70: 500–6.

Grimm, Herman Nicolas (1683). "Vermes rari lucentes." *Miscellanea Curiosa, sive Ephemiridum Medico-Physicarum Germanicarum Academiae Naturae Curiosorum* 2, 1: 406–7.

Grinnell, A.D., P.M. Narins, F.T. Awbrey, W.M. Hamner, and P.P. Hamner (1988). "Eye/photophore coordination and light-following in krill, *Euphausia superba*." *Journal of Experimental Biology* 134: 61–77.

Groeben, Christiane (1985). "Anton Dohrn – The statesman of Darwinism." *Biological Bulletin* 168 (supplement): 4–25.

– (2001). "La Stazione Zoologica 'Anton Dohrn.'" In *La cultura scientifica e le sue istituzioni: Napoli, 1860–1915*. Napoli: Paparo, Biblioteca universitaria di Napoli. 35–49.

– (2008). "Tourists in science: 19th century research trips to the Mediterranean." *Proceedings of the California Academy of Sciences* 59 (Supplement I): 139–54.

Gudger, Eugene W. (1934). "The five great naturalists of the sixteenth century: Belon, Rondelet, Salviani, Gesner and Aldrovandi: a chapter in the history of ichthyology." *Isis* 22(1): 21–40.

Günther, Albert (1887). *Report on the Deep-Sea Fishes Collected by H.M.S. Challenger During the Years 1873–76*. Edinburgh: Neill and Company.
Guppy, Henry B. (1882). "Note on the pearly organs of Scopelus." *The Annals and Magazine of Natural History*, 5th series, 9: 202–4.
Haddock, Steven H.D. and James F. Case (1994). "A bioluminescent chaetognath." *Nature* 367: 225–6.
– (1999). "Bioluminescence spectra of shallow and deep-sea gelatinous zooplankton: ctenophores, medusae and siphonophores." *Marine Biology* 133: 571–82.
Haddock, Steven H.D. and Casey W. Dunn (2015). "Fluorescent proteins function as a prey attractant: experimental evidence from the hydromedusa *Olindias formosus* and other marine organisms." *Biology Open* 4: 1094–1104.
Haddock, Steven H.D., Mark A. Moline, and James F. Case (2010). "Bioluminescence in the sea." *Annual Reviews of Marine Science* 2: 443–93.
Haddock, Steven H. D., Trevor J. Rivers, and Bruce H. Robison (2001). "Can coelenterates make coelenterazine? Dietary requirement for luciferin in cnidarian bioluminescence." *Proceedings of the National Academy of Sciences* 98: 11148–51.
Haddock, Steven H.D., Douglas J. Neilson, Edith A. Widder, and James F. Case (1998). "Feasibility of using in situ measurements of bioluminescence spectra to determine the vertical distribution of plankton." In A. C. Pierrot-Bults and S.van der Spoel, eds., *Pelagic Biogeography ICoPB II: Proceedings of the 2nd International Conference: Final Report of SCOR/IOC Working Group 93*. Paris, France: UNESCO. 137–42.
Haddock, Steven H.D., Casey W. Dunn, Philip R. Pugh, and Christine E. Schnitzler (2005). "Bioluminescent and red-fluorescent lures in a deep-sea siphonophore." *Science* 309: 263.
Haffer, Jürgen (2007). *Ornithology, Evolution, and Philosophy: The Life and Science of Ernst Mayr 1904–2005*. San Francisco: Ignatius Press.
Halverson, Roger C., James F. Case, and Darwin Tiemann (1973). "Control of luminescence in phengodid beetles." *Journal of Insect Physiology* 19: 1327–39.
Hamy, Ernest-Théodore (1894). *Vie et travaux de M. de Quatrefages*. Paris: F. Alcan.
Handrick, Kurt (1901). "Zur Kenntnis des Nervensystems und der Leuchtorgane des Argyropelecus hemigymnus." *Zoologica* (Stuttgart) 13, 32: 1–68.
Haneda, Yata (1938). "On the luminescence of the deep-sea fish *Malacocephalus laevis* (Lowe)." *Japanese Journal of Physiology* 3: 318–26.
– (1939a). "A few observations on the luminescent fungi of Micronesia." *Kagaku Nanyo* 1: 116–28 (in Japanese).
– (1939b). "Luminosity of *Rocellaria grandis* (Deshayes) (Lamellibranchia)." *Kagaku Nanyo* 2: 36–9 (in Japanese).
– (1940). "On the luminescence of the fishes belonging to the family Leiognathidae of the tropical Pacific." Palao: *Tropical Biological Station Studies* 2: 29–39.
– (1941). "Trip to Rabaul, New Britain (Observations on synchronous flashing of fireflies)." *Kagaku Nanyo* 3: 66–9 (in Japanese).
– (1942). "Further investigation upon luminescent fungi." *Kagaku Nanyo* 4: 49–63 (in Japanese).
– (1943). "On the photogenous organs of *Anomalops katoptron*, a luminous fish." *Kagaku Nanyo* 5: 81–8 (in Japanese).

– (1946). "A luminous land snail, *Dyakia striata*, found in Malaya." *Seibutsu* 1, 5/6: 294–8 (in Japanese).
– (1950a). "Star-worm in Singapore." *Shin Konchu* (New Insect) 3: 2–5 (in Japanese).
– (1950b). "Luminous organs of fish which emit light indirectly." *Pacific Science* 4: 214–27.
– (1951). "The luminescence of some deep-sea fishes of the families Gadidae and Macrouridae." *Pacific Science* 5: 372–8.
– (1952). "Some luminous fishes of the genera *Yarrella* and *Polyipnus*." *Pacific Science* 6: 13–15.
– (1953). "Observations on luminous organisms of the Miura Peninsula, Japan." *Yokosukashi-Shi* 5: 1–49 (in Japanese).
– (1954). "A luminous land snail (Planaxidae)" (in Japanese).
– (1955). "Luminous organisms of Japan and the Far East." In Frank H. Johnson, ed., *The Luminescence of Biological Systems*. Washington, DC: American Association for the Advancement of Science. 335–85.
– (1966). "Synchronous flashing of fireflies in New Guinea." *Science Report of the Yokosuka City Museum* 12: 1–8.
– (1985). *Hakko seibutsu (Luminous Organisms)*. Tokyo: Tokyo: Koseisha Koseikaku.
Haneda, Yata and Frank H. Johnson (1958). "The luciferin-luciferase reaction in a fish, *Parapriacanthus beryciformis*, of newly discovered luminescence." *Proceedings of the National Academy of Science* 44: 127–9.
– (1962). "The photogenic organs of *Parapriacanthus beryciformes* Franz and other fish with the indirect type of luminescent system." *Journal of Morphology* 110: 187–98.
Haneda, Yata and N. Kumagai (1939). "The luminous material of *Pontodrilus matsushimensis*, Izuka." *Japanese Journal of Physiology* 4: 328–33 (in Japanese).
Haneda, Yata and Frederick I. Tsuji (1971). "Light production in the luminous fishes *Photoblepharon* and *Anomalops* from the Banda Islands." *Science* 173: 143–5.
Haneda, Yata, Frank H. Johnson, and Osamu Shimomura (1966). "The origin of luciferin in the luminous ducts of *Parapriacanthus ransonneti*, *Pempheris klunzingeri*, and *Apogon ellioti*." In Frank H. Johnson and Yata Haneda, eds., *Bioluminescence in Progress*. 533–45
Haneda, Yata, Frank H. Johnson, and Edward H.-C. Sie (1958). "Luciferin and luciferase extracts of a fish, *Apogon marginatus*, and their luminescent cross-reactions with those of a crustacean, *Cypridina hilgendorfii*." *Biological Bulletin* 115: 336.
Haneda, Yata, Frederick I. Tsuji, and Noboru Sugiyama (1969). "Luminescent systems in apogonid fishes from the Philippines." *Science* 165: 188–90.
Hanna, Charles H., Thomas A. Hopkins, and John Buck (1976) "Peroxisomes of the firefly lantern." *Journal of Ultrastructure Research* 57: 150–62.
Hansen, Hans J. (1903). "On the crustaceans of the gernera *Petalidium* and *Sergestes* from the 'Challenger,' with an account of luminous organs in *Sergestes challengeri*, n. sp." *Proceedings of the Zoological Society of London* 1: 52–77.
Hansen, Kjeld and Peter J. Herring (1977). "Dual bioluminescent systems in the anglefish genus *Linophryne* (Pisces: Ceratioidea)." *Journal of Zoology* 182: 103–24.
Hanson, Frank E., James F. Case, Elizabeth Buck, and John Buck (1971). "Synchrony and flash entrainment in a New Guinea Firefly." *Science* 174: 161–4.

Harker, Allen (1887). "On a luminous oligochaete." *Report of the British Association for the Advancement of Science*, 57th meeting, 767.

Harvey, E. Newton (1913). "The temperature limits of phosphorescence of luminous bacteria." *Biochemical Bulletin* 2: 456–7.

– (1914). "On the chemical nature of the luminous material of the fire-fly." *Science* 40: 33–4.

– (1915). "Experiments on the nature of the photogenic substance in the fire-fly." *Journal of the American Chemical Society* 37: 396–401.

– (1916a). "The mechanism of light production in animals." *Science* 44: 208–9.

– (1916b). "Studies on bioluminescence. II. On the presence of luciferin in luminous bacteria." *American Journal of Physiology* 41: 449–53.

– (1916c). "The light-producing substances, photogenin and photophelein, of luminous animals." *Science* 44: 652–4.

– (1917a). "The chemistry of light-production in luminous organisms." *Publications of the Carnegie Institution of Washington* 251: 171–234.

– (1917b). "Studies on bioluminescence. IV. The chemistry of light production in a Japanese ostracod crustacean, Cypridina hilgendorfii, Müller." *American Journal of Physiology* 42: 318–41.

– (1918). "Studies on bioluminescence. VII. Reversibility of the photogenic reaction in Cypridina." *Journal of General Physiology* 1: 133–45.

– (1919a). "Studies on bioluminescence. IX. Chemical nature of Cypridina luciferin and Cypridina luciferase." *Journal of General Physiology* 1: 269–93.

– (1919b). "Studies on bioluminescence. X. Carbon dioxide production during luminescence of Cypridina luciferin." *Journal of General Physiology* 2: 133–6.

– (1919c). "Studies on bioluminescence XI. Heat production during luminescence of Cypridina luciferin." *Journal of General Physiology* 2: 137–45.

– (1920a). "Studies on bioluminescence XII. The action of acid and of light in the reduction of Cypridina oxyluciferin." *Journal of General Physiology* 2: 207–13.

– (1920b). *The Nature of Animal Light*. Philadelphia: J.B. Lippincott.

– (1921a). "A fish with a luminous organ designed for the growth of luminous bacteria." *Science* 53: 314–15.

– (1921b). "Studies on bioluminescence. XIII. Luminescence in the coelenterates." *Biological Bulletin* 41: 280–7.

– (1922a). "Studies on bioluminescence. XIV. The specificity of luciferin and luciferase." *Journal of General Physiology* 4: 285–95.

– (1922b). "The production of light by the fishes, Photoblepharon and Anomalops." *Publications of the Carnegie Institution of Washington* 312: 43–60.

– (1924a). "Studies on bioluminescence. XVI. What determines the color of the light of luminous animals?" *American Journal of Physiology* 70: 619–23.

– (1924b). "Recent advances in bioluminescence." *Physiological Review* 4: 639–71.

– (1925a). "Studies on bioluminescence. XVII. Fluorescence and inhibition of luminescence in ctenophores by ultraviolet light." *Journal of General Physiology* 7: 331–9.

– (1925b). "The inhibition of Cypridina luminescence by light." *Journal of General Physiology* 7: 679–85.

– (1927). "On the quanta of light produced and the molecules of oxygen utilized during Cypridina luminescence." *Journal of General Physiology* 10: 875–81.

– (1928). "The oxygen consumption of luminous bacteria." *Journal of General Physiology* 11: 469–75.
– (1931). "Stimulation by adrenalin of the luminescence of deep sea fish." *Zoologica*, NY, 12: 67–70.
– (1932). "The evolution of bioluminescence and its relation to cell respiration." *Proceedings of the American Philosophical Society* 71: 135–41.
– (1939). "Deep-sea photography." *Science* 90: 187.
– (1940). *Living Light*. Princeton: Princeton University Press.
– (1941). "Review of bioluminescence." *Annual Review of Biochemistry* 10: 531–52.
– (1944). "The nature of the red and green luminescence of the South American 'railroad worm,' Phryxothrix." *Journal of Cellular and Comparative Physiology* 23, 2: 31–8.
– (1952). *Bioluminescence*. New York: Academic Press.
– (1956). "Evolution and bioluminescence." *Quarterly Review of Biology* 31, 4: 270–87.
– (1957). *A History of Luminescence: From the Earliest Times Until 1900*. Philadelphia: American Philosophical Society.
Harvey, E. Newton and Irvin M. Korr (1928). "Luminescence in absence of oxygen in the ctenophore, Mnemiopsis leidyi." *Journal of Cellular and Comparative Physiology* 12: 319–23.
Harvey, E. Newton and Alfred L. Loomis (1928). "High Frequency Sound Waves of Small Intensity and their Biological Effects." *Nature* 121: 622–4.
– (1929). "The destruction of luminous bacteria by high frequency sound waves." *Journal of Bacteriology* 17: 373–6.
Harvey, E. Newton and Peter A. Snell (1931). "The analysis of bioluminescences of short duration, recorded with photoelectric cell and string galvanometer." *Journal of General Physiology* 14: 529–45.
Harvey, E. Newton and Kenneth P. Stevens (1928). "The brightness of the light of the West Indian elaterid beetle, Pyrophorus." *Journal of General Physiology* 12: 269–72.
Hastings, J. Woodland (1952). "Oxygen concentration and bioluminescence intensity. II. Cypridina hilgendorfii." *Journal of Cellular and Comparative Physiology* 40, 1: 1–9.
– (1971). "Light to hide by: ventral luminescence to camouflage the silhouette." *Science* 173: 1016–17.
– (1978). "The chemistry and biology of bacterial light emission." *Photochemistry and Photobiology* 27: 397–404.
– (1983). "Biological diversity, chemical mechanisms, and the evolutionary origins of bioluminescent systems." *Journal of Molecular Evolution* 19: 309–21.
– (1995). "Bioluminescence: similar chemistries but different evolutionary origins." *Photochemistry and Photobiology* 62: 599–600.
– (2001a). "Fifty years of fun." *Journal of Biological Rhythms* 16: 5–18.
– (2001b). "Cellular and molecular mechanisms of circadian regulation in the unicellular dinoflagellate *Gonyaulax polyedra*." In Joseph S. Takahashi, Fred W. Turek, and R.Y. Moore, eds., *Circadian Clocks*. New York: Kluwer Academic/Plenum Publishers. 321–34.

– (2004). "William David McElroy January 22, 1917–February 17, 1999." *Biographical Memoirs of the National Academy of Sciences* 85: 165–82.
Hastings, J. Woodland and George Mitchell (1971). "Endosymbiotic bioluminescent bacteria from the light organ of pony fish." *Biological Bulletin* 141: 261–8.
Hastings, J. Woodland and Beatrice M. Sweeney (1956). "The luminescence reaction in extracts of the marine dinoflagellate, *Gonyaulax polyedra*." *Journal of Cellular and Comparative Physiology* 49: 209–26.
– (1958). "A persistant diurnal rhythm of luminescence in Gonyaulax polyedra." *Biological Bulletin* 115: 440–58.
Haswell, William A. (1882). "On the structure and functions of the elytra of the aphroditacean annelids." *Annals and Magazine of Natural History*, 5th series, 10: 238–42.
Hayashi, Sotoo (1927). "Studies on the luminous organs of Watasenia scintillans (Berry)." *Folia Anatomica Japonica* 5: 417–27.
Hedgpeth, Joel W. (1946). "The voyage of the Challenger." *Scientific Monthly* 63, 3: 194–202.
Hedley, Charles (1919). "Henry Suter, 1841–1918." *Transactions and Proceedings of the Royal Society of New Zealand* 51: ix–x.
Heinemann, Carl (1872). "Untersuchungen über die Leuchtorgane der bei Vera-Cruz vorkommenden Leuchtkäfer." *Archiv für mikroskopische Anatomie* 8: 461–71.
– (1886). "Zur Anatomie und Physiologie der Leuchtorgane mexikanischer Cucuyo's." *Archiv für mikroskopische Anatomie* 27: 296–382.
Henderson, Thomas F. (1885) "Baker, Henry (1698–1774)." In Leslie Stephen, ed., *Dictionary of National Biography*, Vol. 3,. London: Elder Smith. 9–10.
Henry, Jean-Pierre and A. Michael Michelson (1978). "Bioluminescence: physiological control and regulation at the molecular level." *Photochemistry and Photobiology* 28: 293–310.
Henry, Jean-Pierre, C. Monny, and A. Michelson (1975). "Characterization and properties of *Pholas* luciferase as a metalloglycoprotein." *Biochemistry* 14: 3458–66.
Herdman, William A. (1921). "Edinburgh and the rise of oceanography." *Nature* 108: 308–11.
– (1923). *Founders of Oceanography and Their Work. An Introduction to the Science of the Sea.* London: Edward Arnold & Co.
Herren, Christen M., Steven H.D. Haddock, Cyril Johnson, Cristina M. Orrico, Mark A. Moline, and James F. Case (2005). "A multi-platform bathyphotometer for fine-scale, coastal bioluminescence research." *Limnology and Oceanography Methods* 3: 247–62.
Herrera, Albert A. (1979). "Electrophysiology of bioluminescent excitable epithelial cells in a polynoid polychaete worm." *Journal of Comparative Physiology* 129: 67–78.
Herring, Peter J. (1972). "Bioluminescence of searsid fishes." *Journal of the Marine Biological Association of the United Kingdom* 52: 879–87.
– (1974). "New observations on the bioluminescence of echinoderms." *Journal of Zoology* 172: 401–18.
– (1977). "Bioluminescence in an evermannellid fish." *Journal of Zoology* 181: 297–307.

– (1978). "Bioluminescence of invertebrates other than insects." In Peter J. Herring, ed., *Bioluminescence in Action*. 199–240.

– (1983). "The spectral characteristics of luminous marine organisms." *Proceedings of the Royal Society of London* B220: 183–217.

– (1987). "Systematic distribution of bioluminescence in living organisms." *Journal of Bioluminescence and Chemiluminescence* 1: 147–63.

– (1988). "Copepod luminescence." *Hydrobiologia* 167/168: 183–95.

– (2000). "Bioluminescent signals and the role of reflectors." *Journal of Optics, Part A: Pure and Applied Optics*, 2: R29–R38.

– (2002). *The Biology of the Deep Ocean*. Oxford, UK: Oxford University Press.

– (2007). "Sex with the lights on? A review of bioluminescent sexual dimorphism in the sea." *Journal of the Marine Biological Association of the United Kingdom* 87: 829–42.

Herring, Peter J., ed. (1978). *Bioluminescence in Action*. New York and London: Academic Press.

Herring, Peter J. and N.A. Locket (1978). "The luminescence and photophores of euphausiid crustaceans." *Journal of Zoology* 186: 431–62.

Herring, Peter J. and Julian C. Partridge (1992). "Light and life on RRS 'Discovery.'" *Journal of the Marine Biological Association of the United Kingdom* 72: 1–4.

Herring, Peter J. and M. Watson (1993). "Milky seas: a bioluminescent puzzle." *The Marine Observer* 63: 22–30.

Hess, Walter N. (1922). "Origin and development of the light-organs of Photurus pennsylvanica De Geer." *Journal of Morphology* 36: 245–77.

Hess, Wilhelm (1904). "Greeff, Richard." *Allgemeine Deutsche Biographie* 49: 523–4.

Hestmark, Geir (2009). "G.O. Sars, zoolog." *Norsk Biografisk Leksikon* (23 March 2009).

Hickling, Charles F. (1928). "The luminescence of the dogfish, Spinax niger, Cloquet." *Nature* 121: 280–1.

Hilgendorf, Franz Martin (1887). "Peters: Wilhelm Karl Hartwig."*Allgemeine Deutsche Biographie* 25: 489–93.

Holberg, Ludvig (1845). *Niels Klim's Journey Under the Ground* (translated by John Gierlow). Boston: Saxton, Pierce & Co.

Holmes, Richard (2009). *The Age of Wonder: How the Romantic Generation Discovered the Beauty and Terror of Science*. London: HarperCollins.

Hoyle, William E. (1902). "The luminous organs of Pterygioteuthis margaritifera, a Mediterranean cephalopod." *Memoirs and Proceedings of the Manchester Literary & Philosophical Society* 46: 2–14.

– (1904). "The luminous organs of certain Cephalopoda." *Bulletin of the Museum of Comparative Zoology at Harvard College* 43: 51–64.

Hudson, George V. (1886). "On New Zealand glowworms." *Transactions of the New Zealand Institute* 29: 62–4.

– (1898). "On seasonal time." *Transactions and Proceedings of the Royal Society of New Zealand* 31: 577–88.

– (1926). "Observations made on the New Zealand glow-worm (*Arachnocampa luminosa*) during 1926." *Annals and Magazine of Natural History*, 9th series, 18, 108: 667–70.

Hull, Gillian (2007). "John MacCullough, MD (1773–1835): A dedicated geologist." *Journal of Medical Biography* 15, 4: 235–40.

Huxley, Thomas (1854). "Professor Edward Forbes, F.R.S." *Journal of Science and Literary Gazette*, 1854: 1016–18.

Inaba, Kazuo (2015). "Japanese marine biological stations: Preface to the special issue." *Regional Studies in Marine Science* 2: 154–7.

Inoue, Shoji and Hisae Kakoi (1976). "*Oplophorus* luciferin, bioluminescent substance of the decapod shrimps, *Oplophorus spinosus* and *Heterocarpus laevigatus*." *Journal of the Chemical Society, Chemical Communications*, 1976: 1056–7.

Inoue, Shoji, Sumi Sugiura, Hisae Kakoi, and Kiyomatsu Hasizume (1975). "Squid bioluminescence. II. Isolation from *Watasenia scintillans* and synthesis of 2-(p-Hydroxybenzyl)-6-(p-Hydroxyphenyl)-3,7-Dihydroimidazo[1,2-a)-Pyrazin-3-One." *Chemistry Letters*, 1975: 141–4.

Izumi, Masanami (1992). "The Palao Tropical Biological Station and marine resource studies in Micronesia in the 1930s." *South Pacific Commission Fisheries Newsletter* 63: 26–8.

Jackson, J.W. (1926). "Obituary Notice: William Evans Hoyle." *Journal of Conchology* 18: 33–4.

Jin, K., J.C. Klima, G. Deane, Dale M. Stokes, and Michael I. Latz (2013). "Pharmacological investigation of the bioluminescence signaling pathway of the dinoflagellate *Lingulodinium polyedrum*: evidence for the role of stretch-activated ion channels." *Journal of Phycology* 49: 733–45.

Johann, Leopold (1899). "Über eigenthümliche epitheliale Gebilde (Leuchtorgane) bei Spinax niger." *Zeitschrift für wissenschaftliche Zoologie* 66: 136–60.

Johnson, Frank H. (1967). "Edmund Newton Harvey 1887–1959." *Biographical Memoirs of the National Academy of Sciences* 39: 193–266.

Johnson, Frank H. and E. Newton Harvey (1937). "The osmotic and surface properties of marine luminous bacteria." *Journal of Cellular and Comparative Physiology* 9: 363–80.

– (1938). "Bacterial luminescence, respiration and viability in relation to osmotic pressure and specific salts of sea water." *Journal of Cellular and Comparative Physiology* 11: 213–32.

Johnson, Frank H. and Henry Eyring (1948). "The fundamental action of pressure, temperature and drugs on enzymes as revealed by bacterial luminescence." *Annals of the New York Academy of Sciences* 49: 376–96.

Johnson, Frank H., Henry Eyring, and Milton J. Polissar (1954). *The Kinetic Basis of Molecular Biology*. New York: John Wiley & Sons.

Johnson, Frank. H. and Yata Haneda, eds., (1966). *Bioluminescence in Progress*. Princeton: Princeton University Press.

Johnson, Frank H., Yata Haneda, and Edward H.-C. Sie (1960). "An interphylum luciferin-luciferase reaction." *Science* 132: 422–3.

Johnson, Frank H., Osamu Shimomura, Yo Saiga, Lewis C. Gershman, George T. Reynolds, and John R. Waters (1962). "Quantum efficiency of *Cypridina* luminescence, with a note on that of *Aequorea*." *Journal of Cellular and Comparative Physiology* 60: 85–103.

Johnson, G. David and Richard H. Rosenblatt (1988). "Mechanisms of light organ occlusion in flashlight fishes, family Anomalopidae (Teleostei: Beryciformes), and the evolution of the group." *Zoological Journal of the Linnean Society 94*: 65–96.

Johnston, George (1847). *A History of the British Zoophytes*. London: J. van Voorst.

Johnston, T.C., R.B. Thompson, and T.O. Baldwin (1986). "Nucleotide sequence of the luxB gene of *Vibrio harveyi* and the complete amino acid sequence of the beta subunit of bacterial luciferase." *Journal of Biological Chemistry* 261: 4805–11.

Jordan, Maria (1987). "Skowron Stanislaw Teofil (1900–1976)." *Dictionary of Polish Biologists (Słownik biologów polskich)*, 490–1. Warszawa (Warsaw): Pa stwowe Wydawnictwo Naukowe.

Joubin, Louis (1893a). "Recherches sur l'appareil lumineux d'un céphalopode: *Histioteuthis ruppellii*, Verany." *Bulletin de la Société scientifique et médicale de l'Ouest* 2: 49–78.

– (1893b). "Note sur une adaptation particulière des chromatophores chez un céphalopode." *Bulletin de la Société zoologique de France* 18: 146–51.

– (1893c). "Quelques organes colorés de la peau chez deux céphalopodes du genre *Chiroteuthis*. *Mémoires de la Societe zoologique de France* 6: 331–43.

– (1895). *Résultats des campagnes scientifiques accomplies sur son yacht par Albert 1er, Prince de Monaco. Fascicule IX: Contribution à l'étude des Céphalopodes de l'Atlantique Nord*. Monaco: Imprimerie de Monaco.

– (1924). *Résultats des campagnes scientifiques accomplies sur son yacht par Albert 1er, Prince de Monaco. Fascicule LXVII: Contribution à l'étude des Céphalopodes de l'Atlantique Nord (4ème série)*. Monaco: Imprimerie de Monaco.

Jourdan, Étienne (1885). "Structure des élytres de quelques Polynoës." *Zoologischer Anzeiger* 8: 128–34.

Julin, Charles (1909). "Les embryons de Pyrosoma sont phosphorescents: les cellules du testa (calymnocytes de Salensky) constituent les organes lumineux du cyathozoïde." *Comptes rendus des séances de la Société de biologie* 66, 1: 80–2.

Kampa, Elizabeth M. and Brian P. Boden (1953). "Light generation in a sonic-scattering layer." *Deep Sea Research* 4: 73-92.

Kanda, Sakyo (1910). "The religious development of a modern Japanese." *The Open Court* 1910, 12: 753–61.

– (1920a). "Physico-chemical studies on bioluminescence. I. On the luciférine and luciférase of Cypridina hilgendorfii." *American Journal of Physiology* 50: 544–60.

– (1920b). "Physico-chemical studies on bioluminescence. II. The production of light by Cypridina hilgendorfii is not an oxidation." *American Journal of Physiology* 50: 561–73.

– (1920c). "Physico-chemical studies on bioluminescence. III. The production of light by Luciola vitticollis is an oxidation." *American Journal of Physiology* 53: 137–49.

– (1924). "Physico-chemical studies on bioluminescence. V. The physical and chemical nature of the luciferine of Cypridina hilgendorfii." *American Journal of Physiology* 68: 435–43.

Kastle, Joseph H. and F. Alex McDermott (1910). "Some observations on the production of light by the firefly." *American Journal of Physiology* 27: 122–51.

Kay, R.H. (1965). "Light-stimulated and light-inhibited bioluminescence of the euphausiid *Meganyctiphanes norvegica* (G.O. Sars)." *Proceedings of the Royal Society of London* B162: 365–86.

Kemp, Stanley (1910). "Notes on photophores of decapod Crustacea." *Proceedings of the Zoological Society of London* 1910: 639–51.

Kingsley, John S. (1879) "Scientific News." *American Naturalist* 13, 9: 596–602.

Kingston, Ralph (2007). "A not so Pacific voyage: the "floating laboratory" of Nicolas Baudin." *Endeavour* 31, 4: 145–51.

Kircher, Anasthasius (1671). *Ars magna lucis et umbrae.* Amsderdam: Joannes Jansson van Waesberghe.

Kishi, Yoshito, Toshio Goto, Yoshimasa Hirata, Osamu Shimomura, and Frank H. Johnson (1966). "The structure of *Cypridina* luciferin." In Frank H. Johnson and Yata Haneda, eds., *Bioluminescence in Progress.* 89–113.

Kishitani, Teijiro (1928a). "Ueber das Leuchtorgan von *Euprymna morsei* Verrill and die symbiotischen Leuchtbakterien." *Proceedings of the Imperial Academy of Japan* 4: 306–9.

– (1928b). "Preliminary report on the luminous symbiosis in *Sepiola birostrata* Sasaki." *Proceedings of the Imperial Academy of Japan* 4: 393–6.

– (1928c). "L'étude de l'organe photogène du *Loligo edulis* Hoyle. (Notes préliminaires)." *Proceedings of the Imperial Academy of Japan* 4: 609–12.

– (1932). "Studien über Leuchtsymbiose von japanischen Sepien." *Folia Anatomica Japonica* 10, 3: 315–418.

Kleinschmidt, Adolf (1955). "Brandes, Gustav Philipp Hermann." *Neue Deutsche Biographie* 2: 520.

Klewitz, Andreas von (2013). *Carl Chun, die Valdivia und die Entdeckung der Tiefsee.* Berlin: Parthas Verlag.

Kohler, Robert E. (1982). *From Medical Chemistry to Biochemistry: The Making of a Biomedical Discipline.* Cambridge, UK: Cambridge University Press.

Komárek, Julius (1934). "Luminescence of Casparthian worms and its causes." *Bulletin international de l'Académie des Sciences de Bohême* 35: 63–4.

Kraemer, Fabian and Helmut Zedelmaier (2014). "Instruments of invention in Renaissance Europe: The cases of Conrad Gesner and Ulisse Aldrovandi." *Intellectual History Review* 24: 321–41.

Krebs, Hans and Fritz Lipmann (1988). "Dahlem in the later nineteen twenties." In Horst Kleinhauf and Lothat Jaenicke, eds., *The Roots of Modern Biochemistry: Fritz Lipmann's Squiggle and its Consequences.* Berlin and New York: Walter de Gruyter. 111–24.

Krekel, Anna (1921). "Die Leuchtorgane von Chaetopterus variopedatus Clap." *Zeitschrift für wissenschaftliche Zoologie* 118: 480–509.

Kritsky, Gene and Ron Cherry (2000). *Insect Mythology.* Bloomington, Indiana: iUniverse.

Krönström, Jenny, Susanne Holmgren, Fernand Baguet, Lorenza Salpietro, and Jérôme Mallefet (2005). "Nitric oxide in control of luminescence from hatchetfish (*Argyropelecus hemigymnus*) photophores." *Journal of Experimental Biology* 208: 2951–61.

Krönström, Jenny, Samuel Dupont, Jérôme Mallefet, Michael Thorndyke, and Susanne Holmgren (2007). "Serotonin and nitric oxide interaction in the control of bioluminescence in northern krill, *Meganyctiphanes norvegica* (M. Sars)." *Journal of Experimental Biology* 209: 3179–87.

Krönström, Jenny, Waldemar Karlsson, Bengt R. Johansson, and Susanne Holmgren (2009). "Involvement of contractile elements in control of bioluminescence in Northern krill, *Meganyctiphanes norvegica* (M. Sars)." *Cell and Tissue Research* 336: 299–308.

Krusenstern, Adam Johann von (1813). *Voyage Round the World in the Years 1803, 1804, 1805 & 1806*. London: John Murray.

Langley, Samuel P. and F.W. Very (1890). "On the cheapest form of light, from studies at the Allegheny Observatory." *American Journal of Science*, 3rd Series, 40: 97–113.

Lapota, David (2012). "Seasonal changes of bioluminescence in photosynthetic and heterotrophic dinoflagellates at San Clemente Island." In David Lapota, ed., *Bioluminescence – Recent Advances in Oceanic Measurements and Laboratory Applications*. Croatia: InTech. 29–46.

Lapota, David, Jon R. Losee, and Mark L. Geiger (1986). "Bioluminescence displays induced by pulsed light." *Limnology and Oceanography* 31: 886–9.

Lapota, David, Charles Galt, Jon R. Losee, Howard D. Huddell, James K. Orzech, and Kenneth H. Nealson (1988). "Observations and measurements of planktonic bioluminescence in and around a milky sea." *Journal of Experimental Marine Biology and Ecology* 119: 55–81.

Latz, M.I., A.R. Juhl, A.M. Ahmed, S.E. Elghobashi, and J. Rohr (2004). "Hydrodynamic stimulation of dinoflagellate bioluminescence: a computational and experimental study." *Journal of Experimental Biology* 207: 1941–51.

Latz, Michael I. and James F. Case (1982). "Light organ and eyestalk compensation to body tilt in the luminescent midwater shrimp, *Sergestes similis*." *Journal of Experimental Biology* 98: 83–104.

Latz, Michael I., Tamara M. Frank, Mark R. Bowlby, Edith A. Widder, and James F. Case (1987a). "Variability in flash characteristics of a bioluminescent copepod." *Biological Bulletin* 173: 489–503.

Latz, Michael I., Tamara M. Frank, James F. Case, Elijah Swift, and Robert R. Bidigare (1987b). "Bioluminescence of colonial radiolaria in the western Sargasso Sea." *Journal of Experimental Marine Biology and Ecology* 109: 25–38.

Latz, Michael I., Tamara M. Frank, and James F. Case (1988). "Spectral composition of bioluminescence of epipelagic organisms from the Sargasso Sea." *Marine Biology* 98: 441–6.

Latz, Michael I., James F. Case, and R.L. Gran (1994). "Excitation of bioluminescence by laminar fluid shear associated with simple Couette flow." *Limnology and Oceanography* 39: 1424–39.

Lawry, James V. Jr (1974). "Lantern fish compare downwelling light and bioluminescence." *Nature* 247: 155–7.

Legocki, Roman P., Misuk Legocki, Thomas O. Baldwin, and Aladar A. Szalay (1986). "Bioluminescence in soybean root nodules: demonstration of a general approach to assay gene expression in vivo by using bacterial luciferase." *Proceedings of the National Academy of Science USA* 83: 9080–4.

Leitch, Alexander (2015). *A Princeton Companion*. Princeton: Princeton University Press.

Lendenfeld, Robert von (1887). "Report on the structure of the phosphorescent organs

of fishes." In Albert Günther, *Report on the Deep-Sea Fishes Collected by H.M.S. Challenger During the Years 1873–76*. Edinburgh: Neill and Company. 277–329.
– (1905). "The radiating organs of the deep sea fishes." *Memoirs of the Museum of Comparative Zoology at Harvard College* 30, 2: 169–208.
Lenhoff, Howard M. (1991). "Ethel Browne, Hans Spemann, and the discovery of the organizer phenomenon." *Biological Bulletin* 181: 72–80.
Leuckart, Rudolf (1865). "Ueber mutmassliche Nebenaugen bei einem Fisch, Chauliodus Sloani." *Amtlicher Bericht über die Versammlung Deutscher Naturforscher und Aerzte* 39: 153–5.
Lewis, Sara M. (2016). *Silent Sparks: The Wondrous World of Fireflies*. Princeton and Oxford: Princeton University Press.
Lewis, Sara M. and Christopher K. Cratsley (2008). "Flash signal evolution, mate choice, and predation in fireflies." *Annual Review of Entomology* 53: 293–321.
Leydig, Franz (1881). *Die augenähnlichen Organe der Fische*. Bonn: Verlag von Emil Strauss.
Li, Hou-Feng, Nan-Yao Su, and Wen-Jer Wu (2010). "Solving the hundred-year controversy of Coptotermes taxonomy in Taiwan." *American Entomologist* 56, 4: 222–7.
Lieberman, Stephen H., David Lapota, Jon R. Losee, and Alberto Zirino (1987). "Planktonic bioluminescence in the surface waters of the Gulf of California." *Biological Oceanography* 4: 25–46.
Liebersohn, Harry (2009). *The Travelers' World: Europe to the Pacific*. Cambridge, MA: Harvard University Press.
Linklater, Eric (1972). *The Voyage of the Challenger*. London: John Murray.
Lloyd, James E. (1965). "Aggressive mimicry in Photuris: firefly femmes fatales." *Science* 149: 653–4.
– (1966). "Studies on the flash communication system in *Photinus* fireflies." *Miscellaneous Publications of the Museum of Zoology, University of Michigan* 130: 1–95.
– (1973a). "Model for the mating protocol of synchronously flashing fireflies." *Nature* 245: 268–70.
– (1973b). "Fireflies of Melanesia: bioluminescence, mating behavior, and synchronous flashing (Coleoptera: Lampyridae)." *Environmental Entomology* 2: 991–1008.
Lohmann, Hans T. (1899) "Das Gehäuse der Appendicularien, sein Bau, seine Funktion und Entstehung." *Schriften des Naturwissenschaftlichen Vereins für Schleswig-Holstein* 11: 348–406.
Long, Tulley (2009). "William McElroy, the McCollum-Pratt Institute, and the transformation of biology at Johns Hopkins, 1945–1960." *Journal of the History of Biology* 42: 765–809.
Lurie, Edward (1960). *Louis Agassiz: A Life in Science*. Chicago: University of Chicago Press.
Macaire, Jean-François (1821). "Mémoire sur la phosphorescence des lampyres." *Annales de chimie et de physique* 17: 251–67.
MacAlister, Alexander (1900). *James Macartney: A Memoir*. London: Hodder and Stoughton.
Macartney, James (1810). "Observations upon luminous animals." *Philosophical Transactions of the Royal Society of London* 100: 258–93.

MacCullough, John (1819). "Remarks on luminous marine animals." *Quarterly Journal of Science, Literature and the Arts* 11: 248–60.
Mackie, George O. and Quentin Bone (1978). "Luminescence and associated effector activity in Pyrosoma (Tunicata: Pyrosomida)." *Proceedings of the Royal Society of London* B202: 483–95.
Mackie, George O. and G.V. Mackie (1963)."Systematics and biological notes on living hydromedusae from Puget Sound." *Contributions in Zoology of the National Museum of Canada*, 199: 63–84.
Maddock, Linda (2006). "Joseph Arthur Colin Nicol. 5 December 1915–20 December 2004." *Biographical Memoirs of Fellows of the Royal Society* 52: 203–18.
Maher, Charles-Adolphe (1869). *Eloge de J.R.C. Quoy*. Rochefort: Imprimerie Ch. Thèze. 28 pages.
Mallefet, Jérôme (2009). "Echinoderm bioluminescence: where, how and why do so many ophiuroids glow?" In Victor B. Meyer-Rochow, ed., *Bioluminescence in Focus: A Collection of Illuminating Essays*. Kerala, India: Research Signpost. 67–81.
Mallefet, Jérôme and Michel Anctil (1992). "Inmunohistochemical detection of biogenic amines in the photophores of the midshipman fish *Porichthys notatus*." *Canadian Journal of Zoology* 70: 1968–75.
Mallefet, Jérôme and Fernand Baguet (1984). "Oxygen consumption and luminescence of *Porichthys* photophores stimulated by potassium cyanide." *Journal of Experimental Biology* 109: 341–52.
Mallefet, Jérôme and Fernand Baguet (1985). "Effects of adrenaline on the oxygen consumption and luminescence of the photophores of the mesopelagic fish *Argyropelecus hemigymnus*." *Journal of Experimental Biology* 118: 341–9.
Mallefet, Jérôme and Osamu Shimomura (1995). "Presence of coelenterazine in mesopelagic fishes from the Strait of Messina." *Marine Biology* 124: 381–5.
Mallefet, Jérôme, P. Vanhoutte, and Fernand Baguet (1992). "Study of *Amphipholis squamata* luminescence." In L. Scalera-Liaci and C. Canicatti, eds., *Echinoderm Research 1991*. Rotterdam: A.A. Balkema. 125–30.
Mangold, Ernst (1907). "Ueber das Leuchten der Tiefseefische." *Pflügers Archiv* 119: 583–601.
– (1908). "Ueber das Leuchten und Klettern der Schlangensterne." *Biologische Zentralblatt* 28: 169–76.
– (1910). "Die Produktion von Licht." In Hans Winterstein, ed., *Handbuch der vergleichenden Physiologie*, Vol. 3. Jena: Gustav Fischer. 225–392.
Marshall, Norman B. (1960). "Swimbladder structure of deep-sea fishes in relation to their systematics and biology." *Discovery Reports* 31: 1–121.
Martin, Normand and Michel Anctil (1984). "Luminescence control in the tubeworm *Chaetopterus variopedatus*: role of nerve cord and photogenic gland." *Biological Bulletin* 166: 583–93.
Mason, Howard S. and E.F. Davis (1952). "Cypridina luciferin. Partition chromatography." *Journal of Biological Chemistry* 197: 41–5.
Matsen, Bradford (2005). *Descent: The Heroic Discovery of the Abyss*. New York: Pantheon Books.
Matteucci, Carlo (1843). "Lettre à M. Dumas sur la phosphorescence des vers luisants." *Annales de chimie et de physique* 9 (série 3): 71–89.

– (1844). *Traité des phénomènes électro-physiologiques des animaux*. Paris: Fortin, Masson & Cie.

Mazliak, Paul (2006). *La biologie au siècle des Lumières*. Paris: Vuibert – Adapt Editions.

McAllister, Donald E. (1961). "A collection of oceanic fishes from off British Columbia with a discussion of the evolution of black peritoneum." *Bulletin of the National Museum of Canada* 172: 39–43.

McCullough, David (2015). *The Wright Brothers*. New York: Simon & Schuster.

McDermott, F. Alex and Charles G. Crane (1911). "A comparative study of the structure of the photogenic organs of certain American Lampyridae." *The American Naturalist* 45: 306–13.

McElroy, William D. (1944). "The effect of narcotics and other inhibitors on the oxidation and assimilation of glucose by the luminous bacterium Achromobacter fischeri." *Journal of Cellular and Comparative Physiology* 23, 3: 171–92.

– (1947). "The energy source for bioluminescence in an isolated system." *Proceedings of the National Academy of Sciences* 33: 342–3.

– (1964). "Insect bioluminescence." In Morris Rockstein, ed., *The Physiology of Insecta*, Vol. 1. New York: Academic Press. 463–508.

– (1976). "From the precise to the ambiguous: light, bonding, and administration." *Annual Review of Microbiology* 30: 1–20.

McElroy, William D. and Bernard H. Strehler (1949). "Factors influencing the response of the bioluminescent reaction to adenosine triphosphate." *Archives of Biochemistry* 22: 420–33.

McElroy, William D. and Howard H. Seliger (1962). "Origin and evolution of bioluminescence." In Michael Kasha and Bernard Pullman, eds., *Horizons in Biochemistry*. New York and London: Academic Press. 91–101.

– (1966). "Firefly bioluminescence." In Frank H. Johnson and Yata Haneda, eds., *Bioluminescence in Progress*. 427–58.

McFall-Ngai, Margaret J. (1983). "Adaptations for reflection of bioluminescent light in the gas bladder of *Leiognathus equulus* (Perciformes: Leiognathidae)." *Journal of Experimental Zoology* 227: 23–33.

McFall-Ngai, Margaret J. and Paul V. Dunlap (1984). "External and internal sexual dimorphism in leiognathid fishes: morphological evidence for sex-specific bioluminescent signaling." *Journal of Morphology* 182: 71–83.

McFall-Ngai, Margaret and Mary K. Montgomery (1990). "The anatomy and morphology of the adult bacterial light organ of *Euprymna scolopes* Berry (Cephalopoda: Sepiolidae)." *Biological Bulletin* 179: 332–9.

McFall-Ngai, Margaret and Edward G. Ruby (1991). "Symbiont recognition and subsequent morphogenesis as early events in an animal-bacterial mutualism." *Science* 254: 1491–4.

McMeekin, Sean (2010). *The Berlin-Bagdad Express: The Ottoman Empire and Germany's Bid for World Power*. Cambridge, MA: Harvard University Press.

M'Intosh, William C. (1872). "On the abyssal theory of light, the protozoic-absorption theory, and the azoic-mud theory." *The Annals and Magazine of Natural History*, 4th series, 9: 1–13.

– (1915). "Obituary notices of Fellows deseased: A.C.L.G. Günther." *Proceedings of the Royal Society of London, Section B*, 88: xi–xxvi.

Menzel, Randolf (2004). In Larry A. Squire, ed., *The History of Neuroscience as Autobiography*, Vol. 4. San Diego and London: Elsevier Academic Press. 455–7.
Merian, Maria Sybilla (1705). *Metamorphosis Insectorium Surinamensium. Ofte Verandering der Surinaamsche Insecten*. Amsterdam: Gerard Valck.
Merrett, Nigel R., Julian Badcock, and Peter J. Herring (1971). "Observations on bioluminescence in a scopelarchid fish, *Benthalbella*." *Deep Sea Research* 18: 1265–7.
– (1973). "The status of *Benthalbella infans* (Pisces: Myctophoidei), its development, bioluminescence, general biology and distribution in the eastern North Atlantic." *Journal of Zoology, London*, 170: 1–48.
Merriman, Daniel (1972). "Challengers of Neptune: the 'Philosophers.'" *Proceedings of the Royal Society of Edinburgh. Section B. Biology*, 72: 15–45.
Mertens, Robert (1957). "Carl Chun." *Neue Deutsche Biographie* 3: 252.
Meyen, Franz Julius Ferdinand (1834). "Über das Leuchten des Meeres und Beschreibung einiger Polypen und anderer niederer Thiere." *Nova acta physico-medica Academiae Caesareae Leopoldino-Carolinae* 8 (Supplement): 125–216.
Meyer, Werner T. (1906). "Über das Leuchtorgan der Sepiolini." *Zoologischer Anzeiger* 30: 388–92.
– (1908). "Über das Leuchtorgan der Sepiolini: II. Das Leuchtorgan von Heteroteuthis." *Zoologischer Anzeiger* 32: 505–8.
Meyer-Rochow, Victor B. (1976a). "Some observations on spawning and fecundity in the luminescent fish *Photoblepharon palpebratus*." *Marine Biology* 37: 325–8.
– (1976b). "Loss of bioluminescence in *Anomalops katoptron* due to starvation." *Experientia* 32: 1175–6.
– (2007). "Glowworms: a review of *Arachnocampa* spp. and kin." *Luminescence* 22: 251–65.
– (2009). "Bioluminescence through the ages: a brief, but enlightening survey." In Victor B. Meyer-Rochow, ed., *Bioluminescence in Focus – A Collection of Illuminating Essays*. Kerala, India: Research Signpost. 1–8.
– (2016). "Eyes of two keroplatid dipterans: the luminescent *Arachnocampa luminosa* and the non–luminescent *Neoditomyia farri* plus comments on luminescent and non-luminescent beetles and gastropods." *Entomologie heute* 28: 117–26.
Meyer-Rochow, Victor B. and Stephen Moore (1988). "Biology of *Latia neritoides* GRAY 1850 (Gastropoda, Pulmonata, Basomniatophora): the only light-producing freshwater snail in the World." *Internationale Revue der gesamten Hydrobiologie* 73: 21–42.
– (2009). "Hitherto unreported aspects of the ecology and anatomy of a unique gastropod: the bioluminescent freshwater pulmonate *Latia neritoides*." In Victor B. Meyer-Rochow, *Bioluminescence in Focus – A Collection of Illuminating Essays*. Kerala, India: Research Signpost. 85–104.
Miller, David (1952). "Sir William Blaxland Benham, K.B.E., F.R.S., (1860–1950)." *Transactions and Proceedings of the Royal Society of New Zealand* 80: 103–12.
Miller, Robert (1984). *An Interview with Frank H. Johnson*. Salt Lake City: University of Utah Marriott Library (Everett L. Cooley Oral History Project).
Miller, Steven D., Steven H.D. Haddock, Christopher D. Elvidge, and Thomas F. Lee (2005). "Detection of a bioluminescent milky sea from space." *Proceedings of the National Academy of Science* 102: 14181–4.

– (2006). "Twenty thousand leagues over the seas: the first satellite perspective on bioluminescent 'milky seas.'" *International Journal of Remote Sensing* 27: 5131–43.

Mills, Eric L. (1980). "Alexander Agassiz, Carl Chun and the problem of the intermediate fauna." In Mary Sears and Daniel Merriman, eds., *Oceanography: The Past.* New York, Springer-Verlag. 360–72.

Miron, Marie-José, Luc LaRivière, Jean-Marie Bassot, and Michel Anctil (1987). "Immunohistochemical and radioautographic evidence of monoamine-containing cells in bioluminescent elytra of the scale-worm *Harmothoe imbricata* (Polychaeta)." *Cell and Tissue Research* 249: 547–56.

Miyamori, Asatarō, ed. (1932). *Classic Haiku: An Anthology of Poems by Bashō and His Followers.* Tokyo: Maruzen Company, Ltd.

Mizuno, Hiromi (2009). *Science for the Empire: Scientific Nationalism in Modern Japan.* Stanford, California: Stanford University Press.

Mliner, Josef (2010). "200 Jahre Firma Zugmayer." *Waldegger Chronik* 4: 13–14.

Monod, Théodore (1954). "Sur un premier essai d'utilisation scientifique du bathyscaphe F.N.R.S. 3." *Comptes rendus des séances de l'Académie des sciences* 238: 1951–3.

– (1991). *Plongées profondes: Bathyfolages.* Paris: Actes Sud.

Montgomery, Mary K. and Margaret McFall-Ngai (1993). "Embryonic development of the light organ of the sepiolid squid *Euprymna scolopes* Berry." *Biological Bulletin* 184: 296–308.

Moore, Arthur R. (1924). "Luminescence in Mnemiopsis." *Journal of General Physiology* 6, 4: 403–12.

Moore, Hilary B. (1950). "The relation between the scattering layer and the euphausiid." *Biological Bulletin* 99: 181–212.

Morange, Michel (1998). *A History of Molecular Biology.* Cambridge, MA: Harvard University Press.

Morgese, Francesca and Viviana Vinci (2010). *Performascienza. Laboratori teatrali di storia della scienza a scuola.* Milano: FrancoAngeli.

Morin, James G. (1974). "Coelenterate bioluminescence." In Leonard Muscatine and Howard M. Lenhoff, eds., *Coelenterate Biology: Reviews and Perspectives.* New York: Academic Press. 397–438.

Morin, James F. (1983). "Coastal bioluminescence: patterns and functions." *Bulletin of Marine Science* 33: 787–817.

– (1986). "Firefleas of the sea: luminescent signaling in marine ostracode crustaceans." *Florida Entomologist* 69: 105–20.

Morin, James G. and Anne C. Cohen (1991). "Bioluminescent displays, courtship, and reproduction in ostracodes." In R. Bauer and J. Martin, eds., *Crustacean Sexual Biology.* New York: Columbia University Press. 1–16.

– (2010). "It's all about sex: bioluminescent courtship displays, morphological variation and sexual selection in two new genera of Caribbean ostracodes." *Journal of Crustacean Biology* 30: 56–67.

Morin, James G. and Ian M. Cooke (1971a). "Behavioural physiology of the colonial hydroid *Obelia.* II. Stimulus-initiated electrical activity and bioluminescence." *Journal of Experimental Biology* 54: 707–21.

– (1971b). "Behavioural physiology of the colonial hydroid *Obelia.* III. Characteristics of the bioluminescent system." *Journal of Experimental Biology* 54: 723–35.

Morin, James G. and J. Woodland Hastings (1971a). "Biochemistry of the bioluminescence of colonial hydroids and other coelenterates." *Journal of Cell Physiology* 77: 305–12.

– (1971b). "Energy transfer in a bioluminescent system." *Journal of Cell Physiology* 77: 313–18.

Morin, James G. and George T. Reynolds (1974). "The cellular origin of bioluminescence in the colonial hydroid *Obelia*." *Biological Bulletin* 147: 397–410.

Morin, James G., Anne Harrington, Kenneth Nealson, Neil Krieger, Thomas O. Baldwin, and J. Woodland Hastings (1975). "Light for all reasons: versatility in the behavioral repertoire of the flashlight fish." *Science* 190: 74–6.

Morrison, Thomas F. (1929). "Observations on the synchronous flashing of fireflies in Siam." *Science* 69: 400–1.

Moseley, Henry N. (1879). *Notes of a Naturalist on the "Challenger."* London: Macmillan and Co.

Müller, Gustav W. (1891). "Neue Cypridiniden." *Zoologische Jahrbücher* 5: 211–52.

Murray, John (1821). "On the luminosity of the sea." *Memoirs of the Wernerian Natural History Society of Edinburgh* 3: 466–71.

Murray, Sir John (1885). *Report on the Scientific Results of the Voyage of H.M.S. Challenger During the Years 1873–1876.* Narrative in two volumes. London: Longmans & Co.

– (1899). "Address to the geographical section of the British association, 1899." *Scottish Geographical Magazine* 15, 10: 505–22.

Murray, Sir John and Johan Hjort (1912). *The Depths of the Ocean.* London: Macmillan and Co.

Nakamura, Hideshi, Yoshito Kishi, Osamu Shimomura, David Morse, and J. Woodland Hastings (1989). "Structure of dinoflagellate luciferin and its enzymatic and nonenzymatic air-oxidation products." *Journal of the American Chemical Society* 111: 7607–11.

Nansen, Fridtjof (1887). *The Structure and Combination of the Histological Elements of the Central Nervous System.* Bergen: John Grieg.

Naranjo, David, Charles Plant, Kathleen Dunlap, and Paul Brehm (1994). "Two subcellular mechanisms underlie calcium-dependent facilitation of bioluminescence." *Neuron* 13: 1293–1301.

Naumann, Eva A., Adam R. Kampff, David A. Prober, Alexander F. Schier, and Florian Engert (2010). "Monitoring neural activity with bioluminescence during natural behavior." *Nature Neuroscience* 13: 513–20.

Nealson, Kenneth H. (1977). "Autoinduction of bacterial luciferase: occurrence, mechanism and significance." *Archives of Microbiology* 112: 73–9.

Nealson, Kenneth H. and J. Woodland Hastings (1979). "Bacterial bioluminescence: its control and ecological significance." *Microbiological Reviews* 43: 496–518.

– (2006). "Quorum sensing on a global scale: massive numbers of bioluminescent bacteria make milky seas." *Applied and Environmental Microbiology* 72: 2295–7.

Nealson, Kenneth H., Terry Platt, and J. Woodland Hastings (1970). "Cellular control of the synthesis and activity of the bacterial luminescent system." *Journal of Bacteriology* 104: 313–22.

Neumann, Dirk (2006). "Type catalogue of the ichthyological collection of the Zo-

ologische Staatssammlung München. Part I: Historic type material from the 'Old Collection' destroyed in the night 24/25 April 1944." *Spixiana* 29, 3: 259–85.
Nicol, J.A. Colin (1945). "The homing ability of the carrier pigeon: its value in warfare." *Auk* 62: 286–98.
– (1952a). "Studies on *Chaetopterus variopedatus* (Renier). I. The light-producing glands." *Journal of the Marine Biological Association of the United Kingdom* 30: 417–30.
– (1952b). "Studies on *Chaetopterus variopedatus* (Renier). II. Nervous control of light production." *Journal of the Marine Biological Association of the United Kingdom* 30: 433–52.
– (1952c). "Studies on *Chaetopterus variopedatus* (Renier). III. Factors affecting the light response." *Journal of the Marine Biological Association of the United Kingdom* 31: 113–44.
– (1953). "Luminescence in polynoid worms." *Journal of the Marine Biological Association of the United Kingdom* 32: 65–84.
– (1954a). "Effect of external milieu on luminescence in *Chaetopterus variopedatus*." *Journal of the Marine Biological Association of the United Kingdom* 33: 173–5.
– (1954b). "Fatigue of the luminescent response of *Chaetopterus*." *Journal of the Marine Biological Association of the United Kingdom* 33: 177–86.
– (1954c). "The nervous control of luminescent responses in polynoid worms." *Journal of the Marine Biological Association of the United Kingdom* 33: 225–55.
– (1955a). "Observations on luminescence in *Renilla* (Pennatulacea)." *Journal of Experimental Biology* 32: 299–320.
– (1955b). "Nervous regulation of luminescence in the sea pansy *Renilla köllikeri*." *Journal of Experimental Biology* 32: 619–35.
– (1955c). "Physiological control of luminescence in animals." In Frank H. Johnson, ed., *The Luminescence of Biological Systems*. Washington, DC: American Association for the Advancement of Science. 299–319.
– (1957). "Observations on photophores and luminescence in the teleost *Porichthys*." *Quarterly Journal of Microscopical Science* 98: 179–88.
– (1958a). "Observations on luminescence in *Noctiluca*." *Journal of the Marine Biological Association of the United Kingdom* 37: 535–49.
– (1958b). "Observations on luminescence in pelagic animals." *Journal of the Marine Biological Association of the United Kingdom* 37: 705–52.
– (1960a). "Studies on luminescence. On the subocular light-organs of stomiatoid fishes." *Journal of the Marine Biological Association of the United Kingdom* 39: 529–48.
– (1960b). "The regulation of light emission in animals." *Biological Reviews* 35: 1–40.
– (1960c). *The Biology of Marine Animals*. New York: Interscience Publishers.
– (1967). "The luminescence of fishes." *Symposia of the Zoological Society of London* 19: 27–55.
– (1978). "Bioluminescence and vision." In Peter J. Herring, ed., *Bioluminescence in Action*. 367–98.
Nicol, J.A. Colin and Hiraoki Somiya (1989). *The Eyes of Fishes*. Oxford: Clarendon Press.
Nicolas, Marie-Thérèse, Jean-Marie Bassot, and Gisèle Nicolas (1989). "Immunogold

labeling of luciferase in the luminous bacterium *Vibrio harveyi* after fast-freeze fixation and different freeze-substitution and embedding procedures." *Journal of Histochemistry and Cytochemistry* 37: 663–74.

Nicolas, Marie-Thérèse, Jean-Marie Bassot, and Osamu Shimomura (1982). "Polynoidin: a membrane photoprotein isolated from the bioluminescent system of scaleworms." *Photochemistry and Photobiology* 35: 201–7.

Nicolas, Marie-Thérèse, Gisèle Nicolas, Carl H. Johnson, Jean-Marie Bassot, and J. Woodland Hastings (1987). "Characterization of the bioluminescent organelles in *Gonyaulax polyedra* (dinoflagellates) after fast-freeze fixation and antiluciferase immunogold staining." *Journal of Cell Biology* 105: 723–35.

Nusbaum-Hilarowicz, Joseph (1912). "Notes préliminaires sur l'anatomie comparée des poissons provenant des campagnes de S.A.S. le Prince de Monaco." *Bulletin de l'Institut océanographique de Monaco* 246: 1–13.

– (1916). *Memoirs of a naturalist* (in Polish). Lviv, Poland: H. Altenberg Publishing House.

– (1920). *Résultats des campagnes scientifiques accomplies sur son yacht par Albert Ier, Prince Souverain de Monaco. Fascicule LVIII: Etudes d'anatomie comparée sur les Poissons provenant des campagnes scientifiques de S. A. S. le Prince de Monaco*. Imprimerie de Monaco.

– (1923). *Résultats des campagnes scientifiques accomplies sur son yacht par Albert Ier, Prince Souverain de Monaco. Fascicule LXV: Etudes d'anatomie comparée sur les Poissons provenant des campagnes scientifiques de S. A. S. le Prince de Monaco. Deuxième partie*. Imprimerie de Monaco.

Nutting, Charles C. (1899). "The utility of phosphorescence in deep-sea animals." *American Naturalist* 33: 793–9.

Nyhart, Lynn K. (1995). *Biology Takes Form: Animal Morphology and the German Universities*. Chicago and London: University of Chicago Press.

Oba, Yuichi and Darrin T. Schultz (2014). "Eco–Evo bioluminescence on land and in the sea." In Gérald Thouand and Robert Marks, eds., *Bioluminescence: Fundamentals and Applications in Biotechnology*. Berlin-Heidelberg: Springer-Verlag. 3–36.

Oba, Yuichi, Shin-ichi Kato, Makoto Ojika, and Satoshi Inouye (2009). "Biosynthesis of coelenterazine in the deep-sea copepod, *Metridia pacifica*." *Biochemical and Biophysical Research Communications* 390: 684–8.

O'Day, William T. (1972). *The Histology and Fine Structure of Some Bioluminescent Organs in Deep-Sea Fishes*. PhD dissertation, University of Southern California.

Oertel, D., Kenneth A. Linberg, and James F. Case (1975). "Ultrastructure of the larval firefly light organ as related to control of light emission." *Cell and Tissue Research* 164: 27–44.

Oertel, Donata and James F. Case (1976). "Neural excitation of the larval firefly photocyte: slow depolarization possibly mediated by a cyclic nucleotide." *Journal of Experimental Biology* 65: 213–27.

Ogilvie, Brian W. (2006). *The Science of Describing. Natural History in Renaissance Europe*. Chicago and London: University of Chicago Press.

Ohba, Nobuyoshi (2004). "Flash communication systems of Japanese fireflies." *Integrative and Comparative Biology* 44: 225–33.

Ohba, Nobuyoshi and Victor B. Meyer-Rochow (2012). "Insect species co-existing with the Papua New Guinea firefly *Pteroptyx effulgens* share aspects of appearance and behavior." *Lampyrid* 2: 127–37.

Ohmiya, Yoshihiro, Satoshi Kojima, Mitsuhiro Nakamura and Haruki Niwa (2005). "Bioluminescence in the limpet-like snail *Latia neritoides*." *Bulletin of the Chemical Society of Japan* 78: 1197–1205.

Ohshima, Hiroshi (1911). "Some observations on the luminous organs of fishes." *Journal of the College of Science, Imperial University, Tokyo*, 27, 15: 1–25.

Ohtsuka, Hiroko, Noel G. Rudie and John E. Wampler (1976). "Structural identification and synthesis of luciferin from the bioluminescent earthworm, *Diplocardia longa*." *Biochemistry* 15: 1001–4.

Okada, Tokindo S. (1994). "Experimental Embryology in Japan, 1930–1960. A historical background of developmental biology in Japan." *Internation Journal of Developmental Biology* 38: 135–54.

Okada, Yô K. (1926). "On the photogenic organ of the knight-fish (*Monocentris japonicus* (Houttuyn))." *Biological Bulletin* 50: 365–73.

– (1927a). "Luminescence et organe photogène des ostracodes." *Bulletin de la Société zoologique de France* 51: 478–86.

– (1927b). "Luminescence chez les mollusques lamellibranches." *Bulletin de la Société zoologique de France* 52: 95–9.

– (1966). "Observations on rod-like contents in the photogenic tissue of *Watasenia scintillans* through the electron microscope." In Frank H. Johnson and Yata Haneda, eds., *Bioluminescence in Progress*. 611–25.

O'Kane, Dennis J. and Douglas C. Prasher (1992). "Evolutionary origins of bacterial bioluminescence." *Molecular Microbiology* 6: 443–9.

Oliveira, Anderson G., Dennis E. Desjardin, Brian A. Perry and Cassius V. Stevani (2012). "Evidence that a single bioluminescent system is shared by all known bioluminescent fungal lineages." *Photochemical and Photobiological Sciences* 11: 848–52.

Ormö, Mats, Andrew B. Cubitt, Karen Kallio, Larry A. Gross, Roger Y. Tsien, and S. James Remington (1996). "Crystal structure of the *Aequoria victoria* green fluorescent protein." *Science* 273: 1392–5.

Osawa, Kotaro, Toyo Sasaki and Victor B. Meyer-Rochow (2014). "New observations on the biology of *Keroplatus nipponicus* Okada, 1938 (Diptera: Mycetophiloidea; Keroplatidae), a bioluminescent fungivorous insect." *Entomologie heute* 26: 139–49.

Osborn, Karen J., Steven H.D. Haddock, Fredrik Pleijel, Laurence P. Madin and Greg W. Rouse (2009). "Deep-sea, swimming worms with luminescent 'bombs.'" *Science* 325: 964.

Osorio, Balthazar (1912). "Une propriété singulière d'une bactérie phosphorescente." *Comptes rendus des séances de la Société de biologie* 72: 432–3.

Oviedo, Gonzalo Fernández de (1851). *Historia General y Natural de las Indias*. Primera Parte. Edited by José Amador De Los Rios. Madrid: Imprenta de la Real Academia de la Historia.

Packard, Alpheus S. (1896). "The phosphorescent organs of insects." *Journal of the New York Entomological Society* 4, 2: 61–6.

Panceri, Paolo (1871). "Intorno alla sede del movimento luminoso nelle meduse." *Rendiconti dell'Accademia delle Scienze fisiche, matematische e naturali in Napoli* 10: 140–6.

– (1872a). "Études sur la phosphorescence des animaux marins." *Annales des sciences naturelles, Paris,* Série 5, 16: 1–67.

– (1872b). "The luminous organs and light of the Pennatulae." *Quarterly Journal of Microscopical Science,* Series 2: 12, 4: 248–54.

– (1872c). "The luminous organs and the light of the Pholades." *Quarterly Journal of Microscopical Science,* Series 2: 12, 47: 254–60.

– (1873a). "Gli organi luminosi e la luce della pennatule." *Atti dell'Accademia di Scienze Fisiche e Matematiche Napoli* 5, 10: 1–38.

– (1873b). "Gli organi luminosi e la luce dei Pirosomi e della Foladi." *Atti dell'Accademia di Scienze Fisiche e Matematiche Napoli* 5, 13: 1–51.

– (1873c). "Intorno alla luce che emana delle cellule nervosi della Phyllirrhoe bucephala." *Atti dell'Accademia di Scienze Fisiche e Matematiche Napoli* 5, 14: 1–12.

– (1873d). "La luce e gli organi luminosi dei Beroidei." *Atti dell'Accademia di Scienze Fisiche e Matematiche Napoli* 5, 20: 1–15.

– (1873e). "The luminous organs and light of *Pyrosoma.*" *Quarterly Journal of Microscopical Science,* Series 2: 13, 49: 45–51.

– (1873f). "On the light emanating from the nerve-cells of *Phyllorhoë bucephala.*" *Quarterly Journal of Microscopical Science,* Series 2: 13, 50: 109–16.

– (1878a) "La luce e gli Organi luminosi di alcuni annelidi." *Atti dell'Accademia di Scienze Fisiche e Matematiche Napoli* 7, 1: 1–20.

– (1878b). "Intorno alla sede del movimento luminoso nelle Campanularie." *Atti dell'Accademia di Scienze Fisiche e Matematiche Napoli* 7, 9: 1–6.

Parker, George H. (1919). "The organization of Renilla." *Journal of Experimental Zoology* 27: 499–507.

– (1920). "Activities of colonial animals. II. Neuromuscular movements and phosphorescence in Renilla." *Journal of Experimental Zoology* 31: 474–515.

– (1938). "William Morton Wheeler 1865–1937." *Biographical Memoirs of the National Academy of Sciences* 19: 203–41.

Partridge, Julian C. and Ron H. Douglas (1995). "Far-red sensitivity of dragon fish." *Nature* 375: 21–2.

Pauly, Philip J. (1987). *Controlling Life. Jacques Loeb and the Engineering Ideal in Biology.* New York and Oxford: Oxford University Press.

Pavans de Ceccatty, Max, André Bilbaut and Jean-Marie Bassot (1972). « Correlations entre les signaux électriques spontanés, la motricité et la luminescence chez *Acholoe astericola* delle Ch." *Comptes Rendus des séances de l'Académie des Sciences* D275: 2523–6.

Paxton, John R. (1972). "Osteology and relationships of the lanternfishes (Family Myctophidae)." *Bulletin of the Natural History Museum of Los Angeles County* 13: 1–81.

Pellerano, Alessandro (2013). "L'Istituto di zoologia della Università degli studi di Genova – Cenni storici e recordi." *Bollettino dei musei e degli istituti biologici dell'Università di Genova* 75: 1–159.

Péron, François (1804a). "Mémoire sur le nouveau genre *Pyrosoma.*" *Annales du Muséum national d'histoire naturelle* 4: 437–46.

– (1804b). "Précis d'un mémoire lu à l'Institut national, sur la température des eaux de la mer, soit à sa surface, soit à diverses profondeurs." *Annales du Muséum national d'histoire naturelle* 5: 123–48.

Peters, Amos W. (1905). "Phosphorescence in ctenophores." *Journal of Experimental Zoology* 2: 103–16.

Peters, Wilhelm (1841). "Ueber das Leuchten der Lampyris italica." *Archiv für Anatomie, Physiologie und wissenschaftliche Medicin* (Jahrgang 1841): 229–33.

Pettitt, George A. (2015) "Conrad Gesner." *Encyclopedia Britannica Online*, 23 September 2015.

Petushkov, Valentin N., Maxim A. Dubinnyi, Aleksandra S. Tsarkova, Natalja S. Rodionova, Mikhail S. Baranov, Vadim S. Kublitski, Osamu Shimomura, and Ilia V. Yampolsky (2014). "A novel type of luciferin from the Siberian luminous earthworm *Fridericia heliota*: structure elucidation by spectral studies and total synthesis." *Angewandte Chemie International Edition* 53: 5566–8.

Phipson, Thomas Lamb (1862). *Phosphorescence or, the Emission of Light by Minerals, Plants and Animals*. London: Lovell Reeve & Co.

Pickford, Grace E. (1949). "*Vampyroteuthis infernalis* (Chun) an archaic dibranchiate cephalopod. II. External anatomy." *Dana Reports* 32: 1–132.

Pierantoni, Umberto (1918). "Organi luminosi, organi simbiotici e glandola nidamentale accessoria nei Cefalopodi." *Bollettino della Società dei naturalisti in Napoli* 30: 30–6.

– (1921). "Gli organi luminosi simbiotici ed il loro ciclo ereditario in *Pyrosoma giganteum*." *Pubblicazioni della Stazione Zoologica di Napoli* 3: 191–221.

Pieters, Florence F.J.M. and Jaap de Visser (1993). "The scientific career of the zoologist Max Wilhelm Carl Weber (1852–1937)." *Bijdragen tot de Dierkunde* 62, 4: 193–214.

Pitelka, Morgan (2011). "Art, agency, and networks in the career of Tokugawa Ieyasu (1543–1616)." In Rebecca M. Brown and Deborah S. Hutton, eds., *Companion to Asian Art and Architecture*. Chichester, West Sussex, UK: Wiley-Blackwell. 447–70.

Plinius Secundus, C. (1634). *The Historie of the World*. Translated into English by Philemon Holland. London: Adam Islip.

Polimanti, Osvaldo (1911). "Über das Leuchten von *Pyrosoma elegans* Les." *Zeitschrift für Biologie* 55: 505–29.

Portier, Paul Jules (1918). *Les symbiotes*. Paris: Masson et cie.

Pourcel, Léon (1929). "Notice biographique du Professeur Raphaël Dubois." *Annales de la Société d'histoire naturelle de Toulon* 14: 9–11.

Powers, Meghan L., Amy G. McDermott, Nathan C. Shaner, and Steven H.D. Haddock (2013). "Expression and characterization of the calcium-activated photoprotein from the ctenophore *Bathocyroe fosteri*: insights into light-sensitive photoproteins." *Biochemical and Biophysical Research Communications* 431: 360–6.

Prasher, Douglas, Richard O. McCann and Milton J. Cormier (1985). "Cloning and expression of the cDNA coding for aequorin, a bioluminescent calcium-binding protein." *Biochemical and Biophysical Research Communications* 126: 1259–68.

Prasher, Douglas C., Virginia K. Eckenrode, William W. Ward, Frank G. Prendergast, and Milton J. Cormier (1992). "Primary structure of the *Aequorea victoria* green-fluorescent protein." *Gene* 111: 229–33.

Purtov, Kinstantin V. and 16 colleagues (2015). "The chemical basis of fungal bioluminescence." *Angewandte Chemie International Edition* 54: 8124–8.

Pütter, August (1905). "Sammelreferate: Leuchtende Organismen." *Zeitschrift für allgemeine Physiologie* 5: 17–53.

Quatrefages, Jean Louis Armand de (1843). "Sur un mode nouveau de phosphorescence observé chez quelques Annelides et Ophyures." *Annales des sciences naturelles* 19 (ser. 2): 183–92.
– (1850). "Mémoire sur la phosphorescence de quelques invertébrés marins." *Annales des sciences naturelles* 14 (ser. 3): 236–81.
Querner, Hans (2008). "Meyen, Franz Julius Ferdinand." *Complete Dictionary of Scientific Biography*. Encyclopedia.com, 13 August 2015.
Quoy, Jean René Constant and Joseph Paul Gaimard (1825). "Observations sur quesques Mollusques et Zoophytes, envisagés comme causes de la Phosphorescence de la mer." *Annales des sciences naturelles* 4: 4–13.
Rawitz, Bernhard (1892). "Der Mantelrand der Acephalen. Dritte Teil." *Jenaische Zeitschrift für Naturwissenschaft* 27: 1–232.
Réaumur, René-Antoine Ferchault de (1712). "Observations sur le mouvement progressif de quelques coquillages de mer, sur celuy des hérissons de mer, et sur celuy d'une espèce d'étoile." *Mémoires de l'Académie Royale des Sciences* (1712): 115–45.
– (1723). "Des merveilles des dails ou de la lumière qu'ils répandent."*Mémoires de l'Académie royale des sciences* (1723): 198–204.
Rechnitzer, Andreas B. (1962). *Summary of the bathyscaph Trieste research program results: (1958–1960)*. San Diego, California: U.S. Navy Electronics Laboratory (Research Report 1095).
Rees, Jean-François, Bertrand de Wergifosse, Olivier Noiset, Marlene Dubuisson, Bernadette Janssens, and Eric M. Thompson (1998). "The origins of marine bioluminescence: turning oxygen defence mechanisms into deep-sea communication tools." *Journal of Experimental Biology* 201: 1211–21.
Regnard, Paul (1891). *Recherches expérimentales sur les conditions physiques de la vie dans les eaux*. Paris: G. Masson, éditeur.
Renwart, Marie, Jérôme Delroisse, Julien M. Claes, and Jérôme Mallefet (2014). "Ultrastructural organization of lantern shark (*Etmopterus spinax* Linnaeus, 1758) photophores." *Zoomorphology* 133: 405–16.
Renwart, Marie, Jérôme Delroisse, Patrick Flammang, Julien M. Claes, and Jérôme Mallefet (2015). "Cytological changes during luminescence production in lanternshark (*Etmopterus spinax* Linnaeus, 1758) photophores." *Zoomorphology* 134: 107–16.
Reynolds, George T. (1978). "Application of photosensitive devices to bioluminescence studies." *Photochemistry and Photobiology* 27: 405–21.
Richard, Jules (1885). "Un mot sur la phosphorescence des myriapodes." *Annales de la Société entomologique de Belgique* 29: 15–25.
Ridgway, Ellis B. and Christopher C. Ashley (1967). "Calcium transients in single muscle fibers." *Biochemical and Biophysical Research Communications* 29, 2: 229–34.
Rigby, Lisa M. and David J. Merritt (2011). "Roles of biogenic amines in regulating bioluminescence in the Australian glowworm *Arachnocampa flava*." *Journal of Experimental Biology* 214: 3286–93.
Rivers, Trevor J. and James G. Morin (2008). "Complex sexual courtship displays by luminescent male marine ostracods." *Journal of Experimental Biology* 211: 2252–62.
– (2012). "The relative cost of using luminescence for sex and defense: light budgets in cypridinid ostracods." *Journal of Experimental Biology* 215: 2860–8.
Roberts, P. Ann, Jan Knight, and Anthony K. Campbell (1987). "Pholasin – a biolu-

minescent indicator for detecting activation of single neutrophils." *Analytical Biochemistry* 160: 139–48.

Robertson, Harold A. and Albert D. Carlson (1976). "Octopamine: presence in firefly lantern suggests a transmitter role." *Journal of Experimental Zoology* 195: 159–64.

Robison, Bruce H. (2004). "Deep pelagic biology." *Journal of Experimental Marine Biology and Ecology* 300: 253–72.

Robison, Bruce H., Kim R. Reisenbichler, James C. Hunt, and Steven H.D. Haddock (2003). "Light production by the arm tips of the deep-sea cephalopod *Vampyroteuthis infernalis*." *Biological Bulletin* 205: 102–9.

Roda, Aldo (2011). "A history of bioluminescence and chemiluminescence from ancient times to the present." In Aldo Roda, ed., *Chemiluminescence and Bioluminescence: Past, Present and Future*. Cambridge, UK: Royal Society of Chemistry. 1–50.

Rohr, James, Michael I. Latz, S. Fallon, J.C. Nauen, and E. Hendricks (1998). "Experimental approaches towards interpreting dolphin-stimulated bioluminescence." *Journal of Experimental Biology* 201: 1447–60.

Rooney, Jennifer and Sara M. Lewis (2002). "Fitness advantage from nuptial gifts in female fireflies." *Ecological Entomology* 27: 373–7.

Root, Michael (1988). "Glow-in-the-dark biotechnology. The future of bioluminescence monitoring is bright." *Bioscience* 38: 745–7.

Roper, Clyde F.A. (1984). "S. Stillman Berry (1887–1984): a tribute through glimpses and reflections." *American Malacological Bulletin* 3, 1: 55–61.

Rosenberg, Jörg and Victor B. Meyer-Rochow (2009). "Luminescent myriapoda: a brief review." In Victor B. Meyer-Rochow, ed., *Bioluminescence in Focus – A Collection of Illuminating Essays*. Kerala, India: Research Signpost. 139–46.

Rostand, Jean (1951). *Les origines de la biologie expérimentale et l'abbé Spallanzani*. Paris: Fasquelle.

Rota, Emilia (2003). "Redescription of *Fridericia heliota* (Annelida, Clitellata: Enchytraeidae), a luminous worm from the Siberian taiga, with a review of bioluminescence in the Oligochaeta." *Journal of Zoology, London*, 260: 291–9.

– (2009). "Lights on the ground: a historical survey of light production in the Oligochaeta." In Victor Benno Meyer-Rochow, ed., *Bioluminescence in Focus – A Collection of Illuminating Essays*, Kerala, India: Research Signpost. 105–38.

Rowe, Laura, Emre Dikici, and Sylvia Daunert (2009). "Engineering bioluminescent proteins: expanding their analytical potential." *Analytical Chemistry* 81: 8662–8.

Ruby, Edward G. (1996). "Lessons from a cooperative, bacterial-animal association: the *Vibrio fischeri-Euprymna scolopes* light organ symbiosis." *Annual Review of Microbiology* 50: 591–624.

Ruby, Edward G. and James G. Morin (1978). "Specificity of symbiosis between deep-sea fishes and psychrotrophic luminous bacteria." *Deep Sea Research* 25: 161–7.

Ruby, Edward G. and Kenneth H. Nealson (1976). "Symbiotic association of *Photobacterium fischeri* with the marine luminous fish *Monocentris japonica*: a model of symbiosis based on bacterial studies." *Biological Bulletin* 151: 574–86.

Sala-Newby, Graciela B., Catherine M. Thomson, and Anthony K. Campbell (1996). "Sequence and biochemical similarities between the luciferases of the glow-worm *Lampyris noctiluca* and the firefly *Photinus pyralis*." *Biochemical Journal* 313: 761–7.

Salfi, Mario (1960). "Umberto Pierantoni." *Bollettino di zoologia* 27: 21–33.

Salmon, John T. (1947). "George Vernon Hudson, F.R.S.N.Z. (1867–1946)." *Transactions and Proceedings of the Royal Society of New Zealand* 76: 264–6.

Salpietro, L., A. Donato, F. Baguet, J. Mallefet, M.A. Gargano, and R. Marino (1998). "Determination by HPLC of adrenalin (E) and of noradrenalin (NE) in certain species of mesopelagic fish in the Strait of Messina." *Journal of Bioluminescence and Chemiluminescence* 13: 311–14.

Sapolsky, Harvey M. (1990). *Science and the Navy. The History of the Office of Naval Research.* Princeton: Princeton University Press.

Sapp, Jan (1994). *Evolution by Association: A History of Symbiosis.* New York: Oxford University Press.

Sargent, Rose-Mary (1995). *The Diffident Naturalist: Robert Boyle and the Philosophy of Experiment.* Chicago and London: University of Chicago Press.

Sars, Georg O. (1875). *Report on the Schizopoda collected by H.M.S. Challenger during the Years 1873–76.* Edinburgh: Neill and Company.

Satterlie, Richard A., Peter A.V. Anderson, and James F. Case (1980). "Colonial coordination in anthozoans: Pennatulacea." *Marine Behavior and Physiology* 7: 25–46.

Schnitzler, Christine E. and colleagues (2012). "Genomic organization, evolution, and expression of photoprotein and opsin genes in *Mnemiopsis leidyi*: a new view of ctenophore photocytes." *BMC Biology* 10: 107.

Schram, Frederick R. (1986). "Martin Davis Burkenroad (20 March 1910–12 January 1986)." *Journal of Crustacean Biology* 6: 302–7.

Seliger, Howard H. (1975). "The origin of bioluminescence." *Photochemistry and Photobiology* 21: 355–61.

Seliger, Howard H. and William D. McElroy (1959). "Quantum yield in the oxidation of firefly luciferin." *Biochemical and Biophysical Research Communications* 1: 21–4.

Sherwell, Philip (2008). "The scientist, the jellyfish protein and the Nobel prize that got away." *The Telegraph*, 11 October 2008.

Shimomura, Osamu (1987). "Presence of coelenterazine in non-bioluminescent marine organisms." *Comparative Biochemistry and Physiology* 86B: 361–3.

Shimomura, Osamu (2006). *Bioluminescence: Chemical Principles and Methods.* Singapore: World Scientific Publishing Co.

– (2008). *Discovery of Green Fluorescent Protein, GFP.* Nobel Lecture, 8 December 2008.

Shimomura, Osamu and Frank H. Johnson (1966). "Partial purification and properties of the *Chaetopterus* luminescence system." In Frank H. Johnson and Yata Haneda, eds., *Bioluminescence in Progress.* 495–521.

– (1968a). "*Chaetopterus* photoprotein: crystallization and cofactor requirements for bioluminescence." *Science* 159: 1239–40.

– (1968b). "Light-emitting molecule in a new photoprotein type of luminescence system from the euphausiid shrimp *Meganyctiphanes norvegica.*" *Proceedings of the National Academy of Science* 59: 475–7.

– (1975). "Chemical nature of bioluminescence systems in coelenterates." *Proceedings of the National Academy of Science* 72: 1546–9.

Shimomura, Osamu, Toshio Goto and Yoshimasa Hirata (1957). "Crystalline Cypridina luciferin." *Bulletin of the Chemical Society of Japan* 30: 929–33.

Shimomura, Osamu, Frank H. Johnson, and Hiroshi Morise (1974). "The aldehyde content of luminous bacteria and of an 'aldehydeless' dark mutant." *Proceedings of the National Academy of Science* 71: 4666–9.

Shimomura, Osamu, Frank H. Johnson and Yo Saiga (1961). "Pufirication and properties of *Cypridina* luciferase." *Journal of Cellular and Comparative Physiology* 58: 113–23.

– "Extraction, purification and properties of aequorin, a bioluminescent protein from the luminous hydromedusan, *Aequorea*." *Journal of Cellular and Comparative Physiology* 59: 223–39.

Shimomura, Osamu, Shoji Inoue, Frank H. Johnson, and Yata Haneda (1980). "Widespread occurrence of coelenterazine in matine bioluminescence." *Comparative Biochemistry and Physiology* 65B: 435–7.

Shrope, Mark (2007). "Marine biology: lights in the deep." *Nature* 450: 472–4.

Siesser, William G. (1981). "Christian Gottfried Ehrenberg: founder of micropaleontology." *Centaurus* 25: 166–88.

Silvestre, Baron Augustin-François de (1829). *Notice biographique sur M. Louis-Augustin-Guillaume Bosc*. Paris: Madame Huzard. 28 pages.

Skowron, Stanislaw (1926). "On the luminescence of *Microscolex phosphoreus* Dug." *Biological Bulletin* 51: 199–208.

– (1928). "The luminous material of *Microscolex phosphoreus* Dug." *Biological Bulletin* 54: 191–5.

Slabber, Martinus (1778) *Natuurkundige verlustigingen: behelzende microscopise waarneemingen van in- en uitlandse water- en land-dieren*. Te Haarlem: J. Bosch.

Smalley, Katherine N. (1965). "Adrenergic transmission in the light organ of the firefly, *Photinus pyralis*." *Comparative Biochemistry and Physiology* 16: 467–70.

Smith, David S. (1963). "The organization and innervation of the luminescent organ in a firefly, *Photuris pennsylvanica* (Coleoptera)." *Journal of Cell Biology* 16: 323–59.

Smith, Hugh M. (1935). "Synchronous flashing of fireflies." *Science* 82: 151–2.

Snell, Peter A. (1931). "The neuro-muscular mechanism controlling flashing in the lampyrid fireflies." *Science* 73: 372–3.

– (1932). "The control of luminescence in the male lampyrid firefly, Photuris pennsylvanica, with special reference to the effect of oxygen tension on flashing." *Journal of Cellular and Comparative Physiology* 1: 37–51.

Southward, Alan J. and Elizabeth K. Roberts (1987). "One hundred years of marine research at Plymouth." *Journal of the Marine Biological Association of the United Kingdom* 67: 465–506.

Spallanzani, Lazaro (1784). "Lettera prima relativa a diverse produzioni marine." *Memorie di matematica e di fisica della Società italiana delle scienze, Verona*, 2: 603–61.

– (1794). "Memoria sopra le meduse fosforiche." *Memorie di matematica e di fisica della Società italiana delle scienze, Verona*, 7: 271–90.

– (1796–97). "Chimico esame degli experimenti del Signor Goettling professore a Jena, sopra la luce del fosforo di Kunkel osservata nell'aria comune ed in diversi fluidi aeriformi." Published in Modena. Also: "Examen chimique des expériences de M. Goettling sur la lumière du phosphore dans les différents gaz." *Annales de chimie* 22: 246–54.

Stammer, Hans-Jürgen (1932). "Zur Biologie und Anatomie der leuchtenden Pilzmückenlarve von Ceroplatus testaceus Dalm. (Diptera, Fungivoridae)." *Zeitschrift für Morphologie und Ökologie der Tiere* 26: 135–46.

Stanley, Philip E. (1989). "Marlene A. DeLuca (1936–1987)." *Journal of Bioluminescence and Chemiluminescence* 4: 7–11.

Staples, Robert F. (1966). *The Distribution and Characteristics of Surface Bioluminescence in the Oceans.* Technical Report, U.S. Naval Oceanographic Office, Washington, DC.

Stead, David G. (1906). *Fishes of Australia: a popular and systematic guide to the study of the wealth within our waters.* Sydney: W. Brooks & Co.

Steche, Otto H. (1909). "Die Leuchtorgane von Anomalops katoptron und Photoblepharon palpebratus, zwei Oberflächenfischen aus dem Malaiischen Archipel. Ein Beitrag zur Morphiologie und Physiologie der Leuchtorgaue der Fische." *Zeitschrift für wissenschaftliche Zoologie* 93: 349–408.

Stewart, Charles (1802). *Elements of the Natural History of the Animal Kingdom.* Edinburgh: Bell & Bradfute.

Strehler, Bernard H. (1955). "Firefly luciferin." In Frank H. Johnson, ed., *The Luminescence of Biological Systems.* Washington, DC: The Americal Association for the Advancement of Science. 199–208.

Strum, Judy M. (1969a). "Fine structure of the dermal luminescent organs, photophores, in the fish, *Porichthys notatus.*" *Anatomical Record* 164: 433–62.

– (1969b). "Photophores of *Porichthys notatus*: ultrastructure of innervation." *Anatomical Record* 164: 463–78.

Summers, Adam P. and Thomas J. Koob (1997). "A biographical sketch of Samuel Walton Garman." In Adam P. Summers, ed., *Plagiostomia: The Sharks, Skates and Rays.* Los Angeles: Benthic Press. l–lxxi.

Suter, Henry (1890). "Miscellaneous communications on New Zealand land and freshwater molluscs." *Transactions and Proceedings of the New Zealand Institute* 23: 93–6.

Sweeney, Beatrice M. (1971). "Laboratory studies of a green *Noctiluca* from New Guinea." *Journal of Phycology* 7: 53–8.

– (1978). "Ultrastructure of *Noctiluca miliaris* (Pyrrophyta) with green flagellate symbionts." *Journal of Phycology* 14: 116–20.

– (1980). "Intracellular source of bioluminescence." *International Review of Cytology* 68: 173–95.

Sweeney, Beatrice M., Francis T. Haxo, and J. Woodland Hastings (1959). "Action spectra for two effects of light on luminescence in *Gonyaulax polyedra.*" *Journal of General Physiology* 43: 285–99.

Takasuna, Miki (2005). "Six researchers' contributions to comparative psychology in Japan: 1900–1945." *Japanese Psychological Research* 47: 88–94.

Terao, Arata (1917). "Notes on the photophores of *Sergestes prehensilis* Bate." *Annotationes zoologicae japonenses* 9: 299–316.

Thomson, Catherine M., Peter J. Herring, and Anthony K. Campbell (1995). "Coelenterazine distribution and luciferase characteristics of oceanic decapod crustaceans." *Marine Biology* 124: 197–207.

– (1997). "The widespread occurrence and tissue distribution of the imidazolopyrazine luciferins." *Journal of Bioluminescence and Chemiluminescence* 12: 87–91.

Thompson, D'Arcy Wentworth (translator) (1910). *Aristotle's A History of Animals.* Oxford: Clarendon Press.

Thompson, Eric M., Yoshiaki Toya, Basil G. Nafpaktitis, Toshio Goto, and Frederick I. Tsuji (1988). "Induction of bioluminescence capability in the marine fish, *Porichthys notatus,* by *Vargula* (crustacean) [14C] luciferin and unlabelled analogues." *Journal of Experimental Biology* 137: 39–51.

Thust, Kurt A. (1916). "Zur Anatomie und Histologie der Brisinga coronata G.O. Sars unter besonderer Berückstichtigung der Luminescenz der Brisingiden." *Mitteilungen aus der Zoologischen Station zu Neapel* 22: 368–432.

Tilesius von Tilenau, Wilhelm Gottlieb (1819). "Des Hofrath Tilesius Resultate seiner, während der drei Jahre der Krusenstern'schen Entdeckungsreise angestellten Untersuchungen über das Leuchten des Meeres." *Annalen der Physik* 61, 1: 36–44.

Todd, Kim (2007). *Chrysalis. Maria Sybilla Merian and the Secrets of Metamorphosis.* London and New York: I.B. Tauris.

Tompkins, Tom (1981). *Yokosuka: Base of an Empire*. Novato, California: Presidio Press.

Tong, Deyan, Natalia S. Rozas, Todd H. Oakley, Jane Mitchell, Nansi J. Colley, Margaret J. McFall-Ngai, and J. Woodland Hastings (2009). "Evidence for light perception in a bioluminescent organ." *Proceedings of the National Academy of Sciences of the United States of America* 106: 9836–41.

Torlais, Jean (1961). *Un esprit encyclopédique en dehors de "L'encyclopédie": Réaumur.* Paris: Blachard.

Townsend, Anne B. (1904). "The histology of the light organs of Photinus marginellus." *American Naturalist* 38: 127–51.

Trimmer, Barry A., June R. Aprille, D.M. Dudzinski, Christopher J. Lagace, S.M. Lewis, T. Michel, S. Qazi, and R.M. Zayas (2001). "Nitric oxide and the control of firefly flashing." *Science* 292: 2486–8.

Trojan, Emanuel (1907). "Zur Lichtentwicklung in den Photosphärien der Euphausien." *Archiv für mikroskopische Anatomie* 70: 177–89.

– (1909). "Leuchtende Ophiopsilen." *Archiv für mikroskopische Anatomie* 73: 883–912.

– (1910). "Die Leuchtorgane von *Cyclothone signata* Garman." *Sitzungsberichte der Kaiserlichen Akademie der Wissenschaften in Wien* 124: 291–316.

– (1913). "Über Hautdrüsen des Chaetopterus variopedatus Clap." *Sitzungsberichte der Kaiserlichen Akademie der Wissenschaften in Wien* 122: 565–96.

Tsuji, Frederick I. (2010). "Early history, discovery, and expression of *Aequorea* green fluorescent protein, with a note on an unfinished experiment." *Microscopy Research and Technique* 73: 785–96.

Tsuji, Frederick I. and Yata Haneda (1966). "Chemistry of the luciferases of *Cypridina hilgendorfii* and *Apogon ellioti*." In Frank H. Johnson and Yata Haneda, eds., *Bioluminescence in Progress*. 137–49.

– (1971). "Luminescent system in a myctophid fish, *Diaphus elucens* Brauer." *Nature* 233: 623–4.

Tsuji, Frederick I. and Raymond Sowincki (1961). "Purification and molecular weight of *Cypridina* luciferase." *Journal of Cellular and Comparative Physiology* 58, 2: 125–9.

Tsuji, Frederick I., Anthony T. Barnes, and James F. Case (1972). "Bioluminescence in the marine teleost, *Porichthys notatus*, and its induction in a non–luminescent form by *Cypridina* (ostracod) luciferin." *Nature* 237: 515–16.

Tsuji, Frederick I., Aurin M. Chase, and E. Newton Harvey (1955). "Recent studies on the chemistry of *Cypridin*a luciferin." In Frank H. Johnson, ed., *The Luminescence of Biological Systems*. Washington, DC: American Association for the Advancement of Science. 127–56.

Tucker, Gordon H. (1951). "Relation of fishes and other organisms to the scattering of underwater sound." *Journal of Marine Research* 10: 215–38.

Turnbull, Constance M. (2009). *A History of Modern Singapore, 1819–2005*. Singapore: National University of Singapore Press.

Turner, John R., E.M. White, Martin A. Collins, Julian C. Partridge, and Ron H. Douglas (2009). "Vision in lanternfish (Myctophidae): adaptations for viewing bioluminescence in the deep-sea." *Deep Sea Research* 56: 1003–17.

Ussow, Mikhail M. (1879). "Ueber den Bau der sogenannten augenaehnlichen Flecken einiger Knochenfische." *Bulletin de la Société impériale des naturalistes de Moscou* 54: 79–115.

Valenstein, Elliot S. (2005). *The War of the Soups and the Sparks*. New York: Columbia University Press.

Valiadi, Martha and Debora Iglesias-Rodriguez (2013). "Understanding bioluminescence in dinoflagellates – How far have we come?" *Microorganisms* 1: 3–25.

Vallentin, Rupert and John T. Cunningham (1888). "The Photospheria of Nyctiphanes Norvegica, G.O. Sars." *Quarterly Journal of Microscopical Science* 28: 319–40.

Van Aken, Hendrik (2005). "Dutch oceanographic research in Indonesia in colonial times." *Oceanography* 18, 4: 30–41.

Van Der Reijden, E.D., Jeff D. Monchamp, and Sara M. Lewis (1997). "The formation, transfer, and fate of spermatophores in *Photinus* fireflies (Coleoptera: Lampyridae)." *Canadian Journal of Zoology* 75: 1202–7.

Vandiver, J. Kim and Pagan Kennedy (2005). "Harold Eugene Edgerton April 6, 1903–January 4, 1990." *Biographical Memoirs of the National Academy of Sciences* 86: 97–118.

Vanhöffen, Ernst (1918). "Zur Erinnerung an August Brauer." *Mitteilungen aus dem Zoologischen Museum in Berlin* 9: 1–12.

Vérany, Jean-Baptiste (1851). *Mollusques méditerranéens. 1ère Partie: Céphalopodes méditerranéens*. Gènes: Beuf Libraire.

Verworn, Max (1892). "Ein automatisches Centrum für die Lichtproduction bei Luciola italica L." *Centralblatt für Physiologie* 6: 69–74.

Vianelli, Giuseppe (1749). *Nueve scoperte intorno le luci notturne dell'aqua marina*. Venice: Pitteri.

Visick, Laren L., Jamie Foster, Judith Doino, Margaret McFall–Ngai, and Edward G. Ruby (2000). "Vibrio fischeri *lux* genes play an important role in colonization and development of the host light organ." *Journal of Bacteriology* 182: 4578–86.

Viviani, Domenico (1805). *Phosphorescentia maris quatuordecim lucescentium animalculorum*. Genoa: Tipis Joannis Giossi.

Viviani, Vadim R. (2002). "The origin, diversity, and structure-function relationships of insect luciferases." *Cellular and Molecular Life Sciences* 59: 1833–50.
Vorderman, Adolphe G. (1900). "Twee lichtgewende visschen van Banda. Ikan lewéri batoe (*Heterophthalmus palpebratus*) en Ikan lewéri ajer (*Heterophthalmus katoptron*)." *Natuurkundig Tijdschrift voor Nederlandsch Indië* 59: 72–7.
Walsh, Don (2010). "The Last Seven Miles: The Bathyscaph *Trieste* Story." *Diver* magazine, post retrieved 18 August 2016.
Ward, William W. and Howard H. Seliger (1974). "Extraction and purification of calcium-activated photoproteins from the ctenophores *Mnemiopsis* sp. and *Beroë ovata.*" *Biochemistry* 13: 1491–9.
– (1976). "Action spectrum and quantum yield for the photoinactivation of mnemiopsin, a bioluminescent photoprotein from the ctenophore *Mnemiopsis* sp." *Photochemistry and Photobiology* 23: 351–63.
Warner, Jon A., Michael I. Latz, and James F. Case (1979). "Cryptic bioluminescence in a midwater shrimp." *Science* 203: 1109–10.
Wass, John (2015). "The fantastical world of hormones." *The Endocrinologist* 115: 6–7.
Wassink, Eline C. and Seishi Kuwabara (1966). "Some notes on semi-large-scale cultivation of luminous fungi." In Frank H. Johnson and Yata Haneda, eds., *Bioluminescence in Progress*, 247–64.
Watanabe, Hiroshi (1897). "The phosphorescence of *Cypridina hilgendorfii*, Müller." *Annotationes zoologicae japonenses* 1: 69–70.
Watasé, Shozaburo (1890). *On the morphology of the compound eyes of arthropods*. Baltimore: Press of I. Friedenwald.
– (1895). *On the physical basis of animal phosphorescence*. Biological lectures delivered at the Marine Biological Laboratory of Wood's Hole. Boston: Ginn & Co.
– (1898). *Protoplasmic contractility and phosphorescence*. Biological lectures delivered at the Marine Biological Laboratory of Wood's Hole. Boston: Ginn & Co.
Weber, Max W.C. (1902). *Siboga-Expeditie: Introduction et Description de l'Expédition*. Leiden: E.J. Brill.
Welsh, John H. and Fenner A. Chace, Jr. (1937). "Eyes of deep-sea crustaceans. I. Acanthephyridae." *Biological Bulletin* 72: 57–74.
Wheeler, William M. (1925). "Obituary: Carlo Emery." *Entomological News* 36: 318–20.
Wheeler, William M. and Francis X. Williams (1915). "The luminous organ of the New Zealand glow-worm." *Psyche* 22: 36–43.
White, Emil H., Frank McCapra, George F. Field, and William D. McElroy (1961). "The structure and synthesis of firefly luciferin." *Journal of the American Chemical Society* 83: 2402–3.
White, Harris H. (1979). "Effects of dinoflagellate bioluminescence on the ingestion rates of herbivorous zooplankton." *Journal of Experimental Marine Biology and Ecology* 36: 217–24.
Widder, Edith A. (2010). "Bioluminescence in the ocean: origins of biological, chemical, and ecological diversity." *Science* 328: 704–8.
Widder, Edith A. and James F. Case (1981). "Two flash forms in the bioluminescent dinoflagellate *Pyrocystis fusiformis*." *Journal of Comparative Physiology* 143: 43–52.

– (1982). "Distribution of subcellular bioluminescent sources in a dinoflagellate, *Pyrocystis fusiformis*." *Biological Bulletin* 162: 423–48.

Widder, Edith A., Michael I. Latz and James F. Case (1983). "Marine bioluminescence spectra measured with an Optical Multichannel Detection System." *Biological Bulletin* 165: 791–810.

Widder, Edith A., Michael I. Latz, Peter J. Herring, and James F. Case (1984)."Far red bioluminescence from two deep-sea fishes." *Science* 225: 512–14.

Widder, Edith A., S.A. Bernstein, D.F. Bracher, J.F. Case, K.R. Reisenbichler, J.J. Torres, and B.H. Robison (1989). "Bioluminescence in the Monterey Submarine Canyon: image analysis of video recordings from a midwater submersible." *Marine Biology* 100: 541–51.

Wiedermann, Eilhard (1888). "Ueber Fluorescenz und Phosphorescenz. I. Abhandlung." *Annalen der Physik* 34: 446–63.

Wielowiejski, Heinrich Ritter von (1882). "Studien über die Lampyriden." *Zeitschrift für wissenschaftliche Zoologie* 37: 354–428.

Will, Johann Friedrich (1844). "Ueber das Leuchten einiger Seethiere." *Archiv für Naturgeschichte* 10: 328–37.

Williams, Francis X. (1916). "Photogenic organs and embryology of lampyrids." *Journal of Morphology* 28: 145–206.

Williams, Paul, Klaus Winzer, Weng C. Chan, and Miguel Cámara (2007)."Look who's talking: communication and quorum sensing in the bacterial world."*Philosophical Transactions of the Royal Society of London* B362: 1119–34.

Wilson, Thérèse and J. Woodland Hastings (1998). "Bioluminescence." *Annual Review of Cell and Developmental Biology* 14: 197–230.

Winter, F.H. (1914). "Carl Chun." *Bericht der Senckenbergischen Naturforschenden Gesellschaft in Frankfurt am Main* 45: 176–83.

Wolf, Mark J.P. (2012). *Building Imaginary Worlds: The Theory and History of Subcreation*. New York: Routledge.

Wolff, Torben (1993). "More than 200 years of crustacean research in Denmark." In Frank Truesdale, ed., *History of Carcinology*. Boca Raton, Florida: CRC Press. 207–24.

Wood, Keith V. (1995). "The chemical mechanism and evolutionary development of beetle bioluminescence." *Photochemistry and Photobiology* 62: 662–73.

Wood, Keith V., Y. Amy Lam, Howard H. Seliger, and William D. McElroy (1989). "Complementary DNA coding click beetle luciferases can elicit bioluminescence of different colors." *Science* 244: 700–2.

Wormer, Eberhard (1990). "Mangold Ernst." *Deutsche Biographie* 16: 29–30.

Wyville Thomson, Charles (1873). *The Depths of the Sea*. London: Macmillan and Co.

– (1878). *The Voyage of the "Challenger." The Atlantic* (in two volumes). New York: Harper & Brothers.

Yajima, Michiko (2007). "Franz Hilgendorf (1839–1904): introducer of evolutionary theory in Japan around 1873." In Patrick W. Jackson, ed., *Four Centuries of Geological Travel: The Search for Knowledge on Foot, Bicycle, Sledge and Camel*. London: Geological Society of London. 389–93.

Yasaki, Yoshiwo (1928). "On the nature of the luminescence of the knight-fish (Monocentris japonicus (Houttuyn))." *Journal of Experimental Zoology* 50: 495–505.

Yasaki, Yoshiwo and Yata Haneda (1935). "Bioluminescence of the deep-sea fish, Macrouridae." *Journal of Applied Zoology of Tokyo* 7: 165–76 (in Japanese).

Yasaki, Yoshiwo and Yata Haneda (1936). "Uber einen neuen Typus von Leuchtorgan in Fische (*Acropoma japonicum*). *Proceedings of the Imperial Academy of Japan* 12, 2: 55–7.

Yatsu, Naohide (1917). "Note on the structure of the maxillary gland of Cypridina hilgendorfii." *Journal of Morphology* 29: 435–40.

Yonge, Charles Maurice (1972). "The inception and significance of the *Challenger* Expedition." *Proceedings of the Royal Society of Edinburgh, Section B. Biology*, 72: 1–13.

Young, Richard E. (1973). "Information feedback from photophores and ventral countershading in mid-water squid." *Pacific Science* 27: 1–7.

Zarubin, Margarita, Shimshon Belkin, Michael Ionescu and Amatzia Genin (2012. "Bacterial bioluminescence as a lure for marine zooplankton and fish." *Proceedings of the National Academy of Science* 109: 853–7.

Zirpolo, Giuseppe (1918). "I batteri fotogeni degli organi luminosi di Sepiola Intermedia NAEF (*Bacillus pierantonni* n. sp.)." *Bollettino della Società dei naturalisti in Napoli* 30: 206–19.

– (1919). "Micrococcus pierantonii. Nuova specie di batterio fotogeno dell'organo luminoso di *Rondeletia minor*." *Bollettino della Società dei naturalisti in Napoli* 32: 75–87.

Zottoli, Steven J. (2001). "The origins of the Grass Foundation." *Biological Bulletin* 201: 218–26.

Zugmayer, Erich (1911a). "Diagnoses des poissons nouveaux provenant des campagnes du yacht 'Princesse-Alice' (1901 à 1910)." *Bulletin de l'Institut océanographique de Monaco* 193: 1–14.

– (1911b). *Résultats des campagnes scientifiques du Prince de Monaco. Poissons provenant des campagnes du yacht Princesse-Alice (1901–1910)*. Monaco: Imprimerie de Monaco.

– (1913). "Diagnoses des Stomiatidés nouveaux provenant des campagnes du yacht 'Hirondelle II' (1911 et 1912)." *Bulletin de l'Institut océanographique de Monaco* 253: 1–7.

– (1914). "Diagnoses de quelques poissons nouveaux provenant des campagnes du yacht *Hirondelle II* (1911–1913)." *Bulletin de l'Institut océanographique de Monaco* 288: 1–4.

Index